NORTH-HOLLAND
PERSONAL LIBRARY

PAUL EHRENFEST

VOLUME 1

THE MAKING OF A THEORETICAL PHYSICIST

PAUL EHRENFEST

VOLUME 1

THE MAKING OF A THEORETICAL PHYSICIST

MARTIN J. KLEIN

NORTH-HOLLAND
AMSTERDAM · OXFORD · NEW YORK · TOKYO

ISBN: 0 444 86948 4

Published by:

North-Holland Physics Publishing

a division of

Elsevier Science Publishers B.V.
P.O. Box 103
1000 AC Amsterdam
The Netherlands

Sole distributors for the U.S.A. and Canada:

Elsevier Science Publishing Company, Inc.
52 Vanderbilt Avenue
New York, N.Y. 10017
U.S.A.

First edition 1970
Second edition 1972
Third edition as a paperback 1985

Library of Congress Catalog Card Number: 77-102720

Printed in The Netherlands

To my mother and father

Preface

The thought of a book on Paul Ehrenfest first crossed my mind in the spring of 1953. I was spending the year at the Dublin Institute for Advanced Studies and I had gone to the library of the Royal Irish Academy to look up some work of Ehrenfest's referred to by Professor Erwin Schrödinger in a conversation. I was struck by Ehrenfest's unique style: the man himself seemed to be present in some of his scientific papers, a most unusual situation. When I also read Einstein's moving and personal memorial notice of his friend Ehrenfest I decided that I wanted to know more about this extraordinary physicist.

My rather vague initial idea was strengthened by the information and encouragement I received the next year, after my return to Cleveland and the Case Institute of Technology. These came from a new Dutch friend, Paul H. E. Meijer, who had studied at Leyden with Ehrenfest's own student and successor, H. A. Kramers. Later that year, when Ehrenfest was already much in my thoughts, I was startled by being introduced to a Paul Ehrenfest – nephew and namesake of the subject of this book, who provided more information and encouragement.

I did not decide to go ahead until I received the first indication of approval from Ehrenfest's widow, Mrs. Tatyana Ehrenfest-Afanassjewa, and also the enthusiastic and whole-hearted support of Professor George E. Uhlenbeck, who has been a tower of strength to me in this entire endeavor.

I certainly had no idea then of either the magnitude of the project or the effect it would have on my own work.

The kindness and help extended to me by virtually everyone I ever approached concerning Ehrenfest has made my work pleasurable as well as possible. I cannot properly thank all those who helped me, but I do want to mention many of them here.

The Ehrenfest family must come first. I met the late Mrs. Ehrenfest on a brief visit to Leyden in 1956 and came to know her during the year I spent there in 1958–59. She eventually made all her husband's papers, notebooks, manuscripts and extensive correspondence available to me. In my many visits and discussions of scientific questions with her, I was exposed to at least part of the unique intellectual atmosphere of the Ehrenfests' study-living room in the house on Witterozenstraat. We corresponded until her death in 1964. It is a privilege to have known her. The Ehrenfest daughters, Mrs. T. van Aardenne-Ehrenfest of Dordrecht and Mrs. A. van Bommel-Ehrenfest of Gronsveld, have been extremely helpful to me in a great many ways. It is a pleasure to thank them here.

I received correspondence, information and reminiscences from a great many people including Ehrenfest's friends and former students: P. M. S. Blackett, N. Bohr, M. Born, G. Breit, J. M. Burgers, H. B. G. Casimir, C. Darwin, P. Debye, M. Delbrück, G. H. Dieke, F. Ehrenfest-Egger, W. J. M. v. Eysinga, L. Fermi, V. Fock, A. D. Fokker, J. Franck, A. Gans, H. N. Gilbert, C. J. Gorter, S. A. Goudsmit, G. L. de Haas-Lorentz, A. Joffe, M. Julius, O. Klein, R. Kronig, A. Landé, M. Langeveld, M. v. Laue, L. Meitner, M. Minnaert, R. Oppenheimer, W. Pauli, R. E. Peierls, B. v. d. Pol, L. Rosenfeld, A. J. Rutgers, E. Schrödinger, O. Stern, D. J. Struik, H. Tietze, J. Tinbergen, G. E. Uhlenbeck, and W. v. d. Woude.

I owe a special debt of gratitude to the Estate of Albert Einstein, and to its executor, Dr. Otto Nathan, for permission to quote extensively from Einstein's correspondence. I am, as always, most grateful to Miss Helen Dukas for her great help in all questions concerning Einstein's correspondence and career.

Ehrenfest's correspondence and papers are now in the Ehrenfest Collection at the Museum for the History of Science at Leyden. Both Dr. M. Rooseboom and Mr. P. v. d. Star of the Museum have been very helpful to me. I was also permitted to use the Lorentz collection at the Algemeen Rijksarchief (National Archives) at The Hague.

All of us who work in the history of quantum physics are happy to have the catalogue, *Sources for History of Quantum Physics*, prepared by T. S. Kuhn, J. L. Heilbron, P. Forman, and L. Allen. I am grateful to Professors Kuhn, Heilbron and Forman for particular suggestions.

During the year I spent at Leyden gathering materials for this book I enjoyed the hospitality of the Lorentz Institute for Theoretical Physics. It is a pleasure to thank my friends S. R. de Groot and P. Mazur, the past and present directors of this Institute, for their hospitality, their warm interest and their help.

The first financial support for this work came from the Case Institute of Technology Research Fund. The John Simon Guggenheim Memorial Foundation, that great benefactor of scholarly, artistic and scientific work, awarded me fellowships in 1958–1959 and 1967–68 which made it possible to gather the materials for the book and then to complete the writing of this volume. Case Institute of Technology and Yale University granted me leaves of absence to accompany these Guggenheim Fellowships. The Lorentz Fund of the University of Leyden has made grants to pay for some of my research expenses and to help support the publication of this book. I am deeply grateful to all of them.

When I began working on this book I was a member of the Department of Physics at the Case Institute of Technology. I was remarkably fortunate in having as my department chairmen first Robert S. Shankland, and then Frederick Reines. The understanding, sympathy and support of these two and my other friends and colleagues at Case made it possible for me to follow my own interests, even though I was transformed from a physicist to a historian of science in the process.

Parts of this book have been read in manuscript by George Uhlenbeck, J. M. Burgers, Russell McCormmach, and Helen Dukas, and I am grateful for their criticisms.

I very much appreciate all the secretarial assistance I received with my work in the past at Case Institute of Technology and at the Lorentz Institute. I especially want to thank Mrs. Judith Scudder and Miss Christine Tattersall in the Department of the History of Science and Medicine at Yale.

I am glad to acknowledge the warm interest and cooperation of Mr. M. D. Frank of the North-Holland Publishing Company.

My gratitude to my wife is much too deep for words.

New Haven, Connecticut
April, 1969.

MARTIN J. KLEIN

Contents

List of Illustrations

The page numbers refer to the pages opposite to the illustrations.

Acknowledgement

Figures 2, 6, 8, and 13 are from the collection of the National Museum for the History of Science at Leyden.

Figure 10 is from the General National Archives at the Hague.

The other Figures were supplied by members of the Ehrenfest family.

Introductory Note

Paul Ehrenfest's name, never known to the world at large, is not even so widely known among physicists as it deserves to be. A variety of particular results are referred to as Ehrenfest's, but no great theory bears his name. Nevertheless, Paul Langevin did not exaggerate when he described Ehrenfest in 1933 as having been "at the very heart of the drama of contemporary physics", and even as having personified that drama in his own life. No man was more deeply engaged than he in the struggle to create the concepts of twentieth century physics, and no man was more fully committed to maintaining clarity and intelligibility in the flood of new developments. His efforts were recognized by the informal title his colleagues conferred upon him: the conscience of physics.

Teaching was his true vocation, and he knew it: "The great joy of passing on to others what I believed I understood myself was the real backbone of my life." His unsurpassed lectures, in which only the very essence of a subject was highlighted and made unforgettably clear, were but a part of this teaching. He had a real gift for Socratic dialogue, and practiced it with students and colleagues everywhere, forcing the truth to emerge by his insistent critical questioning. "He was not merely the best teacher in our profession whom I have ever known;" Albert Einstein wrote, "he was also passionately preoccupied with the development and destiny of men, especially his students. To understand others, to gain their friendship and trust, to aid anyone embroiled in outer or inner struggles, to encourage youthful talent – all this was his real element, almost more than immersion in scientific problems."

This "passionate preoccupation" with the development of others grew out of Ehrenfest's profound dissatisfaction with his own development. Despite his achievements in science, despite his many warm friendships, his very close family ties and the devotion of his students, Ehrenfest never lost a sense of his own inadequacy, and this darkened his life.

This first volume deals with Ehrenfest's life and work through the years of the First World War. By the end of this period Ehrenfest was an established physicist, known throughout Europe, with some of his most important scientific contributions already made. The second volume will be concerned with the last fifteen years of his life, when Ehrenfest's personal influence was felt from Pasadena to Leningrad. It was a time when he extended himself up to, and beyond, the limit of his resources trying to cope with the rapid and complex developments in physics, the problems of his own life and the grim world of 1933.

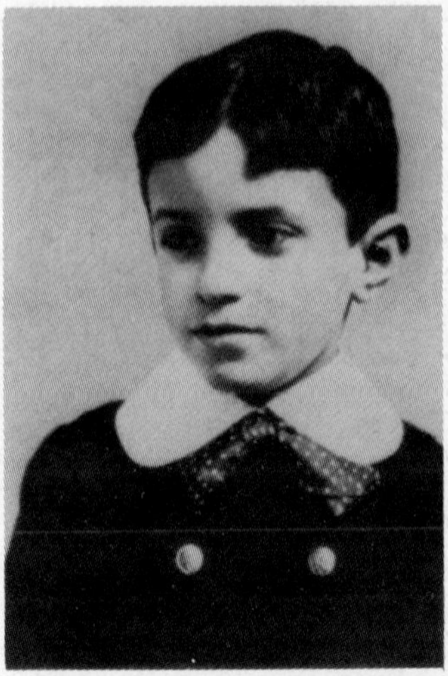

Fig. 1 A very young Paul Ehrenfest.

Fig. 3 "*The inseparable four*": Hans Hahn, Heinrich Tietze, Gustav Herglotz, Paul Ehrenfest. 1900.

Fig. 4 Tatyana and Paul Ehrenfest. 1904.

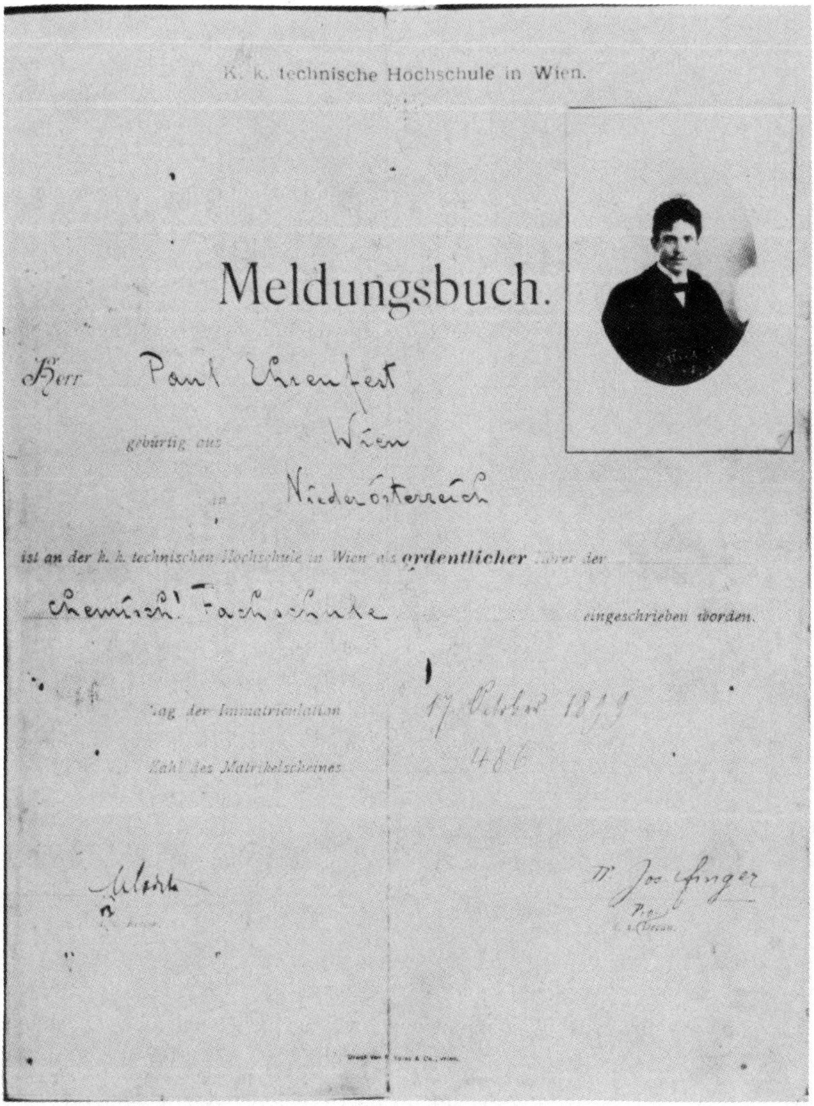

K. k. technische Hochschule in Wien.

Meldungsbuch.

Herr Paul Ehrenfest

gebürtig aus Wien

in Niederösterreich

ist an der k. k. technischen Hochschule in Wien als **ordentlicher** Hörer der

Chemisch. Fachschule eingeschrieben worden.

Tag der Immatriculation 17 October 1899

Zahl des Matrikelscheines 486

Fig. 2 Ehrenfest's registration book at the Technische Hochschule in Vienna, 1899.

CHAPTER 1

The New Professor at Leyden

The audience that gathered for the inaugural lecture of the new professor of theoretical physics in Leyden on December 4, 1912 could not have been prepared for what it heard. Leyden's revered and world renowned H. A. Lorentz, serene master of classical physics, had relinquished his professorship and arranged for the appointment of a youthful successor, Paul Ehrenfest, who was to address them now. Few in the assembled group knew much about the speaker: he was certainly a foreigner, but his Viennese origins were somewhat obscured by the fact that he had come to Leyden from St. Petersburg with his Russian wife. Some, but not many, had seen his published work in one or another of the German journals; a few remembered that he had visited Leyden for some weeks almost a decade earlier to sit at the feet of Lorentz, but for most he was an unknown quantity. It is unlikely that the members of the audience had any great expectation that they would learn very much about the new man, or about much else, from this inaugural lecture; such formal occasions were rarely illuminating. But before the short, black-bearded man on the platform had finished many sentences, it was clear that, formal occasion or not, something exciting was being said and someone exciting was saying it.[1]

1. First hand accounts of Ehrenfest's inaugural lecture can be found in H. A. Kramers, "Physiker als Stilisten", Naturwissenschaften **23** (1935), p. 297, and in W. de Sitter's obituary notice of Ehrenfest in the Leids Dagblad, 26 September 1933. It was also

"Permit me to speak about a crisis", he began, " a crisis which at present seriously threatens a fundamental hypothesis of physics – the aether hypothesis. It seems to me that this crisis gives a vivid picture of the characteristic revolutionary mood which dominates theoretical physics at this moment." The note of immediacy and urgency had been sounded at once. Ehrenfest had chosen a subject that would carry the audience to the center of his science, stressing some of its most problematic features and conveying the excitement generated by the recent, successful, attempts to cope with these problems. His theme was the development leading up to the theory of relativity that Albert Einstein had proposed in 1905 – the development, rather than the new theory itself.[2] The aether of which Ehrenfest spoke was the invisible, intangible substance that had been hypothesized, in one form or another, throughout the preceding century as the medium through which light waves propagate. The properties of this aether had suddenly become self-contradictory as the result of an experiment performed by Albert A. Michelson and Edward W. Morley at Cleveland in 1887. To bring this subtle and delicate experiment alive for his audience Ehrenfest resorted to a device that Leyden would learn to consider absolutely characteristic of him: he introduced what he himself referred to as a "grotesque exaggeration", a fictitious, over-simplified version of the experiment, designed to highlight the principal features of the real experiment as dazzlingly as possible.

"Suppose we have an immense hollow sphere before us. Much larger than the earth, much larger than the earth's orbit. So large that it takes a ray of light about two hours to cross it. An experimenter sits right at the center of the sphere. The sphere is at rest before us. Let the experimenter carry out the following experiment: he allows a very bright lamp to shine for an instant, and then waits for what he will see next. First of all he sees the lamp for an instant. Then darkness for two hours. For it takes the light an hour to travel out from the center to the inner wall of the sphere and another hour to come back to the experimenter

---

described to me by Professor A. D. Fokker. For the lecture itself see P. Ehrenfest, *Zur Krise der Lichtäther-Hypothese* (Leyden: IJdo, 1913; also, Berlin: Springer, 1913). It is reprinted in P. Ehrenfest, *Collected Scientific Papers*, ed. M. J. Klein (Amsterdam: North-Holland, 1959), p. 306. This collection is referred to throughout these notes as *Papers*.

2. The literature on relativity is immense. See, for example, W. Pauli, *The Theory of Relativity* (New York: Pergamon, 1958), and E. T. Whittaker, *A History of the Theories of Aether and Electricity*, 2 vols. (London: Thomas Nelson, 1951, 1953).

after reflection from this wall. And just then the experimenter sees the whole inner wall of the sphere simultaneously illuminated for an instant. Then darkness again. Now suppose we have a second such sphere, just like the first. And again an experimenter sits right at the center. This second sphere, however, is not at rest before us, but instead moves with an enormous velocity, with, for example, a tenth the velocity of light. The experimenter moves with it. Now let this second experimenter, just as the first did, allow a bright lamp to shine for an instant and likewise wait for what he sees next. We ask: does the experimenter in the moving sphere also see the whole inner surface of the sphere at one and the same instant, or does he see something else?"

Using this striking image of the experimenter flashing his lamp in the huge dark sphere hurtling through space, Ehrenfest could readily explain the characteristic differences among the various theories of light by the visibly different answers they provided to his question. In a series of carefully planned strokes, in the same short sharp sentences, working like a caricaturist who captures only the essentials of his subject, Ehrenfest sketched the development of the aether theory. If light is a wave motion in some medium, "the luminiferous aether", as interference and diffraction experiments seemed to prove conclusively early in the nineteenth century, then there were two distinct possibilities for the outcome of the experiment in the huge sphere. For the aether might be carried along by the sphere, as Stokes had assumed, in which case the moving experimenter would see precisely the same thing as his stationary colleague. But if the aether were always at rest, as assumed by Fresnel, so that the sphere moved freely through the aether, then the result would be very different. The propagation of light in the moving sphere would then be analogous to the propagation of water waves when a stone is dropped from a bridge into a rapidly moving river, since there would be an "aether-wind" blowing through the moving sphere at enormous velocity. In this case the moving experimenter, after two hours of darkness, should see "the equator of the sphere" (that is, the circle of greatest radius perpendicular to the direction of motion of the sphere) "light up, then two circles of latitude, symmetrically placed with respect to the equator light up. These illuminated latitudes move symmetrically toward the poles. Finally, the two poles of the sphere light up simultaneously, and then darkness again."

Which of the two suggested answers was one to believe? A real choice had become possible after Maxwell and Hertz established that light was an electromagnetic wave and that the aether was the seat of all electromagnetic phenomena. Lorentz had then shown, in the nineties, that only on the basis of Fresnel's stationary aether could one account exactly for a whole series of experimental results that were simply incompatible with Stokes's assumption of an aether carried along by moving objects. The choice was fixed, the answer was decisive: the experimenter in the moving sphere should see the result predicted from the Fresnel hypothesis. And therefore the crisis, because Michelson and Morley had really seen no trace of the effects of the mighty aether wind. The clash between this "pregnant null result" and the apparently inexorable prediction of the electromagnetic theory produced a variety of new ideas that Ehrenfest went on to sketch in another series of sharp strokes.

Even the least radical of these new ideas was startling enough. This was Lorentz's theory, fully developed in a memoir published in 1904, but first sketched by Lorentz and, independently, by that imaginative Irishman, G. F. Fitzgerald, over a decade earlier. Lorentz postulated that the aether wind affected more than the propagation of light, that it affected even the forces between molecules, and the shape of the basic particles themselves. The miraculous outcome of Lorentz's assumptions was the exact cancellation of these various effects of motion through the aether. In Ehrenfest's words, "The aether wind disturbs the result of the procedures that the moving experimenter employs, but the same aether wind also spoils – if we may put it that way – the experimenter's measuring instruments: it deforms the measuring rods, changes the movement of clocks and the restoring forces in spring balances, and so forth.... And if the experimenter now observes the processes distorted by the aether wind with his instruments that have been spoiled by the same aether wind, he sees exactly what the observer at rest observes for the undistorted processes with his unspoiled instruments". In the case of the moving sphere, Lorentz's theory would require that the aether wind produce a contraction of the radius of the sphere along the line of motion, leaving the equatorial radius undisturbed. The sphere would be flattened just enough so that poles and equator appear to be simultaneously illuminated, just as they were for the experimenter at rest.

From Lorentz's theory Ehrenfest turned to an even bolder solution of the paradox – Einstein's special theory of relativity – which removed the effects of the aether wind by abolishing the aether. In this theory light, and electromagnetic fields generally, were no longer to be considered as depending on the state of a hypothetical aether; light was to be viewed instead as an independent entity traveling through empty space. Einstein had, however, preserved one essential feature of the aether theories by postulating that the velocity of light is independent of the velocity of its source. Ehrenfest did not discuss the basic revision of the concepts of space and time that lay at the heart of Einstein's theory; he limited himself to pointing out that the results of the relativity theory were indistinguishable from those obtained by Lorentz, despite the fundamentally different logical structures of the two theories. Ehrenfest hinted at the novelties implied in Einstein's work as he remarked, with a characteristic touch, that Einstein really required his readers to subscribe to a third statement, in addition to the two postulates of the theory: "We must declare that the combination of these two assertions satisfies us!"

Ehrenfest closed his lecture with a brief description of still another theory that attempted to solve the riddle of the aether by abolishing it. This theory, proposed by his friend Walther Ritz in 1908, differed from Einstein's by making the velocity of light vary with the velocity of its source. Ritz's premature death had prevented the elaboration of his ideas, but his fragmentary theory had not yet been ruled out by experiment and it was still the focus of lively discussion. With a passing reference to the "tangled skein" of problems associated with the quantum theory, Ehrenfest ended his lecture, deliberately leaving his audience with the feeling that much exciting work remained to be done before the last word on the aether problem could be spoken.

Ehrenfest had given his audience much more than they had any right to expect. They had come to hear a professor lecture – they were caught up and captured by a man who taught as though his life depended on it. Over twenty years later some of them remembered the breathless suspense with which they had listened to his words. They remembered, too, the force with which they had been struck by Ehrenfest's personality, by the way in which he managed to throw all of himself into his lecturing. His vivid images and highly unprofessional language, his lively gestures and mobile face, all combined with the particular clarity and structure of his exposition to make a single unforgettable experience. It was something very new and thrilling, especially for the reserved, formal, academic society of the Dutch university, accustomed to the classic polish of Lorentz's lectures.

For Ehrenfest himself it was a great occasion. At the age of thirty-two, after more than a year of discouraging, unsuccessful searching for an academic post, he had suddenly succeeded to the chair of one of the world's foremost physicists. He had been offered a home by a country he had barely visited, whose language he did not know, at a time when, estranged from his native Vienna, he was unable, as a Jew, to follow his vocation in the Russia he loved. His heart was full as he spoke directly to Lorentz, who had made all this possible, in the words of greeting and gratitude that form the traditional coda of a university inaugural in Holland. "When one of us younger men comes in contact with you he must feel one thing above all else: you read our souls like an open book, smiling quietly. You see not only our scientific plans and thoughts spread out before you, with all their merits and defects which we ourselves will only recognize much later; you also see clearly and penetratingly into our purely human feelings, desires, and abilities. And so you also clearly see the tangle of conflicting feelings with which I receive from your hands this office which you desired to give up. But the thought that our joint professorial duties will give me the great privilege of a close personal connection with you fills me with undivided joy."

The chance to succeed Lorentz was a wonderful opportunity but it was also a heavy responsibility, and these inseparable aspects of his new position helped make the "tangle of conflicting feelings" to which Ehrenfest referred. It is never easy for a young man, even a brilliant young man, to replace a mature master of his field, still at the height of his powers; and when that young man is already torn by a lack of full confidence in himself, and when that master is one of the rare ones whose whole life appears to be a serene work of art, the problems may become enormous. But on that December day Paul Ehrenfest was happy, gloriously happy, as he had very rarely been before.

The happiness stayed with him through the following day, when he had to leave the heights of the podium and begin his regular teaching duties. The fifth of December is an odd date for starting a course of lectures in the Netherlands, since that evening is reserved for *Sinterklaas*: it is the eve of the feast of Saint Nicholas. Every man, woman, and child in the country turns poet to write the little verses that will accompany the anonymous gifts they send via the good saint from Spain and his fearsome Moorish helper that evening. Not even a new professor can expect much attention from his class that day. Fortunately Ehrenfest had been made aware of what he could expect, and so he announced to his class that he would not really lecture on his subject matter at all, but only talk to them informally about what the course would try to accomplish. He had not reckoned on the temperament of the Dutch student. He described the outcome in a letter that evening to Lorentz.[3]

"Even my very confused first lecture (it was a chemically pure scandal!) couldn't depress me for more than a little while. . . . What made me so thoroughly confused in the first lecture was this: I began with a few words saying I felt that since today was a holiday it wasn't appropriate to begin lecturing, and so I would therefore use this first hour just to indicate to them what I really wanted to do in the course. And now please imagine my alarm: even while I was saying *this* they were all already writing something (what???!!!) in their notebooks – in other words: instead of their eyes, which I had hoped to see, I saw the tops of about a dozen well barbered heads.

"There is no way I can describe to you how completely confused that made me. And it all grew worse, minute by minute. I became more and more confused and they wrote more and more passionately! After ten minutes I felt as though I were a poor little wounded hare who is desperately fleeing for safety from a dozen pursuing hounds – the more it runs the more vigorously the hounds run too. I don't know just when – but at a certain moment I said to them that I would perhaps try to prepare myself somehow to have notes taken on my next lecture, but that in this *first* hour, when I wanted to speak only very sketchily, I was totally unprepared for that.

"Now the Dutch are really good natured: they *all* immediately put down their pens and now I could finally talk to the people! Instead of twelve to fifteen haircuts (which, naturally, also have their own individuality!) I saw as many faces with eyes. But now the whole lecture was already lost – and I had wanted to show them such a very neat thing – the method for making a continuous transition between two problems that are discretely separated to begin with. I am satisfied with only one single effect – at the end of the hour I saw nothing but gay, 'warm' faces. But they couldn't have understood anything at all, except that I had shown them a few textbooks whose titles they very neatly copied down.

"After a few blunders things will go all right – if they aren't already so terrified today that half of them stay away. I have only one hope:

3. P. Ehrenfest to H. A. Lorentz, 5 December 1912. This incident was also described to me by Dr. J. E. de Vos van Steenwijk who was a student in the class.

it will hurt them too much to let their already spoiled notebooks go unused. *That* will save me."

This comic incident surely underlined the need for what Ehrenfest wanted most – a direct, human contact between himself and the students, and, for that matter, among the students themselves. He had declared his views to them in the coda to his inaugural: "I understand my duties to you in the following way: to the best of my knowledge and ability I shall help each of you to find, with the least possible damage, the way that corresponds to the very essence of his talents. The systematic lectures that I am obliged to give you on the various disciplines and on the individual problems of theoretical physics are a necessary but certainly not a sufficient means for approaching this goal. It is absolutely necessary, in addition, that I make personal contact with individuals. I ask you to look on me as an older colleague in your studies, and not as a person who is at a discontinuously different stage of scientific development." He meant every word of it: the rest of his life would bear witness to the depths from which he spoke.

The customs and traditions of a Dutch university formed a sharp contrast to the ideal that Ehrenfest had in mind, the ideal that he had conceived from his own experiences, positive and negative, in Vienna, Göttingen, and Petersburg. Contacts in Leyden's carefully stratified academic society were at a minimum. The students resident in the small city were divided into noninteracting groups, foremost among them the elite of the Student Corps, "with their noses high and their pockets full of guilders".[4] The many commuters, from the nearby big cities of The Hague and Rotterdam which had no universities of their own, often spent a minimum of time in the university atmosphere. The professors were something of a caste apart, perhaps to be visited on a Sunday afternoon, suitably clad in top hat and striped pants. The University Library was first class, but it offered no reading room set aside for students in the sciences where they could find the books, and especially the journals, that they needed, and where students of the various related disciplines could meet and become acquainted in surroundings conducive to intellectual contact. There was no regular colloquium that could bring the students, research men and professors of physics together frequently to hear a report on new research from

4. These descriptions are based on personal communications to me from Professor D. J. Struik and Professor A. D. Fokker, and on Professor Struik's letter to the Christiaan Huygens Society in Leyden, 22 March 1963.

one of their own group or a visitor from another university. In short, there was no real community of physicists bound together by intellectual and human ties, and Ehrenfest set about to create one.

Within weeks of his arrival in Leyden in October 1912, Ehrenfest made his presence felt in many quarters. Even before his inaugural lecture he instituted the Wednesday evening colloquia that were to become a fixed point in the calendar of Dutch physicists, and were to have an influence on the style of theoretical physics throughout the world. He soon discovered that there was an old society of students in the physical sciences, appropriately named the "Christiaan Huygens", after the Netherlands' great seventeenth century physicist, but this society was no longer very active. Ehrenfest did all he could to revive this organization, seeing it as a focal point where real contact could be made among the science students of various ages, social strata and professional interests. Their fortnightly meetings soon attracted commuters like Dirk Struik, the mathematics student from Rotterdam, Hans Kramers, the brillant and aristocratic student of theoretical physics whom Ehrenfest persuaded to leave the Student Corps for "Huygens", and many others, students of mathematics, physics, chemistry and astronomy. The meetings were held in the rooms of the various members, and they always featured a lecture of some length and depth, as well as lively discussion, shorter talks, and refreshments of tea, coffee and accompanying sweets. Ehrenfest was often present, joining in the discussions, communicating his love for science by example, and his opposition to tobacco and alcohol by precept as well.

Another of Ehrenfest's plans for quickening Leyden's scientific life, especially among the students, was more ambitious, and it required assistance. He felt that there had to be a reading room for the physical sciences. He had a very clear and concrete example of what he had in mind – just such a room existed in Göttingen, established by that university's great professor of mathematics, Felix Klein. Paul Ehrenfest had vivid memories of that room from the years he had spent in Göttingen, and they were fond memories too, for it was in that reading room and at Klein's lectures that he first caught sight of the young Russian mathematician, Tatyana Alexeyevna Afanassjewa, whom he had then married in 1904. The reading room that he wanted must contain the principal texts and treatises, the collected papers of the old and most recent masters of the field, and the important journals; and – most important – all these materials must always be available in that room. Ehrenfest complained that it happened too often in libraries that just the books or journals one needed were always out. The room might be as small or modest as necessary, but it must be a comfortable place to work and read, and it must be open to all students.

Establishing such a room took funds, as well as the good will and cooperation of the University Library, which would have to be willing to transfer some of its books there, on permanent reserve as it were. In this matter, as in all others in which he felt the need for help and guidance, Ehrenfest turned to Lorentz. (Lorentz had given up his professorship, but at fifty-nine, he had by no means retired. He lived close by in Haarlem, as curator of Teyler's Institute, and came regularly to Leyden to deliver his Monday morning lectures as an Adjunct or Special Professor.) Lorentz saw the logic of Ehrenfest's arguments and felt the passion with which his young successor advocated this cause. Through Lorentz's quiet assistance, the reading room was financed and came into being in May 1913, as the "Leeskamer Bosscha", the Bosscha Reading Room. It was named after the recently deceased Dr. Johannes Bosscha (former head of the Institute of Technology at Delft, an older friend and colleague of Lorentz's) whose son donated the necessary funds. "Bosscha", as the reading room was familiarly known, became a major influence on succeeding generations of students at Leyden, and a model for similar arrangements elsewhere.[5]

Ehrenfest gave two courses of lectures in his first year at Leyden. He had begun to plan them in Petersburg, before moving to the Netherlands with his wife and two young daughters. Early in October he wrote Lorentz that he had definite ideas on how to handle the larger of the two courses, on the differential equations of mathematical physics, given to students who were just beginning to work in theoretical physics.[6] "In this course one ought to learn how one can completely permeate the *typical* mathematical apparatus of theoretical physics courses with mechanical and physical intuition, doing this with a severely limited subject matter, but one that is discussed from as many viewpoints as possible." He had done something of the sort in Petersburg, in a course based on the three key ideas of vibrations, waves, and potential fields. "I laid out the course so that, as far as it was at all possible, the mathe-

---

5. See G. L. de Haas-Lorentz, "Reminiscences", in *H. A. Lorentz. Impressions of his Life and Work*, ed. G. L. de Haas-Lorentz (Amsterdam: North-Holland, 1957) pp. 99, 106. Personal communications to the author from Mrs. T. Ehrenfest-Afanassjewa and Mrs. G. L. de Haas-Lorentz. P. Ehrenfest to H. A. Lorentz, 5 December 1912.
6. P. Ehrenfest to H. A. Lorentz, 5 October 1912.

matical operations appeared simply as the more exact form of relationships inferred intuitively". The plan was to proceed step by step from the most trivial special cases to more and more general ones. "What mattered most to me all the way through was how one transforms physical problems into mathematical form, and how one reads the formulas, how one gets a direct feeling for transformations. The really mathematical viewpoint, on the other hand, 'the pleasure in integrating for its own sake', naturally just remained in the background." The notes for the lectures he gave at Leyden that first year show that Ehrenfest followed the plan he had sketched out, adapting it to the needs of his class as they became evident.[7]

He began with the simple oscillator, studying its behavior under various sorts of external forces, making that behavior as vivid and intuitive as possible with the aid of sketches and graphs. After exploring the many, diverse, physical applications of this work he went on to coupled oscillators, gradually developing the theory of normal vibrations, and principal axis transformations in an n dimensional space. The logical next step was the transition to the continuum, to the wave equation and its associated boundary value problems. He emphasized how the discrete theory and the continuous theory threw light on each other, stressing the parallels and structural connections between the principal value problem for matrices and that for differential equations. His students got a kind of insight into the ideas of mathematical physics that was far ahead of its time. Much of what he did was his own reworking and vitalization of the ideas that he had absorbed from David Hilbert in Göttingen.

The second course was a very different sort of thing. This one was aimed at the more advanced group preparing to work at the Kamerlingh Onnes Laboratory, the foremost laboratory in the world for low temperature physics. Heike Kamerlingh Onnes' organizational genius had built a prototype for future large scale research laboratories, based on major equipment on a scale considerably beyond the sealing wax and string of the usual researcher's bench. His staff of glass blowers and machinists was self-perpetuating, since he had incorporated a school for such indispensable, highly skilled technicians into his installation. In 1912 Kamerlingh Onnes, the benevolent despot of this domain, was at the peak of his form. He had just made two major

7. The course was described to me by Professor D. J. Struik. Lecture notes are part of the Ehrenfest collection at the Museum for the History of Science in Leyden.

discoveries that would help set the principal problems of low temperature physics for the next half century: in 1908 he liquefied helium, most recalcitrant of gases, for the first time, and in 1911 he discovered superconductivity, the vanishing of the electrical resistance of certain metals below a definite characteristic temperature. Low temperature physics was full of excitement and promise at just this time, as Walther Nernst's heat theorem and his measurements of specific heats at low temperatures were drawn into closer relationship with the new quantum theory.

Ehrenfest decided to lecture to this group, once a week, on these new turns that theoretical physics had taken, emphasizing the importance of the questions posed by low temperature physics for the future development of basic concepts. His subject matter was very much up to the minute, and so this course had a much less unified character than his other, which was a tightly organized development of a solidly established field.

"There are thirty people at my weekly lecture on the Nernst theorem, (and at my request *all* the Kamerlingh Onnes people have stayed away.) Not one has left yet. Some old people are there too. As you can imagine," Ehrenfest wrote to a close friend,[8] "I really make it completely clear in principle but, naturally, it is absolutely schematized. Not a report on what actually is, but rather on the concepts with which Messrs. A, B, C, ... operate. But what to do? I *can* do this and I can't do the other."

Not all of Ehrenfest's energy went into his teaching, formal and informal. Physics itself – his own drive to understand the questions at issue in his beloved science – made its demands. For a decade Ehrenfest had been struggling to grasp the full significance of what Max Planck had done in 1900 when he introduced energy quanta into physics.[9] Ehrenfest had been among the first to sense that weighty issues were involved, among the first to try to disentangle the implicit assumptions, not only in Planck's work but also in Einstein's much more radical ideas on quanta. The year before he came to Leyden, Ehrenfest had made a major step in this direction when he clarified the constraints enforced upon the quantum theory by the second law of thermodynamics. That work looked as though it could provide the clue for which many were searching, the clue that would permit the quantization of systems other than the simple harmonic oscillator. Ehrenfest

8. P. Ehrenfest to A. F. Joffe, 20 February 1913.
9. See Chapter 10.

had been chewing on this question on and off for a long time, and he came back to it that first winter in Leyden.

During the evening of December twentieth and through the next day he pursued a new connection that he had glimpsed. He saw a relationship between this perplexing problem of the quantum theory and a theorem he had studied in various other contexts years before. Ehrenfest found in this theorem just the approach he needed for extending his 1911 ideas. He had raised as many questions as he answered, but this would be the focus for his greatest constructive work during the next few years. It was the kernel of what Einstein soon named the Ehrenfest Adiabatic Hypothesis, a central theme in the development of the quantum theory for the next decade.[10]

Ehrenfest's idea was not yet worked out that December, but he sensed enough of its importance to describe what he had done in some detail in an important letter to Lorentz[11] written just a day after he had been struck by his new insight. This letter contains a good deal more than the report of Ehrenfest's idea: it reveals his state of mind. And that state of mind is not at all what one would expect from the surface appearance of things. Just two months after his arrival in Leyden, only two weeks after the triumphant success of his inaugural lecture, with his courses going well, the colloquium already in full swing and his plans for the reading room well received by Lorentz, in the very act of reporting that he had found a highly promising line to follow in his research – Paul Ehrenfest was in a mood of despair and self-distrust.

He began the letter in a more formal style than he customarily took in writing to Lorentz. "Please permit me to communicate to you, on the enclosed pages, a little matter concerning the quantum hypothesis that I discovered yesterday, after searching for it in vain for a long time. I should be very grateful to you if you would be willing to make any comments on it, *at your convenience*, (when we meet again)." The several pages that follow report Ehrenfest's work in a remarkably clear, categorical form, as a series of questions, answers and remarks, carefully numbered, with the principal equations set off in boxes and underlined in red. Ehrenfest enumerated the difficulties that he could see confronting him in trying to go further with his idea, (non-trivial difficulties indeed, that took a good many years for their full resolution), and remarked that he had omitted derivations "because a lengthy series

10. See Chapter 11.
11. P. Ehrenfest to H. A. Lorentz, 23 December 1912. See Figure 10.

of complicated formulas looks very repugnant in a letter." Up to this point in the letter Ehrenfest's handwriting, always clear when he was not hurried, is positively meticulous, and the pages are calligraphically elegant. This style persists through one more paragraph as Ehrenfest writes that he will presently send Lorentz a list of the books that he considers desirable for starting off the new reading room.

Suddenly he bursts out in a new vein, writing freely and quickly on a matter close to his heart. "Please have patience with me – with my blunders, my frittering and squandering of my energy. Give me two or three years to decide whether I ought to retain this professorship, or whether I must ask you to give it to someone else. Forgive me! – it was *absolutely* necessary for me to say this to you. Because you ought to be able to keep two things always clearly separate from each other: your interest in my *personal* development and your concern for the future fate of the professorial chair here.... I ask you, *most urgently*, not to answer these last remarks, neither verbally nor in writing. In two or three years I will want to know what you think about the matter *then. But not now*. Only one thing was necessary for me now: to be able to say to you frankly that I look upon these years as being completely a *probationary period* (for myself!). Apart from my wife (and perhaps my two closest friends) it was only you to whom I was obliged to say this. ... I must be able to speak freely to you about whatever interests me, fascinates me, gives me pleasure, and oppresses me at any given time, without having to think with every word that it would give you a cause for concern for the future of the professorship here. Now that I've said this *once and for all*, I will not stop straining to do my best, but I won't feel I have to hide anything from you.

"It had gone so far with me that I felt I couldn't send you the list for the reading room without sending something new in physics (no matter how small). Such a situation would be completely unbearable to me, especially since you would immediately see through the whole deceit. ...You can't imagine how relieved I feel now that I've said everything. Perhaps I should now rewrite the first part of this letter – because there is a great hypocrisy hidden in it."

During the following week, Ehrenfest described his situation and his feelings at greater length and in more detail to one of those "closest friends", Abram Fedorovitch Joffe, in Petersburg.[12]

"Lectures, lectures, lectures – you know how I prepare for them. I

12. P. Ehrenfest to A. F. Joffe, undated.

know nothing clearly enough to be able to teach it – everything must be relearned, recalculated, etc. Discussions with students. Wednesday colloquium. The horribly bad French translation of the Encyclopedia article must be put in shape. A number of reviews for the *Physikalische Zeitschrift* are due. Lectures, lectures, lectures.

"In short I *work* on nothing, read hardly anything... I work on nothing, I am stagnating, while all the others – Einstein, Debye, Laue, you, and all the young fellows go forward, forward.

"Hysterical despair – I know well that it is hysterical, just as you know it. But *you* aren't here to haul me out of all stagnation points with your imperatives and your (seductive) guidance. Tanya does it as best she can. But though I clearly perceive that I have *ridiculously* little to do compared to *everyone* else here and everywhere else, I nevertheless idle away everything, except the lectures. Those at least I carry out *absolutely* conscientiously, if not also satisfactorily.

"Where is the worm? Naturally in the feeling that Lorentz should have taken Debye as his successor and *not* me. And I feel that Lorentz already perceives it. And I know that even you dare not say absolutely anything against it. Thank God that at least I have spoken out quite openly about it with Debye and Lorentz".

He told Joffe about his recent letter to Lorentz and the request it contained, and went on to add: "And see: I seriously want to strive to pull myself together and get to work. But for someone like me that's just the worst background for arriving at good ideas. For everything that I've managed to get out at all has come from the impulse to play, from interest in some paradox, and not from striving to accomplish something 'significant'.

"It is quite possible that this appointment will completely smother me. And don't misunderstand me: I don't have the mad idea that I should be providing an imitation of Lorentz. I know very well: Leyden is a little university in Holland – to which, by chance, Lorentz came as a high school teacher, and then just stayed. His successor had to be a really first rate young fellow; they couldn't get one to come here, and so they had to take a second rater, and I know that I am really in the forefront of the second raters right now. But, but, but – but I know precisely that you understand me *very* well".

When Ehrenfest wrote to Joffe again in February, his spirits had risen.[13] He was full of plans for the meeting he and Joffe were planning

13. P. Ehrenfest to A. F. Joffe, 20 February 1913.

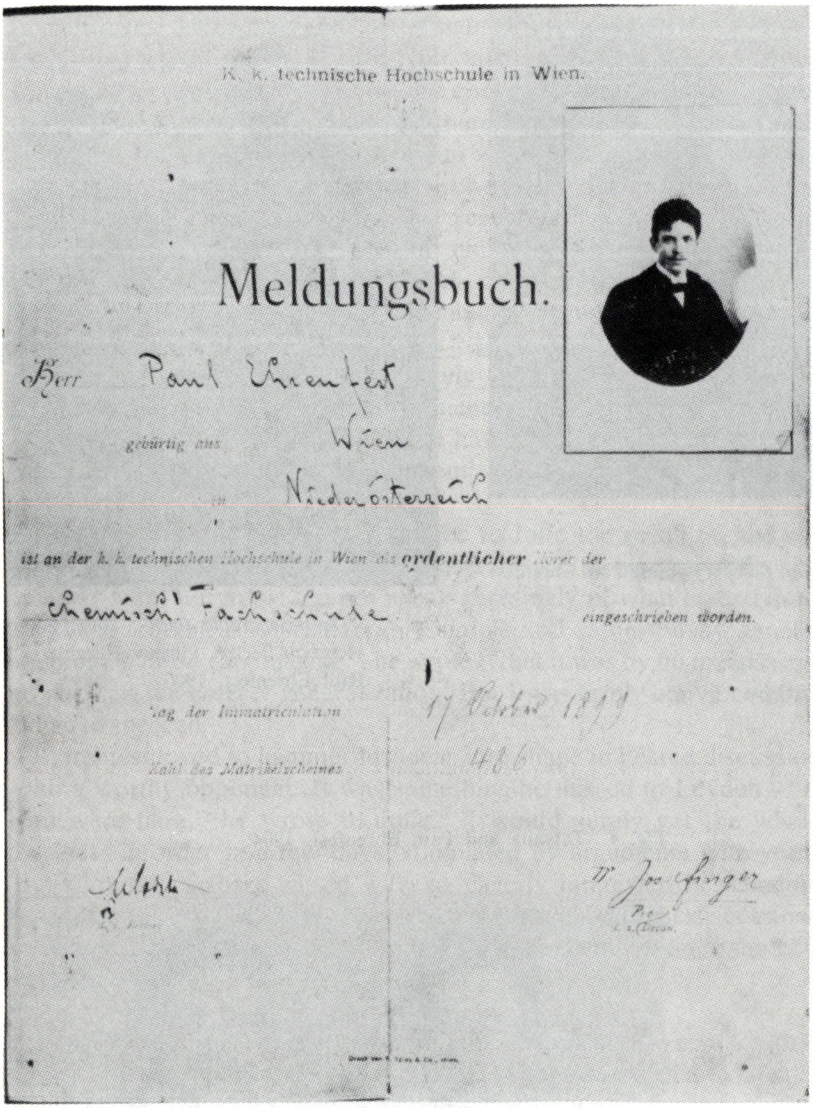

K. k. technische Hochschule in Wien.

Meldungsbuch.

Herr Paul Ehrenfest

gebürtig aus Wien

in Nieder-Österreich

ist an der k. k. technischen Hochschule in Wien als **ordentlicher** Hörer der

Chemisch. Fach-Schule eingeschrieben worden.

Tag der Immatriculation 17 October 1899

Zahl des Matrikelscheines 486

Fig. 2 Ehrenfest's registration book at the Technische Hochschule in Vienna, 1899.

CHAPTER 2

Childhood in Vienna

Paul Ehrenfest was born in Vienna on January 18, 1880. His parents had moved to the capital about twenty years earlier from their home in Loschwitz, a little Jewish village in Moravia.[1] Their move, shortly after their marriage, took place some years before the great influx of Jews into Vienna from Galicia and other regions to the east. Sigmund Ehrenfest, Paul's father, came from a family with no financial means at all, and he had been a poor worker in a weaving mill when he married Johanna Jellinek, the daughter of a merchant in the same village. Her small dowry probably provided the funds for their move and for the beginnings of the grocery business that Sigmund established after they were settled in Vienna's Tenth District, the Favoriten. The reasons for the young couple's choice of the Favoriten rather than the Leopoldstadt, Vienna's Second District, are unknown. The Leopoldstadt, on the north side of the city, had been the home of Vienna's Jewish population since the early seventeenth century, at least. The Favoriten was a

1. Practically all the material given in this chapter about Ehrenfest's childhood years and his family comes from his own detailed reminiscences, written in the fall of 1932 and in the possession of Professor H. B. G. Casimir. (This may account for the somewhat idyllic tone.) I also received information from Mrs. T. Ehrenfest-Afanassjewa and from Ehrenfest's nephew, Fritz Ehrenfest-Egger.

Useful background information was found in the following works:

(a) K. Baedeker, *Austria, Including Hungary, Transylvania, Dalmatia and Bosnia. Handbook for Travellers*, Ninth Edition (Leipzig: Karl Baedeker, 1900).

(b) J. Meurer, *A Handy Illustrated Guide to Vienna and its Environs* (Vienna: A. Hartleben, 1891).

predominantly proletarian quarter a couple of miles south of the tower of St. Stephen's Cathedral which marks the center of the inner city, and it had relatively few Jewish residents.

By the time that Paul, the youngest of their children, was born, the Ehrenfests were fairly well off. Sigmund had built his grocery business into a thriving concern by intensely hard work, side by side with his wife, and by his own business acumen. Although not an educated man, he seems to have had a clear strong mind, considerable imagination, and both the will to drive himself and the temperament to take pleasure in hard work. Most of the inhabitants of the Favoriten were poor working class families of Czech, Yugoslav, or Italian origin, many of them employed in the factories and brick works of the quarter. Sigmund built his business success on the then novel idea of specializing in a small number of staple foods and household items carefully chosen to match the needs of his working class customers. He would stock these products – sugar, coffee, flour, dried peas, rice, eggs, oil, soap, brushes, and so on – carrying only one or at most two varieties of each item, buying in large quantity directly from the producers. Some of his trade also came from the smaller number of Viennese German families living in the Favoriten – minor officials, artisans, and merchants. There was even some business from middle class families of the inner city who found it worth making the trip to this outlying section in order to save money by shopping at the lower prices outside the customs frontier of the city proper, particularly since the shop assistants or Sigmund Ehrenfest's older sons would help transport the bundles.

The Ehrenfest family had four older sons, and had lost a daughter at birth. Arthur had been born in 1862, Emil in 1865, Hugo in 1870, and Otto in 1872. The family lived at Number 17 Himbergerstrasse, a principal street of the Favoriten that carried a tram line running into the city. They occupied one of the few low old houses left on a street that had been built up after 1875 with higher buildings that Paul was to remember as being intensely ugly. The house was a double one with two shops at the street level and two apartments over them. The second shop was occupied by a sausage butcher named Weninger, and this German Catholic family lived above their shop just as the Ehrenfests did. The apartment was not spacious, but in addition to the seven Ehrenfests it accommodated a nursemaid, two maid servants, two shop girls, and four male shop assistants. It was a densely packed world of varied and interwoven lives into which Paul was born, and it set his picture of life for many years to come.

As the youngest by eight years, born when his father was forty-two and his mother about thirty-eight, Paul was very definitely the baby of the family. He became his father's particular favorite, but his mother seems to have exhausted much of her patience for little boys with his four older brothers. Paul remembered many good, quiet times alone with his father. When the first warm spring days appeared, Sigmund Ehrenfest managed to make some free time to take his little son out for a walk, something of an achievement since he was usually busy from six in the morning till ten at night. They would walk out away from town to a "meadow", that is, some empty lots not yet built up, where there were still a few tufts of grass. Father would carefully pick a spot free of broken bricks, glass, or tiles, and Paul could play in the sunshine. At other times the little boy could sit on his father's lap, during a lull in business, on a stool near one of the shop's display windows. One had to look between the wares on display to see the street life beyond, but it was worth it. There was a heavy glass shelf crossing the window above the height of Paul's eyes, so that he had an inexhaustible comic spectacle. Every pair of legs that went by was surmounted – not by a body – but by another pair of legs walking along upside down. The occasional dog that passed added to the fun.

Sigmund Ehrenfest suffered from stomach trouble, often aggravated by his excitable disposition. Some years it was bad enough for him to go off to Karlsbad to take the waters for a few weeks. (A row of dated Karlsbad drinking glasses at home bore witness to these occasions.) This stomach trouble, eventually an ulcer that caused his death in 1896, led to Sigmund's being on a special diet, and taking his meals apart from the rest of the family. The only companion that he chose for these separate meals was his youngest boy, and Paul was generally favored with choice bits of the special foods prepared for his father. Sigmund would also save some of the special pastries baked on Sundays only, and keep a supply of them in his wardrobe to be given secretly to Paul during the week. They smelled of camphor and were dried up, but it was still wonderful to be so uniquely privileged.

Father was the one who would generally yield to Paul if he begged long enough and hard enough to be allowed to do something, and the one who would comfort him if things went hopelessly wrong. One spring day the family and their shop assistants were all out in the yard selecting eggs to be preserved in lime water for that winter's trade. There were thousands of eggs in crates all over the yard, and each egg had to be inspected carefully for cracks and candled in the sunlight.

Paul begged to help, but everyone was against it. He would surely drop the eggs or smash them somehow. Finally Father gave in to his entreaties, against the better judgment of everyone else. Paul was very successful at first, being very careful with the eggs, and Father confirmed that he was sorting them correctly. But somehow the predicted disaster occurred: he toppled backward off his stool right into a crate of eggs. Sigmund Ehrenfest picked up his sticky, yellow and very unhappy little boy and carried him away from the clamor and consternation to wash and console him.

Paul's mother does not appear so often in his memories of his childhood. She died when he was only ten, and it is fair to guess that the shock of her loss, combined perhaps with his complex feelings about having been less close to her than to his father, accounts for this. He remembered her as having an extraordinarily strict sense of duty combined with an impulsiveness and irritability that often led to quick anger. One surmises that it was she who administered the punishments for his childhood escapades.

Her mother, Grandmother Sophie Jellinek, was the only one of his grandparents whom Paul knew. She had been widowed just a year before Paul was born, and was then living in the Favoriten too, on Raaberbahngasse, not far from the Ehrenfests. She had an apartment there with her two unmarried daughters, Mathilde and Josephine, and an unmarried son, Fritz, all much younger than Johanna Ehrenfest. The Jellineks and the Friedmanns, for Grandmother Sophie had been a Friedmann before her marriage, made up a large clan, and they would gather on Sunday afternoons and evenings at the apartment on Raaberbahngasse. Grandmother Sophie held them all together, keeping the connections among the various branches of the family, some still working people, others much more bourgeois. She had herself risen from extreme poverty to a reasonable prosperity as her merchant husband's fortunes had improved, though she was to lose it again and become dependent on her daughters before her death in 1910. Sophie Jellinek was a real Jewish grandmother, unselfishly devoting herself to anyone in the family who needed her, loving all of her own, but loving each in a special way so that the bonds of affection remained strong. She tried to adhere strictly to her religion, rigorously keeping the dietary laws, and attempting to influence the others to keep up their Judaism. Her humor never failed her, even in the midst of suffering.

Paul remembered the Sundays at Grandmother's as crowded, warm occasions, pleasurable for all in various ways. Grandmother, knowing his love for anything mechanical, kept a box of old broken clocks of various sorts for him to "fix", and every Sunday he would rapturously take them all apart and put them together again, though never the same way. He was always sad when it was time to leave these beloved brass clocks. Sometimes, though, there were more exciting pastimes that drew him away from his clock repair. His oldest brother Arthur would organize games for the children – riddles, guessing games, or a lottery with mock prizes. There was always music of one sort or another: many in the family were musically gifted, and especially Uncle Fritz Jellinek who ended up as a night club pianist. On one occasion Arthur and Fritz, both in their early twenties, arranged a hilarious parody of the Jewish marriage ceremony, complete with music and costumes. Paul's young Aunt Josephine, always known as Tante Pepperl, was a high spirited girl who took part in these affairs, doing remarkable impersonations of all the members of the family.

Paul's mother spent much of her time in the shop, where she generally acted as cashier. She sat in the little glass-enclosed cashier's box, keeping her eye on the operations of the business. Paul liked to play there because of all the wonderful things it offered – a telephone, connecting to the apartment upstairs, the little presses for stamping the firm's trademark onto business papers, a magnifying glass for checking forgeries in the bills. And it was also so nice to sit there quietly, half forgotten, and be aware of all the conversations that went on while he played.

Saturday evening was the shop's busiest time, for it was then that the working class wives came, often with their children, to make their purchases. The shop would be crowded, with lines waiting outside for room to come in and buy. Sigmund Ehrenfest and his older sons worked with their half dozen shop helpers serving the customers, marking the purchases on long, narrow slips of paper. It was an exciting time, under the bright, flickering open gas lamps. The work was hard and hurried, but the atmosphere was happy and there was much laughter. One spoke to most of the customers in Slavic, and, over them, to the other Ehrenfests, in German or Yiddish. Mother, in the cashier's box, received payments, and occasionally extended credit. She had a wonderful memory and knew almost all of the hundreds of regular customers by name. Paul was allowed to help too in little ways, so long as he did not get under foot and interfere with the others. He could help even more at quieter times in the shop, when the middle class trade came. The women liked the polite, inquisitive, attentive little boy who so conscientiously tried to select the best eggs for them.

One of the traits that characterized Paul Ehrenfest in his maturity was the joy that he found in his work, a joy and enthusiasm that he radiated and could impart to others. He began to learn this attitude toward work, insofar as such things are really learned, early in his life simply by absorbing the atmosphere that surrounded him at home, where even heavy manual labor was sometimes given the tone of a game. Just that spirit was in the air every time the Ehrenfest grocery received its supply of sugar from the refinery.

Sometimes it was little Paul who first spotted the big freight wagon as it approached, rumbling and rattling, drawn slowly by two stout stallions. The wagon was packed with large, very heavy "loaves" of sugar, well wrapped in blue or black paper. While Sigmund Ehrenfest, excitedly summoned by his youngest son, checked the invoices with the driver, the rest of the men and boys of the establishment gathered around, full of joyful impatience. They formed a human chain from the wagon through the shop to the store-room in the rear, covered by a steep zinc roof. When the papers were found to be in order, and the driver had put the horses' feedbags on, the unloading began – as a rhythmical work-game. The loaves were tossed by the driver to brother Emil in the rear of the wagon, from Emil to Wagner, (one of the clerks), from Wagner to Martin, from Martin to brother Arthur, and so from man to man back to the store-room for stacking. As the loaves flew along they warmed to their work, keeping the rhythm going, so that after two or three hundred heavy loaves had gone in they would boast that they had just gotten into the swing of it and ask when the real work was going to begin!

Sigmund made sure that his little boy was safely out of the path of the flying loaves near the store entrance, stationed there with paper and pencil for his most important share in the job. For Paul was the checker, making a mark on his paper as each sugar loaf sped by, a crucial task. And how gratifying, how reassuring, for Paul to know that his tally agreed exactly with the one that Father made as he stood near the wagon.

The sugar loaves supplied something besides sugar. Sigmund and Johanna Ehrenfest ran their home in a very thrifty style, and writing paper for the children was a luxury they did not indulge in. But under the heavy blue or black wrapping around the loaves was another layer of coarse white paper. This was carefully gathered, folded and cut and then kept to supply the need for paper to write on. Of course it was not much good for this purpose since one could not write on it in ink at all and even the use of a pencil required special care or else the paper would tear at once. Even this paper was not usually available for the youngest boy, who generally got only the very wrinkled pieces that had been around the point of the loaf, which was shaped like an artillery shell. This badly creased paper was a perpetual annoyance, and the regular pieces of the white sugar loaf paper seemed like an unattainable ideal.

As a young child Paul was taken care of by a nursemaid while his mother was occupied in the shop. He was deeply attached to his Nurse Minna, who gave him some of the warm attention that he needed. She often sat quietly by, sewing or knitting, while Paul played. Sometimes she sang Czech songs to herself, or listened lovingly to the results of the little boy's active imagination, conversing with him in her mixture of Czech and German. One of her moving folk songs gave Paul his first encounter with the emotional impact of music as it brought tears to his eyes. Minna loved house plants and cultivated many of them in pots placed near the hall windows. Paul, who liked to watch these grow and observe the new flowers, was inspired to try a little "garden" of his own in the yard, but the dusty clay soil full of bits of tile did not produce anything. He planted beans, watered them incessantly, often looked for roots, but nothing developed.

Paul was a rather sickly child, thin and nervous, subject to heavy nosebleeds and dizzy spells. On every excursion or walk someone always took along wet cloths and a sponge as precautions against the occurrence of one of his nosebleeds; even so grand an occasion as a trip by cab to the Prater, Vienna's famous park and forest, was spoiled by such a nosebleed on the way. If he was sick enough to have to stay in bed, his crib would be moved from its normal nighttime place in his parents' bedroom into the "good parlor" near the grand piano. There he would be, listening to the busy life of the Himbergerstrasse outside while he saw only the still furniture of the quiet room.

His illnesses do not seem to have been major ones, and he did a good deal of playing outdoors. His most frequent playmate was Bertha Weninger, the youngest child of the Weninger family who lived next door, a girl of about his own age. He also played with the Koch boys, the sons of the Ehrenfests' family doctor, who was also a relative, at such things as sailing paper boats in the gutter carrying the run-off from their well.

Bicycles were still new in the eighties, but Paul's oldest brothers, Arthur and Emil, were among Vienna's early cyclists, belonging to the city's first bicycle club. Their cycle had a high front wheel with a very small rear wheel and was as exciting to its owners and their audience as a private plane fifty years later. To Paul it was an object of wonder and delight, a center for his day dreams. The smell of its oil lamps and the magic of their red and green glass reflectors were right out of the Arabian Nights as far as he was concerned. And when he was promised a tricycle of his own, his fantasy took new flights. He would lie in the storeroom dreaming of his bicycle and of the travels he planned with it. No journey seemed too long, though he thought the mountains, and especially the oceans, would offer some obstacles.

When the great day finally came and Arthur, Paul and Mother took the horse-drawn tram into the inner city to buy his long desired tricycle, the reality did not measure up to his dreams. The little poorly made tricycle lacked the elegance and finish of his brothers' wheel, and it had neither lamps nor bell. But it was still a great thing to have his own cycle and to be free to roam the world on it, though that world turned out to be a great deal larger than the range of his tricycle.

The Jews did not have an easy time in the Vienna of the late nineteenth century.[2] It is true that the laws controlling and limiting Jewish life had been greatly liberalized during the previous decades, so that from the eighteen seventies on Jews began to enter and then to dominate Vienna's intellectual and cultural life, and their position in commerce and finance also grew to one of great strength. The old prejudices did not disappear with this success, and the seventies also saw the beginning of organized anti-Semitism, played upon deliberately as a political weapon. The financial crash of 1873 was blamed on the Jews, who were accused of producing it for their own benefit. Both those Jews in positions of power, – in finance and in the press particularly, – and those with none – the steady stream of arrivals from the ghettos of Galicia – were the targets of constant and vicious abuse. This abuse was encouraged by some of the lower Catholic clergy, with a certain Father Sebastian Brunner earning his title as the father of modern Austrian anti-Semitism by writing savage pamphlets on this

2. See A. J. May, *The Hapsburg Monarchy 1867–1914* (Cambridge, Mass.: Harvard University Press, 1951), Chapters 8, 13, Also H. W. Steed, *The Hapsburg Monarchy*, Third Edition (London: Constable, 1914), Chapter 3.

theme. The Christian Socialist Party made anti-Jewish agitation one of its chief political tactics and used it with increasing success, particularly in Vienna under the leadership of Karl Lueger. Lueger was first elected to the Austrian parliament in 1885 on a platform having anti-Semitism as one of its principal planks, and his power in Vienna grew steadily until he became its mayor and "uncrowned king" after a series of stormy elections in 1897. (It may be noted without surprise that Lueger was praised as "the greatest German mayor of all times" by another Austrian, Adolf Hitler.[3])

To a Jewish boy in the predominantly non-Jewish quarter of the Favoriten, anti-Semitism was not a political platform, it was an ugly fact. A child whose looks proclaimed his Jewishness had only to walk down the street to draw vicious shouts, or worse, from the other boys. Respectable adults would readily add their abuse if at a fruit seller's, for example, Paul hesitated, looking from one fruit to another instead of meekly accepting the worst ones pushed in front of him. No wonder that his loving father strictly forbade him to go out on the streets alone before he was about seven years old. Even worse, perhaps, was the fact that his closest non-Jewish friends, his most faithful playmates, like little Bertha Weninger, would remember that he was only a "dirty Jewish pig" whenever there was the slightest disagreement or argument over anything. Paul would experience anti-Semitism in other ways later on, but these early encounters with it in a crude, if non-violent, form left a scar that never healed.

Religion played a large but ambiguous part in his childhood. His parents did not adhere to the ritual or to the requirements of Judaism, though Paul saw these practiced by Grandmother Sophie and in the homes of some of his friends. He attended a Passover Seder in one of these orthodox homes and was deeply impressed by its ceremonies, songs and stories. He liked to read in a German translation of the Old Testament for its stories. Paul had a certain amount of formal instruction in Hebrew and in Judaism on Wednesday afternoons after school. He went to the home of an old man named Strassny for these lessons and was impressed by two things in particular. One was linguistic – the fun of writing German words in the Hebrew script, as is commonly done when Yiddish is printed. The other was his old teacher's passionate and stubborn emphasis on the survival power of the Jews. The

3. See May, *op. cit.*, p. 311.

Egyptians, Babylonians, Persians, Greeks, and so forth, had all been great in their day, but who heard of them now?

The contrast between the precepts of the Jewish religion and their nonobservance at home upset him. Why didn't his parents keep a kosher home? Why did they work on the Sabbath? Why did they answer these questions evasively or with lies? Nothing is so glaringly false and wrong to a child as hypocrisy.

The stories of heaven and angels that he heard stimulated his imagination. As a very small boy he knew where heaven was – right under the roof of the house across the street, right over old Mrs. Schumann's apartment on the fourth floor. This added excitement to the visits to Mrs. Schumann, who also gave him such good pastries to nibble. Later on Paul realized that he had not been looking at quite the correct angle when someone pointed to the sky in answer to his question about where God and his angels lived. Then his fantasies turned to the large balloon tied up in the Prater, the one that could be visited for five gulden. If only he could get the money somehow he could cut the balloon free and it would take him straight up to heaven, away from earth's troubles, lies and sins.

At other times Paul turned his rational mind to more strictly Jewish subjects. There was a Passover song that stayed in his memory about the kid that father bought for two farthings. The kid, representing Israel in its innocence, is killed by a cat, that is bitten by a dog, that is beaten by a stick,... and so it continues until the Holy One, blessed is He, kills the Angel of Death, which is interpreted as marking the establishment of the Kingdom of God on the earth. One day Paul was struck by something peculiar about this song: there was an even number of terms in this series of avengers, so that if the innocent kid was eaten by a wicked cat, the even members of the series were the evil ones – but God appeared in the last, even, place in the series! And, therefore??

Paul could hardly help being aware that Judaism was not the only religion, living as he did in a Catholic district of Vienna. His friends, even his beloved Nurse Minna, were Catholics. Even when he was very small he was familiar with the interior of churches since Nurse Minna would always step in briefly to say a prayer or light a candle when they were out walking. The rich ritual and atmosphere of Catholicism were not lost on the impressionable boy. He went through a period of real longing for Catholicism and for the crucifix. He learned from his Catholic schoolmates, in no uncertain terms, what they were taught about Jesus and the Jews, so that his feelings for Catholicism became a guilty secret to him, "since the Jews were the killers of Christ".

He eventually realized that Catholics were no freer of hypocrisy than Jews, that they deviated from their religion as variously and as much as his own people did. By the age of twelve all of Paul's faith was gone, and he took great pleasure in sharply arguing the absurdity and falseness of all organized religions.

Paul Ehrenfest's brothers played a major role in his early life. They were all much older than he, so that as a youngster of five, he could look up to Otto at thirteen, Hugo at fifteen, Emil at twenty, and Arthur at twenty-two. (Sigmund had named his sons in alphabetical order, giving them short names to avoid nicknames, which he detested, though he always called his youngest by a pet-name.) These four older brothers must have represented the large, free future to him. They knew so many things, did so many things, that went beyond the bounds of his childish existence. The little boy in his crib could see all four working at the large round table in the living room, all busy with studying or work of some sort. Arthur was completing his studies at the Institute of Technology (*Technische Hochschule*), Emil was his father's right hand man in the business, Hugo and Otto were students at academic and technical high schools (*Gymnasium* and *Realschule*). He too would pursue these fascinating things some day.

The greater scope and freedom of his brothers' lives were not limited to matters of the mind. The storeroom of Sigmund Ehrenfest's shop was behind the house and its steep zinc roof was accessible from the living room windows. The older boys were privileged to climb out on it, but this was strictly forbidden to little Paul. They developed a little game out of this situation. When the big boys climbed way up to the highest point of the zinc roof, they could see, or so they told Paul, the ocean. They described it to him in great detail – the sailing ships, the towering waves, sometimes even great storms! When Paul wanted to see the marvels for himself he was reminded that he was too small to climb out: "Maybe after the summer, when you'll be much bigger." And if he wondered why the ocean couldn't be seen from some other place, or how it could be so stormy there when he could see the sun placidly shining, Arthur or Hugo would fabricate some explanation that seemed thoroughly satisfying.

Although much of the wonderful life that his brothers led was beyond the powers of a little boy, they could and did share some of it with him. He learned from them constantly, as one or another was almost always ready to answer the questions that he never stopped asking. The most ready and most gifted of the explainers was Arthur, who had practically finished his education as a mechanical engineer when Paul was still a very small boy. Thanks to Arthur's explanations, Paul was as familiar with certain natural and scientific concepts as he was with his own language. He could never remember when he had first learned about energy and its conservation, or about how an electric bell worked: they were things he had "always" known. Arthur had a great talent for reformulating abstract ideas in forceful everyday language, a talent his youngest brother absorbed and made his own in later years.

Arthur used his gift for explanation, and Paul's infinite readiness to listen to the stories in which he clothed his explanations, as a very effective defensive maneuver in their morning skirmishes. These skirmishes were a result of Paul's morning chore, which consisted of getting his brothers out of bed in time for them to go their respective ways. On cold winter mornings, especially, this was quite a challenge. Arthur would counter his brother's efforts to drive him from bed by offering "to explain something", and the fascinating stories that ensued would hold Paul spellbound enough so that he forgot his mission. Meanwhile Arthur could stay in his warm bed, not even fully awake. And if he did fall asleep during his explaining, Paul would pull the bedclothes off again to get him to resume the story-lecture. On one occasion Paul wanted to know about a coup d'état, a word he had heard used in one of the lively dinner table conversations between Sigmund Ehrenfest and his older sons. Arthur was ready to oblige with a long exciting story about Napoleon's coup d'état, translated into just the sort of boyish escapade that made the whole thing crystal clear to Paul. He could do that kind of thing for almost any of Paul's questions.

Through Arthur, Paul became acquainted with a variety of technical devices, learning to understand them and work with them as a game, a most wonderful form of play. There were electric batteries, bells, telegraph keys; there was Arthur's camera, and his crystal models; and there was the memorable evening when Arthur projected a real image of the oil lamp onto the wall with a lens. It is hardly surprising that to Paul, Arthur was quite simply the wisest person in the world. When Hugo, who was studying the classics in high school, told Paul about Socrates and called him the greatest of all sages, the natural response was, "Even wiser than Arthur?"

Sometimes, to be sure, Arthur could not resist the urge to take advantage of Paul's boundless belief in the stories he told him. When Paul wanted to know about a new black box that Arthur had bought and that looked delightfully electrical to Paul, Arthur convinced him that it was a wireless telephone (in 1886)! The little boy knew enough about electricity and was familiar enough with the real telephone, which Arthur himself had installed between shop and apartment, to be dubious about this new "discovery". Arthur knew how to disarm these suspicions by proposing an experiment. In Paul's presence he would call up Emil, who was below in the shop, and give him any message that Paul wished. Arthur would remain upstairs while Paul ran down to the shop and Emil could then repeat the message to him. It sounded plausible enough and Arthur carried it off in fine style, clicking the little lever on the box to "summon" Emil, becoming impatient because Emil didn't "answer" soon enough, repeating Paul's message several times to be sure Emil had it straight, and "telling" Emil that Paul would be right down to check. Paul dashed through the apartment, down the stairs, through the yard and the shop and, sure enough, received the correct message from Emil's lips. Only long afterward did Paul realize that Arthur had had plenty of time to call Emil on the real phone while he was on his way downstairs, so thoroughly had he been convinced by Arthur's discussion of wireless telephony.

Mathematical puzzles intrigued Paul from an early age. One of his brothers told him the legend about the reward that the inventor of chess was said to have requested: one grain of wheat on the first square of the chess board, two on the second, four on the third, eight on the fourth, and so on, up to the sixty-fourth square. Paul set to work at once to determine the total number of grains, writing on the unused portions of a torn page from a Greek exercise, discarded by Hugo. He calculated and calculated, with the numbers becoming so preposterously large that he finally grew dizzy and almost fainted. Arthur then showed him how to determine the sum of such a geometric progression, without adding all the terms, by a simple formula. That was enormously impressive, particularly since the rationale of this method was a complete mystery to Paul then.

A new intellectual attainment could engross him. When one of his brothers showed him the idea of drawing the floor plan of an apartment, he became absolutely possessed and did nothing but execute such plans for all the apartments he knew, to the exclusion of everything and anything that he should have been doing. He was spanked as a result, but that was a frequent occurrence. Calendars, too, fascinated the boy. At one time everyone in the family saved old calendars to give to Paul on a Sunday at Grandmother's. He especially liked the ones in which the dates fell on the same days of the week as in the current year. Later on he learned to construct his own calendars, taking special pleasure in perpetual calendars. His passion for calendars – Paul's Calendar Craze – became a family byword, used even many years later to describe his intense pursuit of some one interest.

The Ehrenfest family's city life was interrupted almost every summer by a sojourn in the country for a couple of months. Sigmund could not afford elaborate vacations for his family, but he did send them out of the hot, dusty city for July and August. They did not go very far, to be sure, since their summers were usually spent in Brunn am Gebirge, about ten miles southwest of Vienna. One year they went to Hütteldorf, a little to the west of the city, and in the nineties, when they were more prosperous, to Baden, about ten miles beyond Brunn and a real resort. In the summer of 1889, when Paul was nine years old, he and his mother went to Boskowitz – a Jewish-Czech village. But it was Brunn that Paul remembered for the typical summers of his childhood.

Early in July a considerable fraction of the family's belongings were loaded onto a large wagon so that it was piled high with furniture. What was left in the apartment was carefully covered and amply protected with moth balls. The loaded wagon, accompanied by an Ehrenfest brother or two and several servants, went off to Brunn, to be followed in a day or two by most of the other Ehrenfests. Their summer quarters did not provide much of a natural setting for the boy, since they lived in the village, and his memories did not extend much beyond the crowded lodgings that they occupied. These were usually in the extensions built onto the backs of houses on the village street, built by the local residents for the summer trade of the families of Vienna's lower middle class. The rooms behind one such home might house half a dozen or so families for the summer. Paul's only real contact with the countryside came on trips to the place where they swam, though he never really learned to swim, or on excursions in the nearby hills.

The Ehrenfests always took rooms with two families of relatives: the Friedmanns, (Mrs. Friedmann was Johanna Ehrenfest's sister), and the Kochs, (Doctor Koch was the family doctor and a close friend). All three families had children of similar ages, and so they formed their own community, particularly since their various servants were brought along, including two French governesses for the Friedmann girls. The men of the three families could not desert their work in Vienna and appeared in Brunn only late in the evening or for weekends.

Paul's interest in people and their interrelationships had plenty of material to feed on at Brunn as he watched the difficulties that these three families had in getting along smoothly with each other, with the landlord and his wife, and with the other families staying in the same summer quarters. It struck him even then that each of the women had her own particular way of maintaining her self importance, which had a negative effect on the preservation of good neighborly relations. Tante Koch, as the only one whose husband was a real professional man, felt herself in a position to give advice in the form of frequently repeated maxims, often totally inappropriate. "Life is a drama and we all have our parts to play in it", she would say, and Paul would imagine a decked out "artiste" dancing on a few rickety boards spread between two barrels! Tante Lina Friedmann put her whole soul into her pride in her daughters' French lessons and the other graces they received from their governesses. Paul's mother felt her own position to be secured by her oldest son's commanding intelligence. The many large families, each with its entourage of servants, had plenty of material for quarrels, living in such close quarters, and the Herr Landlord could and did profit by these in extorting tips for his services as arbitrator to "the Jewish rabble".

There was a great deal of musical activity during these summers at Brunn. The families brought their pianos with them from Vienna and also their common piano teacher, Ludmilla Gross, whose father had been an acquaintance of Beethoven's! Schubert, Chopin, Mendelssohn and Beethoven were played, there were four hand arrangements of the classical symphonies, some lieder singing, and a lot of current Viennese dance and operetta music. The women often played cards during the afternoons and evenings out on the lawns. Paul learned chess from Doctor Koch, but he was not attracted to it enough to continue playing it for more than a few years. Cards never tempted him at all; his father scorned such a wretched way of killing time, and he found his mother's playing a bore.

Paul learned to read and to do sums at home, more or less by himself, and with his mother's help he learned to write, too, so that when he entered school at the age of six, he was admitted to the second grade.

He did well in elementary school, attending one school from 1886 to 1888 and another from 1888 to 1890, after which he was admitted in due course to secondary school, the Akademisches Gymnasium. So far as one can tell, however, most of his intellectual stimulation continued to come from his brothers, and particularly from Arthur. But once he could read there was a new world of resources opened to him.

The books that first captured his imagination were those that spoke of other ways of life, fantastically and excitingly different from the life of a little boy living over a shop in the proletarian Favoriten. He read *Gulliver's Travels* in a cheap illustrated children's edition, and *Robinson Crusoe*, and then spun out further fantasies on them which he told to patient Nurse Minna. Jules Verne became a passion: Paul could not be torn away from *Around the World in Eighty Days* when he first received a copy of it, and later on he devoured *Twenty Thousand Leagues Under the Sea*. In Grimm's *Tales* and in Wilhelm Busch's *Munich Picture-Sheets* he liked the humorous and marvelous stories rather than the horrible ones. But perhaps the deepest impression of all was made by stories from the *Arabian Nights*, like *Ali Baba* and particularly *Aladdin and his Magic Lamp*. The wonders of Aladdin's garden, with the trees whose fruits were emeralds, rubies and sapphires, were intensely real to Paul, more vivid and alive than the world around him.

The first book on science that he read – secretly – was an old torn elementary school text on physics and chemistry that his brothers had discarded. But his real introduction to science and mathematics came from his brothers directly, in their discussions with him or with each other. (Sometimes there were misunderstandings, of course, as when Paul wrote *a* for 2 in his sums for a while: he had overheard Arthur tutoring Otto in algebra, impatiently shouting, "Don't you understand, *a* is equal to 2.")

This early impression of science as something to be learned with joy, something to be discussed and argued about, was absorbed into Ehrenfest's innermost being.

CHAPTER 3

Student Years

By the time that Paul Ehrenfest began his university studies in 1899, he had lived through a number of critical maturing experiences. His mother had died, of breast cancer, when he was only ten. Not long after her death, Sigmund Ehrenfest married her younger sister, Josephine Jellinek, the vivacious Tante Pepperl who had often joined Arthur Ehrenfest in leading the young people's fun during the family Sunday afternoons at Grandmother Sophie's apartment. Pepperl was much younger than her husband, about the age of his oldest son, in fact, but she managed to preserve the close, warm family relationships among the Ehrenfests. During this period, Arthur Ehrenfest, Paul's mentor and idol in those years, married and left Vienna for a time. Sigmund gave up his grocery business after his wife's death and opened a pawnbroker's shop, an establishment that did not make such demands on the life of the whole family, and the Ehrenfests moved to the busy Kärnthnerstrasse in Vienna's inner city.[1]

Sigmund Ehrenfest outlived his first wife by only six years, and in 1896 he died as a result of the stomach ulcers that had plagued him for many years. By this time all his sons except Paul were grown men, launched on their various careers. Otto was an electrical engineer,

1. Personal communication to the author from Mrs. T. Ehrenfest-Afanassjewa, Mrs. A. van Bommel-Ehrenfest, F. Ehrenfest-Egger, and P. Ehrenfest. Ehrenfest's personal papers include a number of family records.

though not university trained; Hugo was a doctor; Emil, who had been his father's right hand man in business, took over the pawnbroker's shop; and Arthur was a mechanical engineer. At their father's death Arthur was appointed guardian of sixteen year old Paul – and he was sorely needed.

Paul's adolescent years were stormy ones. He was often miserable, deeply depressed and at odds with himself and the world. His father was naturally disturbed and anxious about his youngest and favorite son. The boy, who had been such an eager scholar in elementary school where his fine report cards demonstrated his diligence as well as his intelligence,[2] seemed to move away from his former interests. After Sigmund Ehrenfest's death Arthur had to draw on all his considerable powers of persuasion, and on all the love and respect that his youngest brother felt for him, to convince Paul not to quit school. Paul's grades at secondary school had gone down, his conduct was now only "adequate", and his industry only "uneven"; his mathematics alone continued to be judged as "superior".[3] Despite Paul's lack of interest in school, Arthur felt certain that the boy should go on with his studies, and he somehow managed to convince Paul of this.

Paul was apparently able to work himself out of his depression, which had sometimes been deep enough to make him contemplate suicide. His intellectual interests grew stronger, perhaps as a form of self-protection. He read widely in Schopenhauer and Nietzsche, Wagner and Turgenev. He allowed himself to be fascinated by mathematics and physics, and turned his attention to them, and not to his inner problems. Helped by a perceptive teacher, he began to devour books on mathematics and physics. These outside studies gave a structure to his intellectual life, and they made is possible for him to put up with Latin and Greek and the rest of the required work at school that held no great appeal for him. He successfully completed the final examinations, both written and oral, in the summer of 1899, but his general reaction to all his schooling was bitter: he would later insist on educating his own children at home.

---

2. Paul remembered coming home in tears one day during his first year in school, because his report card had shown one grade that differed from his usual string of 1's (the highest possible grade). His father consoled him quickly by pointing out that the offending number only indicated the number of days absent during that marking period.

3. School records from the Stadtschulrat in Vienna for Ehrenfest's years at the elementary school on Keplergasse. High school records from the Akademisches Gymnasium and the Bundesrealgymnasium, formerly the Kaiser Franz Josef Gymnasium.

Physics and mathematics brought more than intellectual delight to Paul Ehrenfest – they brought friends. He was in his sixth year of the eight year secondary school course at the Akademisches Gymnasium, and preparing to transfer to the Kaiser Franz Josef Gymnasium,[4] when he heard of an outstandingly good mathematics student at the latter school. As Paul was walking in the park one day he spotted a thickset young fellow, wearing glasses and carrying a heavy book that Paul recognized as a familiar work on higher mathematics. Paul surmised that this young fellow must be the Gustav Herglotz whose fame as a mathematics student had extended as far as the Akademisches Gymnasium, and he struck up an acquaintance with the stranger at once. It turned out that Herglotz, who was a class ahead of Paul in school, had progressed even further in his extracurricular study of mathematics. Their acquaintance deepened into friendship, and they corresponded with each other during the summer of 1897 when Paul was away from Vienna.[5] They were together at the Franz Josef Gymnasium the following year and spurred each other on in mathematics, so that both were far in advance of their contemporaries when they finished school.

In October 1899 Paul Ehrenfest enrolled at the Technische Hochschule in Vienna, listing chemistry as his major field. He took courses in a variety of scientific subjects – mechanics, mineralogy, mathematics and, of course, a great deal of chemistry including a laboratory course in analytical chemistry that demanded twenty hours of his time each week.[6] Despite this heavy schedule, Paul was able to keep up his friendship with Herglotz, who was at the University, and to find other friends with common interests and talents, particularly Hans Hahn and Heinrich Tietze. The four mathematical friends, the "inseparable four" as they called themselves for a brief period, found much to discuss together on long walks through the hills of the Wiener Wald.

Ehrenfest and Tietze decided to take part in the great scientific event of November 1899 – observations of the expected Leonid meteor

---

4. After his transfer Paul Ehrenfest did better, receiving good grades in all his subjects during his seventh year. Records for the eighth and final year of gymnasium, 1898-99, were lost during the Second World War. I owe this information to Dr. Ernst Nowotny, director of the Bundesrealgymnasium in 1963.

5. Information from Mrs. T. Ehrenfest-Afanassjewa and from Ehrenfest's record of how he spent his summers during his childhood and his youth.

6. Paul Ehrenfest's *Meldungsbuch* from the Technische Hochschule is dated October 17, 1899, and shows courses for 1899-1900 and 1900-01.

showers that were due to occur after their regular thirty-three year interval. They could not go off to India or any of the other far flung points to which expeditions were being sent, but they fitted out their own little expedition within Austria, preparing to observe and to photograph the meteor showers. The expedition was not very successful: the first night out they had a clear sky, but no meteors were to be seen; the second night brought a snowstorm and ended the expedition.[7]

University students in the German speaking world did not limit themselves to a single institution but travelled about a great deal, going to attend the lectures of particular professors wherever they might be found. It was not long before the group of four friends was broken up. Hahn studied at Strasbourg, Munich and Göttingen; Herglotz went off to Munich to study some astronomy, and Tietze joined him there after doing his military service. Ehrenfest spent two years at Vienna, taking courses at both the Technische Hochschule and the University,[8] but he, too, left in 1901 to study at Göttingen for a while.

At Vienna, however, he had found his métier: it was to be theoretical physics. There is little doubt that his decision was based in large part on the overwhelming example of Vienna's professor in this field, Ludwig Boltzmann, whose lectures on the mechanical theory of heat Ehrenfest attended in 1899–1900.

Boltzmann was a theoretical physicist with all his heart and soul. He was born in Vienna in 1844 and received his doctorate there in 1866.[9] After a few years as assistant to his teacher, Josef Stefan, in Vienna, he taught at Graz, and then moved on to Heidelberg and Berlin for further studies with such notables as Robert Bunsen, Gustav Kirchhoff and Hermann von Helmholtz. He returned to Vienna in 1873 as professor of mathematics, but soon left for Graz again, where he served this time as professor of experimental physics from 1876 to 1889. During these years he attracted gifted foreign students like Svante Arrhenius and Walther Nernst to this provincial Austrian university. From Graz, Boltzmann went to Munich as professor of theoretical physics, after refusing an invitation to succeed Kirchhoff

7. Personal communication to the author from Professor H. Tietze.

8. Ehrenfest's University reports show that he took some work there in 1899–1900 and more in 1900–01.

9. See E. Broda, *Ludwig Boltzmann* (Vienna: Franz Deuticke, 1955) and R. Dugas, *La théorie physique au sens de Boltzmann* (Neuchâtel: Le Griffon, 1959).

in Berlin. He returned to his native Vienna once more in 1894, this time as professor in his own field, but his wanderings were not over yet. He was to leave Vienna for Leipzig in 1900, and then return to his still vacant chair in Vienna in 1902 for the remaining few years of his life. In beginning his inaugural speech in Vienna in 1902[10] Boltzmann remarked that he could spare his audience the conventional hymn of praise for his predecessor since in this case the predecessor and the speaker were identical!

Long before Ehrenfest's university years Boltzmann had won a world-wide reputation for his work in all branches of theoretical physics. Boltzmann had no doubt about the supremacy of mechanics as the branch of science that went deepest in probing the structure of the natural world. The principal line that he followed in his own research was the application of mechanics to the atomic theory, and the one problem to which he devoted his greatest efforts was the mechanical explanation of the second law of thermodynamics, perhaps the grandest of all scientific generalizations.

Boltzmann held up the standard of atomism in a period when the scientific style prevailing on the continent treated atomism, and all attempts to see in detail into the mechanical structure of phenomena, as outmoded, unnecessary and misguided. The phenomenological approach to physical theory, adopted by Kirchhoff, was an effort to avoid structural and mechanical hypotheses in setting up physical theory. Boltzmann could see the virtues of such an effort, though he did not find it really congenial; but when it came to the exaggeration and distortion of this approach in the school of energetics led by Wilhelm Ostwald and Georg Helm, Boltzmann entered the lists to defend the indispensability of atomism and to attack the excesses and errors of his opponents. In his vigorous polemical style, in public lectures and in print, Boltzmann emphasized the insights and successes won by mechanical physics that were simply beyond the reach of the newer approaches.[11] He freely granted that science had passed the era in which mechanical models could be taken as the real and exclusive representation of nature, but he had no doubt that the detailed insights which could be gained only by the older methods should not be scrapped to suit the new fashions.

Boltzmann's often quoted tribute to James Clerk Maxwell's artistry

10. L. Boltzmann, *Populäre Schriften* (Leipzig: Barth, 1905) p. 330.
11. *Ibid.*, pp. 104, 137, 141, 158, 198.

and to the dramatic force of Maxwell's writings is worth repeating here as a typical example of Boltzmann's own style of thinking and writing. He is referring to Maxwell's brilliant device of replacing the hard sphere molecules of the kinetic theory of gases by molecules that repel each other according to a force depending on the inverse fifth power of their separation. Boltzmann, who had made endless calculations using the hard spheres, could fully appreciate the simplifications brought about by Maxwell's maneuver, particularly the elimination of the troublesome relative velocity, V. "Who does not know his [Maxwell's] dynamical theory of gases?", wrote Boltzmann. "At first the variations of the velocities majestically develop; then from one side enter the equations of state, from the other the equations for central motion – the chaos of formulas surges higher and higher. Suddenly the four words resound: 'Put n = 5'. The evil demon V vanishes, just as in music a wild figure in the basses that has been undermining everything up to that point suddenly becomes silent; as if by a stroke of magic all that previously seemed unconquerable is suddenly set in order. There is no time to say why this or that substitution is made; let anyone who does not feel it lay the book down; Maxwell is no program musician who needs to set an explanation over the notes. The formulas obligingly spew forth result after result, until, as a final surprising effect, the thermal equilibrium of a heavy gas is obtained, and the curtain falls."[12]

One is hardly surprised to learn that Boltzmann was considered an incomparable teacher. His mastery of his science and his love for it were combined with a zest for lecturing and an attention to detail and procedure. Even his blackboard technique was carefully developed: the sequence of equations clearly inscribed on the board gave an exact account of the structure of ideas he was building in that lecture. The exposition was not, however, so formal and highly polished as to smooth over the difficulties or make the audience lose interest. Boltzmann never hesitated to point out and correct his own errors; he welcomed questions and discussion; and his lectures drew crowds of eager students.[13]

Boltzmann initiated Ehrenfest into both the substance and the spirit of theoretical physics, and Ehrenfest was to show Boltzmann's influence in his own research and teaching in years to come.

12. *Ibid.*, p. 73. Maxwell also had other reasons for introducing the inverse fifth power force.

13. Broda, *op. cit.*, p. 10.

After Paul Ehrenfest had been at the University for only a year, Boltzmann left Vienna for Leipzig. Ehrenfest stayed on, however, since there were many other younger men to learn from. He studied with Anton Lampa and Fritz Hasenöhrl, who were *Privatdozenten* (instructors); Hasenöhrl had himself been Boltzmann's student and would eventually succeed to his chair in Vienna. There was also Gustav Jäger, who had studied with Josef Loschmidt and Josef Stefan in Vienna in the eighties as Boltzmann had done twenty years earlier, and who was then assistant professor, working in the kinetic theory of gases. Ehrenfest apparently did not study the philosophy of science with Ernst Mach, whose influential ideas on the nature of scientific theory included a rejection of atomism.

By the following year Ehrenfest had had enough of Vienna for a while. He wanted to try another university, and he decided on Göttingen. The initial reason for his choice seems to have been his interest in the work of Johannes Stark, a Bavarian physicist who had come to Göttingen in 1900 as *Privatdozent*.[14] Stark was an extremely active experimentalist who, at the age of twenty-seven, had already written close to forty papers on cathode rays, electric discharges and associated phenomena. Once Ehrenfest arrived at Göttingen, however, he found that there was a much richer scientific life than he had known at Vienna. And for all the physics he studied at Göttingen, it would be the mathematics there that was most exciting.

The eighteenth century university in that quiet old town had a strong and continuous tradition of mathematical excellence dating back to Carl Friedrich Gauss, who had become professor of mathematics there in 1807, and continuing with his successors, Dirichlet, Riemann, and Clebsch. In the early years of the twentieth century Göttingen's mathematicians were led by Felix Klein, professor there since 1886, and a world figure in mathematics. Klein had worked extensively in analysis and in various branches of mechanics, but he made his greatest impact on mathematics with his Erlangen Program for geometry: geometry was to be studied as the theory of the structures left invariant by a group of transformations of space, and each such group generated its own geometry. This program gave a new unity of method to the diverse developments in geometry and helped to form the modern spirit of mathematical research.

Klein was deeply and actively interested in the teaching of mathe-

14. Personal communication from Mrs. T. Ehrenfest-Afanassjewa.

matics at all levels. He lectured regularly to secondary school teachers and prospective teachers, and worked hard to bridge the gap between school and university mathematics. Klein emphasized both the logical structure that forms the skeleton of the subject, and the rich variety of applications that give the mathematical organism its life, to use his own metaphor. His famous books on *Elementary Mathematics from an Advanced Standpoint*[15] show how Klein brought out both the mathematical core and the intuitive meaning of the mathematical materials with which secondary school teachers must work.

His own instruction was carefully organized to initiate young mathematicians into independent study and research. The mathematical institute at Göttingen had its own reading room where all important books and journals were kept, so that they were always available for the use of the students. (This was the room that impressed Ehrenfest so much, as mentioned earlier.[16]) Klein also saw to it that the texts of his own lectures were made available to his students, in the form of handwritten notes kept in the reading room, or in later years, as duplicated notes issued directly to them. An additional and unusual feature of the Göttingen reading room was the collection of mathematical models that Klein assembled, as a concrete way of developing his students' intuition and mathematical insight.[17] He had great organizational talents, too, editing the *Mathematische Annalen* for many years and helping to direct the major enterprise of the *Encyclopädie der mathematischen Wissenschaften*, which was to play a crucial role in Ehrenfest's career a few years later.

The second professor of mathematics at Göttingen was David Hilbert, not quite forty when Ehrenfest arrived there, and already well on the way to his position as one of this century's deepest and most creative mathematicians. His work ranged from studies of the foundations of mathematics to the kinetic theory of gases, from the theory of algebraic numbers to integral equations, and in every field he opened new avenues of research.

Ehrenfest enrolled at Göttingen in November 1901, and signed up

---

15. F. Klein, *Elementary Mathematics from an Advanced Standpoint* Transl. E. R. Hedrick and C. A. Noble, 2 vols. Reprinted (New York: Dover, n.d.)
16. See Chapter 1.
17. See F. Klein, *Lectures on Mathematics* (Delivered from August 28 to September 9, 1893, at Northwestern University) (New York: Macmillan, 1894), p. 96. Also see F. Klein and E. Riecke, *Über angewandte Mathematik und Physik in ihrer Bedeutung für den Unterricht an den höheren Schulen* (Leipzig: Teubner, 1900), p. 22.

for no fewer than fifteen courses![18] Felix Klein was teaching mechanics and directing the mathematical seminar, Hilbert was lecturing on potential theory, and he could not pass up a course in the theory of analytic functions. And then there was physics: W. Voigt, the professor of theoretical physics, was teaching courses on the theory of electricity and on crystal optics, Max Abraham on the electromagnetic theory of light, Stark on ionization phenomena, and the list went on and on. It all seemed too good to pass up, including the chance to work in the physics laboratory, and the courses in applied physics. Ehrenfest stayed with almost all these courses through the semester, but gradually reduced his load in the spring and the following fall. He took work with Walther Nernst, both lectures and laboratory, with the astronomer Karl Schwarzschild and the mathematician Ernst Zermelo, and even attended some of Edmund Husserl's lectures on logic.

Klein and Hilbert and the other widely known scientists on the faculty attracted an international group of students to Göttingen, and Paul Ehrenfest became a part of that group. He soon made a friend in Walther Ritz, a brilliant Swiss theoretical physicist two years older than Ehrenfest, and one year his senior at Göttingen. Ritz was already close to his doctorate, which he completed there in 1902. The fall of 1902 brought new additions to the Göttingen mathematical group and one in particular attracted Paul's attention. This was a young girl whom he saw at Klein's and Hilbert's lectures, and who was often in the reading room studying mathematics. She was a Russian mathematics student, Tatyana Alexeyevna Afanassjewa. She did not appear at the regular weekly meetings of the club for mathematics students, however, and Paul soon discovered the reason: the rules forbade the admission of girls to these meetings. The arbitrariness of such a prohibition, combined with his growing interest in meeting her, led him to propose a change in the rules, which he carried through after a heated discussion. The invitation to attend the club meetings was extended to her through the other Russian students, and Paul soon met Tatyana Alexeyevna.

She was a few years older than Paul, having been born on November 19, 1876 in Kiev.[19] Her background was totally different from his:

---

18. Ehrenfest's *Anmeldungsbuch* from the University of Göttingen is dated 15 November 1901 and shows courses for 1901–02 and for the winter semester of 1902–03. This information is confirmed by letter to me from the Georg-August-Universität, Göttingen of 30 August 1963.

19. Information from Ehrenfest's family records.

she was the only child of a civil engineer who died early in her childhood. She grew up in the home of her childless uncle, Peter Afanassejev, in Petersburg, where he was professor at the Polytechnic Institute. Tatyana showed her mathematical talent at an early age, but, because of her uncle's concern over her somewhat uncertain health, she did not proceed directly from secondary school to the Women's University at Petersburg to pursue her scientific studies. Instead she went to the less demanding Normal School where she completed the mathematics and science curriculum in three years. Only after she was through there did she attend the Women's University, and do outstanding work in mathematics.

Tatyana Alexeyevna came to Göttingen in 1902 to extend and deepen her mathematical and scientific learning. For her, too, mathematics and science were not just subjects to be studied, disciplines leading to a professional career: they were, rather, the center of her vigorous, exciting intellectual life. In this matter she and Paul Ehrenfest were in complete agreement, different as they were in their pasts and their personalities. Her sharp mind was combined with an absolutely independent spirit and a strong and stable approach to life that could and would withstand all that time would bring.[20] Her personal code was a strict one, demanding the rigorous fulfillment of all obligations, and excluding indulgences like alcohol and tobacco. But this strong will and strict sense of duty were coupled with a lively interest in others and a real willingness to help them. She made friends easily and readily – among those whose intellectual and moral outlooks were compatible with her own. Her rooms were often the center for student gatherings and lively discussions. Her Aunt Sonya, who had brought her up, and who accompanied her to Germany, knew how to make young people at home and set them at ease.

Paul Ehrenfest soon found that there were deeper bonds joining him to Tatyana Alexeyevna than their joint admiration for Felix Klein's artistic approach to mathematics, and sometime during that winter of 1902-1903 they decided on marriage.

During the year and a half he spent in Göttingen Paul Ehrenfest was developing his scientific independence and preparing to start his own investigations. Perhaps the fact that he was not under the direct

---

20. See C. Visser, "In Memoriam T. Ehrenfest-Afanassjewa." Vernieuwing van Opvoeding en Onderwijs **22** (1964) p. 217 (June).

influence and guidance of Boltzmann accelerated this development. In any event one can trace his studies, his reasoning, and his intense scientific curiosity in his notebooks. He began at Göttingen to record his ideas systematically in the first of what was to be a long series of small pocket notebooks.[21] These notebooks are distinctly different from those kept by most scientists: of the thousands of entries made over the thirty years of his active life, by far the larger part consists of questions. He carried such a notebook around, and entered questions as they occurred to him, questions suggested by his reading, often on points of detail in papers he was studying. Sometimes they are recognizable as the germs of ideas later developed in extended articles. Even when a subject is pursued through many entries over a long period of time, it is rare to find really elaborate calculations of the kind that make up most of the work of most theoretical physicists. The small notebook and the sharply pointed question are typical of Ehrenfest.

The motto that he set on the first page of his notebook, dated Christmas, 1902, shows the use for which he originally intended it. He took as this motto some advice contained in a letter written, probably about 1820, by the Hungarian mathematician Wolfgang Bolyai, a close friend of Gauss, to his son Johann, who also became a mathematician and was one of the creators of non-Euclidean geometry: "Trust me, and study now; try to make progress. Make a note of what you don't understand, of places where you discover gaps, and go on; otherwise you can get into a subject that will easily swallow up your whole life."

Almost a decade later, after filling a dozen notebooks with some three thousand entries, largely questions, Ehrenfest explicitly formulated his own views on the value of keeping such records.[22] "One should make a short and precise note of every question one comes upon in reading or in conversation or in free reflection. One after another; and from time to time look through the old questions, and perhaps also systematize them in some way. Above all there is a great satisfaction in seeing how one easily and gradually learns to answer many questions which one hardly knew how to formulate at first, much less to work out. Secondly, in this way one gradually learns how to condense hazy obscurities into sharply defined questions. That is

21. The notebooks are part of the Ehrenfest Collection at the Museum for the History of Science at Leyden. They also include lists of Ehrenfest's reading, addresses, and so forth.

22. From the "Red Book", a family record. This entry is dated 4 August 1912.

enormously important. It is a trivial remark that doing this puts one halfway to the solution. Thirdly, one then has an abundance of questions all ready for asking the right people, in case one comes across them. But that is the true key for opening their minds. Nothing is shabbier than the feeling: now God has granted me the opportunity to meet this man, and I sat before him open-mouthed; how much I might have asked him – but nothing at all occurred to me. It seems to me, however, that the principal value lies in the habit of working over hazy obscurities until they are reduced to sharply defined questions. This is especially so for questions that lie far off from one's own science!"

A selection of some of the questions and problems that Ehrenfest listed during the first few months of 1903 will give some idea of both his systematic studies and the diversity of his scientific curiosity. The themes that frequently recur concern the fields that a young theorist would naturally have been trying to master. His studies of dynamics led to a number of questions and remarks about the relationships among the variational principles that figure so prominently in this subject. (Ehrenfest had evidently profited from Klein's lectures on mechanics at Göttingen.) There are a few questions on electromagnetic theory, and many more on the mathematical methods for its development – properties of spherical harmonics, the concept of a pseudoscalar quantity, and various possible generalizations of ideas in vector analysis to an n-dimensional space. The principal focus of Ehrenfest's interest, however, was already the subject of statistical mechanics. At that time he was trying to work on the theory of dissociation by the methods of statistical mechanics, (a subject already attacked by Boltzmann), thinking about the meaning of phase space, raising the question of a molecular explanation of the anomalous negative thermal expansion of water, and listing obscurities in some of the basic ideas of statistical mechanics. But in addition to these questions along more or less expected lines, all sorts of other problems occurred to Ehrenfest, and were duly recorded in his notebook – problems in electrochemistry, in magnetic hysteresis, in geometry; questions about the stability of a bicycle and the use of series whose convergence is not absolute, about color photography and the value of a repeated factorial operation.

A pocket notebook, even one intended for such specifically scientific use, is a handy place to record many matters. This one contains a few notes to Paul from Tatyana: asking him never to smoke, suggesting that he get a piano to play on, listing Tolstoy's novels and stories for him "to read as soon as possible". There is also a note in her handwriting written in March, setting the first week in August, 1903 as the time when Paul should be sure to return to Göttingen. For he was about to move on to still another university for a while.

He and Walther Ritz, who was now a close friend of both Paul and Tatyana, had decided to go to Holland, to listen to H. A. Lorentz's lectures on theoretical physics at the University of Leyden. Ritz had already received his doctorate at Göttingen; Ehrenfest had decided that he would return to Vienna, after this visit to Leyden, to finish his studies there. They arrived at Leyden at the beginning of April and Paul, at least, remained until the third week in May.[23] Shortly after their arrival they went to visit Kamerlingh Onnes's low temperature laboratory. Their visit came several years before Kamerlingh Onnes's great discoveries, and low temperature physics had still to become an exciting field. In 1903 not even liquid hydrogen, first obtained by Sir James Dewar in 1898, was available yet in Leyden. The two young theorists were shown through the laboratory by C. A. Crommelin,[24] then an equally young assistant to Kamerlingh Onnes. It was he who informed them that their hopes of also attending lectures by J. D. van der Waals, the other world famous Dutch theoretical physicist, could not be fulfilled in Leyden, since he was professor at Amsterdam; they had been misinformed about his location.

During his stay at Leyden Paul Ehrenfest lived in a room on the Apothekersdijk, facing one of the branches of the Rhine that run through the city, and only a short walk from the physics laboratory. He had picked the proper time of year to be in Holland, and he often walked through the tulip fields near Leyden, enjoying the massed colors and the boat loads of flowers on the canals. He went to Katwijk, a little fishing village near Leyden and walked along the sand dunes on the North Sea shore to Noordwijk, a walk he would often take in years to come. He took excursions to Amsterdam, the Hague, Rotterdam and Delft, and noticed characteristic customs and sights – obvious ones like the dog carts and regional costumes that were still common, and less obvious ones like the notices to treat animals gently, the six and seven year old boys smoking, and the spionettje—a mirror set at an appropriate angle on a long support, outside an upstairs sitting room

---

23. Ehrenfest's pocket diary for 1903.
24. C. A. Crommelin in the Algemeen Handelsblad (Amsterdam), 28 September 1933.

window, so that one could easily see who was at the door, or walking by.

There was plenty of time to read during those weeks in Leyden, and he devoured much of the list of Tolstoy's works that Tatyana had prescribed for him, as well as such diversified fare as Strindberg, Hans Christian Andersen, and Theodore Roosevelt on Americanism. But his principal reading was scientific, and he concentrated this on two main themes. The first was again statistical mechanics, and Ehrenfest carefully went through many of the classics of the subject – Clausius', Maxwell's, and Boltzmann's principal papers. He then went on to the new works on the subject – Gibbs's brand new treatise, deceptively named *Elementary Principles in Statistical Mechanics*, and the new articles by an unknown contemporary in Switzerland named Albert Einstein. The second theme was a new one for Ehrenfest, and he was introduced to it by Lorentz's lectures: it was the problem of black-body radiation. He read up the background works, and was soon studying the recent experimental and theoretical papers that had been devoted to this problem, and particularly the many papers written by Max Planck. Lorentz's lectures on radiation had led Ehrenfest to study the work that was to change the whole shape of physics over the following decades.

Even more important to Ehrenfest than this introduction to the radiation problem was his first acquaintance with Lorentz himself. We have already seen something of Ehrenfest's later feelings towards Lorentz, and those feelings would only grow deeper and more intense as the years passed. But Ehrenfest could hardly have dreamed in 1903 that Lorentz and Leyden would move to the center of his life in less than a decade. Even on this first visit, however, Ehrenfest did have some real contact with Lorentz, spending an evening at his home in Leyden before he left to resume his studies in Vienna. It left him with a vivid image of Lorentz in his Leyden home,[25] the first of many such images that he would treasure.

By the end of May Ehrenfest was back in Vienna, and had already paid a call on Boltzmann, who had himself returned from his two year interlude in Leipzig the previous autumn. He set to work at once, continuing his thorough study of all Boltzmann's papers on statistical mechanics, and working out models for understanding various points.

---

25. P. Ehrenfest to H. A. Lorentz, 12 July 1912.

These efforts led to his first publication,[26] a paper dealing with a small point in the theory of gases, which Boltzmann presented for him to the Royal Academy of Sciences in Vienna on July 9, 1903.

Ehrenfest put his work aside that summer and went off on several trips.[27] He visited Schiermonnikoog, one of the Frisian Islands in the North Sea, perhaps to extend his pleasant impressions of the Netherlands. Not surprisingly he spent some time at Göttingen in midsummer with Tatyana Alexeyevna, as he had arranged to do before he left there in the early spring. And then at the end of the summer Ehrenfest went off to Italy for a holiday. He was as overwhelmed by what he saw as generations of tourists have been before and since. His little diary is full of notes on what he saw and rough sketches of landscapes, paintings and buildings. "I will learn to sketch", he wrote to himself. The most impressive scenes were specially marked to be described in his letters to Tatyana. At Florence he noted simply, "I should like to have grown up here!" At Pavia Ehrenfest wrote that he must learn more about architecture so as to appreciate its beauty more thoroughly. "What an infinite source of pleasure that offers." He thought that a year spent in Italy studying mathematics and "otherwise just keeping one's eyes wide open" would be marvelous. But even in Italy Ehrenfest could not sustain this exalted mood. In the midst of the beauties of Lake Como, having just been to Bellaggio, Ehrenfest wrote in his diary: "Isn't the feeling of nervous disgust that is so familiar to me now just another form of the same psychological disposition that used to appear so often to me as 'Sunday boredom'? How to cure it?" Ehrenfest returned to Vienna by way of Switzerland, stopping to visit Walther Ritz in Zürich for a week.

The year that followed was a busy and a significant one for Paul Ehrenfest. He threw himself into his work, eager to finish his formal studies and be free to go on to problems that concerned him more, problems that would take longer to think through and bring to a satisfactory solution. He found the subject for his dissertation in a remark that Boltzmann made in his theoretical seminar concerning Heinrich Hertz's methods in mechanics. Ehrenfest chose the topic himself, and he was sufficiently proud of this fact to mention it several years later when he had to list his credentials in applying for academic positions. He was less proud of the thesis itself, *The Motion of Rigid*

26. P. Ehrenfest, "Zur Berechnung der Volumkorrektion in der Zustandsgleichung von Van der Waals", Wien. Ber. **112**, (1903), p. 1107. *Papers*, p. 77.
27. Family records and Ehrenfest's diary for his 1903 trip to Italy.

Bodies in Fluids and the Mechanics of Hertz,[28] as witnessed by his never publishing it. It has remained in the library of the University of Vienna, written in Ehrenfest's clear, firm hand, its handsomely inscribed equations playing no part in any future developments.

Ehrenfest wrote his thesis very quickly, finishing it on March 25, 1904. He completed his writing at Ragusa (now Dubrovnik), a resort on the Adriatic coast of Dalmatia. The thesis went to Boltzmann, his adviser, who was apparently satisfied with it, referring to it as "very fundamental" and "diligently and cleverly worked out" in his official judgment of the work dated May 13, 1904.[29] Boltzmann did not indulge in any more elaborate praise, however. It is possible that he expected even better things from Ehrenfest; he had certainly recognized the young man's ability. There is even a story that on one occasion, when Ehrenfest cited Boltzmann's works in a particularly complete way, Boltzmann remarked, "If only I knew my own works that well!"[30] Ehrenfest's relationship to Boltzmann does not, however, appear to have been very warm. His intellectual debt to his teacher was enormous, of course, and it left its stamp on his whole career, but one gathers from occasional, somewhat cryptic, remarks in Ehrenfest's notebooks that their personal relationships were not too easy. On one occasion, when Ehrenfest had been pressing him with a series of questions on some difficult point, Boltzmann finally asked with some annoyance whether Ehrenfest thought he was like a lemon that could be squeezed dry for its juice.[31] In any event, Boltzmann accepted the thesis, and Ehrenfest received his doctorate on June 23, 1904, after taking his final examinations.

During the year that he worked on his dissertation in Vienna, Ehrenfest did not lack intellectual companionship. Although his old comrades of the "the inseparable four", Tietze, Hahn, and Herglotz, were no longer in Vienna, there were other students with whom Ehrenfest had contact in varying degrees. There was Philipp Frank, who would eventually become Einstein's successor at Prague and

---

28. See *Papers*, p.1.
29. Boltzmann's opinion, in his own handwriting and signed by him and his colleague, Viktor v. Lang is part of Ehrenfest's file, Philosophischen Rigorosenprotokoll No. 1736, at the University of Vienna. I owe this information and a copy of the opinion to Dr. Franz Gall, University Archivist in 1963.
30. Broda, *op. cit.*, p. 11.
31. Personal communication from Professor Philipp Frank.

Einstein's biographer, as well as a distinguished philosopher of science in his own right. There was Lise Meitner, who soon went into the study of radioactivity and later played a critical role in understanding the discovery of nuclear fission. Ehrenfest looked up Lise Meitner shortly after his return to Vienna from Göttingen, when he heard that she had a detailed set of notes on all of Boltzmann's lectures. He proposed that they study Boltzmann's ideas together, as well as some of the background work by C. G. Jacobi and others in analytical dynamics, a suggestion that she quickly accepted since his gift for clarifying the ideas of theoretical physics was already quite evident. Their joint study was a great stimulus to her scientific development, but she found Ehrenfest's need to probe into the very soul of anyone with whom he came into contact was too much for her, and it prevented a very close acquaintance.[32]

Another fellow student of those years was Felix Ehrenhaft, a Viennese contemporary of Paul Ehrenfest's. Apart from the similarity in their names the two Viennese physicists had little in common. Felix Ehrenhaft became professor of experimental physics in Vienna, and is best known for his long and eventually bitter controversy with Robert A. Millikan over the existence of a natural unit of electric charge. Millikan showed conclusively that Ehrenhaft's "sub-electrons" did not exist, and the aftermath of this battle and defeat seems to have embittered Ehrenhaft's whole career. (In his later years he provoked a new storm with his report that he had discovered free magnetic poles, a discovery that seems to have died the same death as the "sub-electron".) Not surprisingly, the names of Ehrenhaft and Ehrenfest were often confused: even Boltzmann is reported to have asked Ehrenfest, after the latter had made some clever observation, to use his cleverness to devise a way for Boltzmann to tell Ehrenfest and Ehrenhaft apart.[33] Because of the clash of their personalities, and because he did not care to get credit for Ehrenhaft's later achievements, Ehrenfest did not find this confusion of names very amusing. He was never able to escape it completely, (no matter how hard he tried!)

---

32. Letter to the author from Professor Lise Mietner, 12 February 1958.

33. Personal communication from Professor Philipp Frank. I might add that the typed copy of Boltzmann's opinion of Ehren*fest*'s thesis (see note 29) which was sent to me from Vienna actually refers in one place to Paul Ehren*haft*. I am not sure whether this slip was Boltzmann's in 1904 or the typist's in 1963, but in either case it shows how easily the names are confused.

In the summer of 1904, with his doctorate in hand, Paul Ehrenfest was ready to marry, and Tatyana, who had remained in Göttingen, joined him in Vienna. The marriage was not to take place so easily, however: Paul was Jewish and Tatyana was Russian Orthodox. The laws of the Austro-Hungarian Empire did not permit the marriage of a Jew to a Christian; such a marriage could only take place if the couple officially declared themselves "unchurched", and foreswore all religious affiliations. They proceeded to take this step and marry on December 21, 1904, though the consequences were not negligible. It meant that Tatyana took the risk that she might never be permitted to return to Russia, for Czarist Russia was not anxious to welcome unbelievers within its borders. Since the newlywed couple wanted to take a wedding trip to Russia the following spring, where Paul could meet Tatyana's mother and see her beloved homeland, this was a point of some urgency. Fortunately Tatyana was able to presuade the Russian consul in Vienna that the Ehrenfests, even though unbelievers, would not be a threat to Imperial Russia, and the visas were granted. This Easter visit to Russia in 1905 was also intended to be a foretaste of their future life, since they planned to live there, when and if that became possible.

The Ehrenfests' life in Vienna was centered at the University, though Paul held no position there. He continued to take part in Boltzmann's seminar,[34] study with Lise Meitner, and record his constant flow of ideas and questions. Tatyana was at work, too, investigating the basis of dimensional analysis, a theme that she would pursue again years later. Not all of their life was devoted to science; there was time for walking in Vienna's parks, and for endless conversations in coffee houses and dairy restaurants. Tatyana went through a serious illness early that summer – mumps, with high fever and delirium.[35] They went to Switzerland for her convalescence, where they saw their old Göttingen friend Walther Ritz, full of "a thousand and one new ideas" in physics. There they also met Edmond Bauer, a young French physicist who was to be a friend for many years to come.[36] Paul continued to consume books with his usual catholic taste; that summer his reading went from Gibbs to Freud, from Mach to Gorky.

---

34. Ehrenfest's *Meldungsbogen für den ausserordentlichen Hörer* from the University of Vienna shows the courses he attended in the spring term of 1905.
35. Family records in the "Red Book".
36. Personal communication from Professor E. Bauer.

In October 1905 the Ehrenfests' lives took on a new pattern when their first child was born. She was named after her mother, and Paul Ehrenfest proudly recorded her name, complete with the Russian patronymic, as Tatyana Pavlovna. The Russian form would have seemed natural then, with his mother-in-law visiting from Petersburg to help look after her daughter and the new baby. (In later years, when the two Tatyanas had to be distinguished in writing, Paul Ehrenfest would refer to his wife as T and to his daughter as T′, a notation natural to one who constantly read and wrote mathematics.[37])

The fall of 1905 also saw the publication of the first work that shows Ehrenfest's real interests and particular talents and the beginnings of his personal style.[38] This was a study of the physical assumptions underlying Planck's theory of black-body radiation, a theory whose impact was just beginning to be felt in 1905, five years after it had been propounded. During those months Ehrenfest was also digging more and more deeply into statistical mechanics. His notebooks show him raising questions from a variety of approaches – the paradoxes of the subject, new domains for its applicability, and especially questions on the interrelations between the behavior of a gas and the behavior of black-body radiation.

Paul Ehrenfest was not the kind of thinker who develops his ideas slowly in the solitude of his study. He had to talk about them, to work them out by discussing and arguing them with a critical and competent colleague, and Tatyana was willing and able to play this role. Her quick and extraordinarily logical mind was a natural foil for his more inventive one, and her urge to probe to the very bottom of an idea was as deep as his own. Even when the physical problems for which a theory was invented were not known to her, Tatyana Ehrenfest could sometimes strike directly to the nub of the matter and raise a question of logical structure that set her husband off on the right track. Their collaboration first appeared in print early in 1906 in a joint paper point-

---

37. Ehrenfest may have known that William Thomson and Peter Guthrie Tait referred to each other as T and T′ respectively. These descriptions were also used by Maxwell who was a friend of both men. See L. Campbell and W. Garnett, *The Life of James Clerk Maxwell* (London: Macmillan, 1884) or C. G. Knott, *Life and Scientific Work of Peter Guthrie Tait* (Cambridge: University Press, 1911).

38. P. Ehrenfest, "Über die physikalischen Voraussetzungen der Planck' schen Theorie der irreversiblen Strahlungsvorgänge", Wien. Ber. **114** (1905), p. 1301. *Papers*, p. 88. See Chapter 10.

ing out a fundamental error in the *Statistical Mechanics* of J. Willard Gibbs.[39]

In the late spring of 1906 they left Vienna – never to live there again. They spent the summer in Switzerland, at Weesen am Wallensee, not far from Zürich. It was a good summer for Ehrenfest as he developed his ideas on radiation theory, and continued to ponder the problems of statistical mechanics. At the end of the summer they were back in Göttingen. Ehrenfest was hoping for an academic appointment there, but none developed. His future career was nevertheless critically affected by the events that occurred that autumn.

39. T. and P. Ehrenfest, "Bemerkung zur Theorie der Entropiezunahme in der 'Statistischen Mechanik' von W. Gibbs", Wien. Ber. **115** (1906), p. 89. *Papers*, p. 107. See Chapter 6.

CHAPTER 4

A Dissertation on Mechanics

1. Paul Ehrenfest's doctoral dissertation,[1] written in 1904, dealt with a problem in classical mechanics. He showed that it was possible to use the methods of Heinrich Hertz's *Principles of Mechanics*[2] to derive the equations of motion of an incompressible fluid, and also those of a rigid body moving in an incompressible fluid. The dissertation served its purpose. It demonstrated Ehrenfest's ability to do research in theoretical physics, and his doctorate was duly granted by the University of Vienna.

Looking back now, almost two thirds of a century later, we are likely to wonder why Ehrenfest should have chosen to do a dissertation in classical mechanics at all, and why this particular problem should have been of interest. The thesis itself does not fully answer these questions. It is well enough written, with a clear explanation of just what problem is being studied and how Ehrenfest's work is related to other contemporary treatments. But Ehrenfest naturally took for granted his readers' awareness of the major concerns in the theoretical physics of the time. It is just this awareness that we have lost. A physicist today tends to think that his colleagues at the turn of the

1. Paul Ehrenfest, *Collected Scientific Papers*, ed. M. J. Klein, (Amsterdam: North-Holland, 1959), p. 1.

2. Heinrich Hertz, *The Principles of Mechanics Presented in a New Form*, D. E. Jones and J. T. Walley, Trans. Reprinted (New York: Dover, 1956). Originally published in 1894.

century were concerned exclusively with those problems that led into what he would characterize as modern physics. If one imagines that the only subjects of interest in 1900 were X-rays and radioactivity, or the aether and black-body radiation, then Ehrenfest's thesis must seem to be an oddity. It was not.

Mechanics had not yet lost its place as the central subject within physics, although it was on the verge of losing it. The mechanical explanation of phenomena had been the goal of physicists for some three centuries; this goal was now being reinterpreted, reconsidered and even contested strenuously. Ehrenfest's dissertation did not involve him in these weighty issues, but his work was directly related to problems that still concerned much of the scientific community. In the proper context it should become clear why a promising student of Ludwig Boltzmann's might well have chosen to write his thesis on mechanics, and on this particular problem, in 1904. Later on we will see why this same young theorist never went back to the problems of his thesis, but turned his attention to very different matters.

2. When Heinrich Hertz wrote his *Principles of Mechanics* in the early 1890's, it was still possible to assert, as he did in his preface: "All physicists agree that the problem of physics consists in tracing the phenomena of nature back to the simple laws of mechanics."[3] This bold statement of the mechanical ideal summarized the aim that physicists had been pursuing since Newton's time. The great example of what a physical theory could and should be like was Newtonian celestial mechanics, and physicists tried to explain phenomena by reducing them to the motions of masses moving under their mutually exerted forces. This program of mechanization had impressive successes to its credit, ranging from the electrostatics of Cavendish and Coulomb to the various theories of elasticity put forward by Cauchy, Poisson and others.[4] Perhaps the greatest achievement in extending the domain of mechanics came in the middle of the nineteenth century with the mechanical theory of heat. The law of conservation of energy which expressed heat as a form of energy was rapidly followed by the kinetic theory of gases, in which heat was identified still further as a particular mode of energy, the kinetic energy of molecules. In 1866,

3. H. Hertz, *op. cit.*, Author's Preface.

4. See E. T. Whittaker, *A History of the Theories of Aether and Electricity, Volume 1: The Classical Theories*, Reprinted (New York: Harper, 1960), Chapters II, IV, V.

with all the brashness appropriate to a twenty two year old, Ludwig Boltzmann even proposed a "purely analytical, completely general proof of the second law of thermodynamics", considered as a theorem in mechanics.[5]

These mechanical explanations were generally meant to be taken literally. Matter was supposed really to consist of molecules which moved and interacted in the ways described by the theories in order to account for the phenomena. The mechanical explanations portrayed the actual mechanisms by means of which the phenomena took place. Such explanations were also attempted for the variety of electromagnetic phenomena which attracted more and more attention during the century. There was a whole series of attempts to work out the force laws between charges, currents, magnetic poles, etc., from André-Marie Ampère's memoir of 1825 through the work of Franz Neumann and Wilhelm Weber in the 1840's, extending on at least until the 1870's.[6] But the most fruitful development of electromagnetic theory departed from this action-at-a-distance tradition. This was James Clerk Maxwell's theory of the electromagnetic field, whose point of departure was Michael Faraday's work–both Faraday's experimental results and also his novel physical ideas.

Maxwell, following Faraday's lead, abandoned the forces exerted by one charge (or pole or current) on another some distance away, and replaced these by the local action of the field. As Maxwell put it in 1861, "We are dissatisfied with the explanation founded on the hypothesis of attractive and repellent forces directed towards the magnetic poles... and we cannot help thinking that in every place where we find these lines of force, some physical state or action must exist in sufficient energy to produce the actual phenomena."[7] This basic shift in viewpoint did not mean Maxwell gave up the goal of mechanical explanation. No one who has ever glanced through the paper which has just been quoted, "On Physical Lines of Force", could doubt his devotion to mechanics. For in that paper, Maxwell represented the electromagnetic field with the help of an elaborate mechanical construction involving a latticework of vortex rings in

---

5. Ludwig Boltzmann, "Über die mechanische Bedeutung des zweiten Hauptsatzes der Wärmetheorie", Wien. Ber. **53**, (1866), p. 195. *Wiss. Abh.* I, p. 9.

6. See Whittaker, *op. cit.*, Chapter VII.

7. James Clerk Maxwell, "On Physical Lines of Force", Phil. Mag (4) **21**, (1861). *The Scientific Papers of James Clerk Maxwell*, ed. W. D. Niven, Reprinted (New York: Dover, 1965) I, p. 451. (Quote on p. 452.)

an ideal fluid, the vortices being coupled by particles acting like idle wheels. Maxwell, however, adopted a new position with respect to the status of his mechanisms, and his new way of pursuing the mechanical ideal influenced many of his successors, certainly including Hertz.

In his first paper on electromagnetism[8], Maxwell carefully set forth the method he proposed to follow in his work. He would not begin by making "a purely mathematical formula", for then one would "entirely lose sight of the phenomena to be explained". Nor would he prematurely make "a physical hypothesis", that is, an assumption as to the real nature of the phenomena to be explained, for that would lead to "seeing the phenomena only through a medium", making one "liable to that blindness to facts and rashness in assumption which a partial explanation encourages". Instead of either of these Maxwell would exploit physical analogies. He meant something quite specific by a physical analogy, "that partial similarity between the laws of one science and those of another which makes each of them illustrate the other". In other words, to the extent that two physical systems obey laws whose mathematical form is the same, the behavior of one system may be understood by studying the behavior of the other, better known, system. And this could be done without making any hypothesis about the real nature of the system under investigation. Maxwell stated clearly that a dynamical analogy, valuable as it might be, was not a substitute for "a mature theory, in which physical facts will be physically explained". The system of vortices and idle wheel particles of his 1861 papers constituted such a dynamical analogy, offered "not as a mode of connexion actually existing in nature", but only as "a mode of connexion which is mechanically conceivable and easily investigated".[9] Nevertheless, it was the analysis of this dynamical analogy that brought Maxwell to the first formulation of the electromagnetic theory of light.

In the complete version of his theory, published in 1865, the specific mechanical model had disappeared, to be mentioned only in a passing reference halfway through the long memoir.[10] This theory was, however, still what his title implied, "A Dynamical Theory of the

---

8. James Clerk Maxwell, "On Faraday's Lines of Force", Trans. Cambr. Phil. Soc. **10** (1855), *Sci. Papers* I, p. 155.
9. J. C. Maxwell, *op. cit.*, note 7. *Sci. Papers* I, p. 486.
10. J. C. Maxwell, "A Dynamical Theory of the Electromagnetic Field", Trans. Roy. Soc. **155** (1865), p. 459. *Sci. Papers* I, p. 526. (See p. 563)

Electromagnetic Field", "a *Dynamical* Theory, because it assumes that in that space, (in the neighborhood of the electric or magnetic bodies), there is matter in motion, by which the observed electromagnetic phenomena are produced".[11] Maxwell still conceived of the aether, the medium of the field, as "a complicated mechanism capable of a vast variety of motion", and asserted that "such a mechanism must be subject to the general laws of Dynamics".[12] In a letter written to Peter Guthrie Tait a couple of years later Maxwell contrasted his vortex and particle model with the later, more schematic, dynamical analogy. "The former is built up to show that the phenomena (of electromagnetism) are such as can be explained by mechanism. The nature of the mechanism is to the true mechanism what an orrery is to the Solar System. The latter is built on Lagrange's Dynamical Equations and is not wise about vortices."[13]

Maxwell commented again on his new and more subtle concept of dynamical analogy in his *Treatise* of 1873. The only basic assumption he needed was that the system of charges and currents with its fields was a dynamical system. Lagrangian mechanics then allowed him to proceed without detailed knowledge of "the nature of the connexions of the parts of this system".[14] Even this general dynamical treatment was far from a complete theory, however, because of the lack of evidence as to the true nature of the electric current. Only when such questions were answered could a really explanatory theory be created. "A knowledge of these things would amount to at least the beginnings of a complete dynamical theory of electricity in which we should regard electrical action, not, as in this treatise, as a phenomenon due to an unknown cause, subject only to the general laws of dynamics, but as the result of known motions of known portions of matter, in which not only the total effects and final results, but the whole intermediate mechanism and details of the motion, are taken as the objects of study."[15]

3. Maxwell's success with a theory based on dynamical analogies,

---

11. *Ibid.*, p. 527.
12. *Ibid.*, p. 533.
13. J. C. Maxwell to P. G. Tait, 23 December 1867. Quoted in: C. G. Knott, *Life and Scientific Work of Peter Guthrie Tait*, (Cambridge: Cambridge University Press, 1911), p. 215.
14. J. C. Maxwell, *A Treatise on Electricity and Magnetism*, 3rd Ed., 1891 Reprinted (New York: Dover Publications, 1954), Vol. II, p. 213.
15. *Ibid.*, Vol. II, p. 218.

special or general, stimulated a variety of reactions from his contemporaries. Some saw the dynamical analogy as only a mechanical model. To Lord Kelvin this was something to be admired. "It seems to me that the test of 'Do we or do we not understand a particular subject in physics?' is 'Can we make a mechanical model of it?' I have an immense admiration for Maxwell's mechanical model of electromagnetic induction. He makes a model that does all the wonderful things that electricity does in inducing currents, etc...."[16] But when Maxwell's analogy was only to the general structure of the dynamical equations, Kelvin was much less delighted and preferred trying to construct his own models. Others were distressed by Maxwell's models and lamented his failure to provide a proper deductive theory. The most outspoken of these was Pierre Duhem, whose remarks against Maxwell and all the other "ample but weak minds" among English physicists make up one of the liveliest, if not one of the best, chapters in his book *The Aim and Structure of Physical Theory*.[16]

Other physicists recognized the value of the concept of physical analogy in trying to understand the essential features of the natural world. Hermann von Helmholtz, that "intellectual giant", as Maxwell called him, was one of these. Helmholtz, who had independently formulated the law of the conservation of energy while still in his twenties, and who had then spent almost forty years doing major work in just about every one of the sciences, turned his attention to thermodynamics again in the 1880's. After a series of memoirs on the application of thermodynamics to chemistry, he took up the general problem of understanding the second law on the basis of mechanics.[17]

What Helmholtz attempted was not a detailed mechanical explanation in the old sense, where the actual mechanism of thermodynamic

---

16. Quoted in P. Duhem, *The Aim and Structure of Scientific Theory*, trans. P. P. Wiener (Princeton: Princeton University Press, 1954), p. 71.

17. (a) L. Koenigsberger, *Hermann von Helmholtz*, trans. F. A. Welby. Reprinted (New York: Dover, 1965).

(b) H. v. Helmholtz, "Studien zur Statik monocyklischer Systeme", Berliner Berichte (1884), pp. 159, 311.

(c) H. V. Helmholtz, "Principien der Statik monocyklischer Systeme", Crelles Journal **97** (1884), pp. 111, 317.

(d) H. v. Helmholtz, *Vorlesungen über Theoretische Physik, Vol. VI, Vorlesungen über Theorie der Wärme*, ed. F. Richarz (Leipzig: Barth, 1903), p. 338.

(e) J. Larmor and G. H. Bryan, "On our Knowledge of Thermodynamics", B. A. A. S. Report **61** (1891), p. 96.

systems would be exposed; nor was it a statistical mechanical explanation, such as Boltzmann had been developing since 1871, where the second law of thermodynamics became a kind of joint consequence of the laws of mechanics and the laws of chance. Instead it was a physical analogy, precisely in Maxwell's sense of the term (although Helmholtz did not refer explicitly to Maxwell in this matter).

Helmholtz showed that there was a class of simple mechanical systems whose equations of motion could be put in a form that, in many ways, exactly resembled the second law of thermodynamics. These were the systems he called monocyclic. In the simplest case a monocyclic system has one cyclic coordinate; that is, one of the Lagrangian generalized coordinates needed to specify the configuration of the system does not appear in the expression for the energy of the system. If the velocity associated with this cyclic coordinate is large compared to the rates of change of any other coordinates, then Helmholtz could show that the energy dQ added to the system by a change in the cyclic coordinate had the form TdS, where T is the kinetic energy and S is a function of the state of the system. As the notation suggests, dQ represents the heat added to the system, and T and S correspond to temperature and entropy, respectively. The idea is that a cyclic coordinate, such as the angle of rotation for a system rotating uniformly and rapidly about an axis, is a concealed coordinate, in the sense that its changes do not show up in the energy of the system. To this extent it resembles the molecular coordinates of a real thermodynamic system, since the locations of the molecules of a gas, for example, are irrelevant in determining the energy of the gas. To change the energy of the system one can either change the non-cyclic, slowly varying, coordinates, (the parameters of the system), thereby doing work on the system, or one can change the energy by modifying the kinetic energy associated with the cyclic coordinate, a process that corresponds to heat flow into (or out of) the system.

From the outset Helmholtz emphasized that real thermal motion is not monocyclic and that his whole discussion of monocyclic systems was an analogy rather than an explanation. Analysis of the molecular dynamics of real thermodynamic systems was impossibly difficult. "In these circumstances", Helmholtz wrote, "it seems completely rational to me to look for the most general conditions under which the most general physical characteristics of thermal motion can appear in other well known classes of motion. It is in this sense, of course, that I have called special attention to the analogies between the behavior of thermal motion and of the monocyclic motions that I have investigated."[18]

Helmholtz's analogy had its limitations and difficulties, but despite these limitations it made a deep impression on Heinrich Hertz and was a major influence on Hertz's last work, *The Principles of Mechanics Presented in a New Form.*

4. The most characteristic feature of Hertz's new axiomatic version of mechanics is his treatment of the concept of force. Hertz deprived force of its fundamental position and set up his mechanics without it, reintroducing it only later on as a useful but unnecessary auxiliary concept. He was dissatisfied with the "logical obscurity" surrounding the concept of force, which made him comment: "It is exceedingly difficult to expound to thoughtful hearers the very introduction to mechanics without being occasionally embarrassed, without feeling tempted now and again to apologise, without wishing to get as quickly as possible over the rudiments, and on to examples which speak for themselves. I fancy that Newton himself must have felt this embarrassment...."[19] According to Hertz, force was introduced because we cannot account for motion, of a falling stone, for example, by using only what can be directly observed. "If we wish to obtain an image of the universe which shall be well-rounded, complete and conformable to law, we have to presuppose, behind the things which we see, other, invisible things."[20] Force is such an "invisible thing", but Hertz preferred to proceed another way.

"We may admit that there is a hidden something at work, and yet deny that this something belongs to a special category. We are free to assume that this hidden something is nought else than motion and mass again – motion and mass which differ from the visible ones not in themselves but in relation to us and to our usual means of perception." These concealed masses, whose couplings to the visible masses of our experience are to account for the latter's motion, are Hertz's substitute for forces. "What we are accustomed to denote as force and as energy now become nothing more than an action of mass and motion, but not necessarily of mass and motion recognisable by our coarse senses."

18. H. v. Helmholtz, Berliner Berichte (1884), p. 757.
19. H. Hertz, *op. cit.*, p. 6.
20. *Ibid.*, p. 25.

Thus, to cite an example used by Duhem[21], if we pull a mass tied by an invisible thread to a second mass, which will then follow it in motion, we would be tempted to postulate a force of attraction to explain the motion of the second mass, whereas the geometrical constraint of the invisible string can account for the motion without a "real force".

Hertz develops his mechanics by considering a system of particles subject only to such constraints or connections as can be expressed by linear differential forms in the particle coordinates. His single postulated law of motion applies to the motion of a free system, one in which the constraints are independent of time and affect only the relative coordinates of the particles. This postulate requires that the path, defined in the $3n$ dimensional space determined by the coordinates of the n particle system, be as straight as possible subject to the constraints, and be traversed with uniform motion. (Hertz expresses all of this sharply, of course, with the help of a certain amount of technical apparatus taken from differential geometry, but we need not introduce that yet.) The essential point is that, for Hertz, a system which we would normally consider as acted upon by forces is always thought of as part of a free system, the remainder of the system consisting of other masses which may or may not be perceptible to us. All "forces" appear only as Lagrange's undetermined multipliers associated with the geometrical connections between the masses of the observed subsystem and the complementary masses that complete the free system of which it forms a part. Forces are strictly auxiliary constructs in this formulation of mechanics.

The whole Hertzian system is worked out in a closely reasoned analysis, presented in strict deductive style. Hertz's arguments demonstrated that his differential principle of least curvature provided an alternate starting point for mechanics, from which all the usual formulations – Newton's laws, Lagrange's equations, Hamilton's principle – could be deduced as theorems. What he did not do in his *Mechanics* was to show how, or even whether, one could construct the set of concealed masses, and their connections with a given system, in such a way as to account for the observed motions of this system. As Helmholtz wrote in his preface, (a eulogy to Hertz who had died, at the age of 36, before his book appeared): "Unfortunately he has not given examples illustrating the manner in which he supposed such hypothetical mechanism to act; to explain even the simplest cases of physical

21. P. Duhem, *L'Évolution de la mécanique* (Paris: A. Joanin, 1903), p. 160.

forces on these lines will clearly require much scientific insight and imaginative power."[22] The general reaction was that Hertz had offered only an "ideal program", to be achieved in the distant future, if at all.

Quite apart from the question of its usefulness, Hertz's book is remarkably interesting for its treatment of the role of mechanical explanation in theoretical physics. His basic idea of explaining observed motion with the help of the motion of concealed masses rather than forces is obviously related to earlier successes in explaining phenomena mechanically. Hertz referred to these explicitly, giving particular mention to Helmholtz's treatment of "the most important form of concealed motion", (the monocyclic systems), and to "Maxwell's labours.... [showing] that electromagnetic forces are due to the motion of concealed masses".[23] It seems that one of Hertz's goals was to reformulate mechanics in such a way as to facilitate mechanical explanation of the properties of the electromagnetic aether, which he describes as "the field in which the decisive battle between these different fundamental assumptions of mechanics must be fought out".[24] For Hertz, however, mechanical explanation could only be understood in the sense of Maxwell's dynamical analogies.

He was quite explicit on this point in the section of his book entitled "Dynamical Models".[25] On the basis of a sharp definition of dynamical model, Hertz proved that: "In order to determine beforehand the course of the natural motion of a material system, it is sufficient to have a model of that system. The model may be much simpler than the system whose motion it represents." But he then went on to assert that the existence of concealed masses, (essential to his mechanics), means that "it is impossible to carry our knowledge of the connections of material systems further than is involved in specifying models of the actual systems. We can then, in fact, have no knowledge as to whether the systems which we consider in mechanics agree in any other respect with the actual systems of nature which we intend to consider, than in this alone, – that the one set of systems are models of the other." And, finally, Hertz saw "the relation of a dynamical model to the system of which it is regarded as the model" to be "precisely the

22. H. Hertz, *op. cit.*, last page of Helmholtz's Preface. (Unnumbered pages).
23. *Ibid.*, p. 26.
24. *Ibid.*, p. 41.
25. *Ibid.*, pp. 175–177.

same as the relation of the images which our mind forms of things to the things themselves".

5. The various ideas on mechanics and the mechanical interpretation of nature which we have been considering found their most receptive audience in Ludwig Boltzmann, for whom mechanics was always the very core of physics.[26] Throughout his career, Boltzmann admired, developed and expounded Maxwell's ideas. Just those features of Maxwell's papers on electromagnetism that put off so many continental physicists, the dynamical analogies, particularly appealed to him. He translated Maxwell's papers of 1855 and 1861 into German for Ostwald's series of reprints of scientific classics.[27] (Boltzmann's elaborate and detailed editorial notes on these papers still serve as a valuable guide to the reader.) He placed great importance on Maxwell's concept of analogies, describing Maxwell as having been "as much of a pioneer in the theory of knowledge as in theoretical physics",[28] and as having clearly adumbrated in 1855 all the various approaches to the nature of scientific theory that would be developed during the subsequent forty years.

Boltzmann himself found the concept of a scientific theory as an analogy or metaphor of reality a particularly liberating one.[29] It allowed him to continue to develop mechanical explanations without having to assert, for example, that a gas "really" consists of molecules that "really" interact with one another according to a particular force law. If a scientific theory is only an image or picture of nature, one need not worry about developing "the only true theory", and one can be content to portray the phenomena as simply and clearly as possible. From this platform Boltzmann defended the mechanical theories against the claims of those who demanded their replacement by theories of some other, particular, kind. He devoted much time and effort in the nineties to this defense, arguing that neither the believers in energetics nor the

---

26. See the various lectures on mechanics printed in L. Boltzmann, *Populäre Schriften* (Leipzig: J. A. Barth, 1905), pp. 198–227, 253–337. Also see R. Dugas, *La théorie physique au sens de Boltzmann* (Neuchâtel: Éditions du Griffon, 1959), pp. 30–125.
27. Ostwald's Klassiker der exakten Wissenschaften:
(a) No. 69 J. C. Maxwell, *Über Faraday's Kraftlinien* (1895)
(b) No. 102 J. C. Maxwell, *Über physikalische Kraftlinien* (1898) Both edited by L. Boltzmann (Leipzig: Wilhelm Engelmann).
28. See *op. cit.* note 27 (a), pp. 99–100.
29. *Ibid.*, pp. 100–101. Also see L. Boltzmann, *op. cit.* note 26, p. 216.

believers in purely phenomenological theories had a unique passkey to the domain of truth. No approach to physical theory should be excluded so long as it could provide insight, as mechanical theories had in the past and were likely to do in the future.[30]

When Helmholtz's papers on monocycles appeared, Boltzmann immediately seized upon them. He wrote three papers on the subject himself in the years 1884–86[31], giving new proofs of some of the theorems, pointing out a gap in Helmholtz's argument, exploring the relations between the monocycle analogy and his own work on the second law, and discussing many particular examples in great detail. The whole subject exactly suited him. His *Lectures on Maxwell's Theory of Electricity and Light*[32] show how useful he found the idea of cycles. The stationary electric current is treated as a monocycle, coupled circuits as bicycles, etc. Boltzmann not only showed the equivalence of the differential equations describing the electrical and mechanical systems (in the style of Maxwell's *Treatise*, where some of the same analogies are discussed without Helmholtz's terminology), he also worked out elaborately detailed mechanical constructions that would realize these equations. An unwary reader, glancing through the illustrations in Boltzmann's *Lectures*, might well wonder if he had mistakenly picked up a book on the design of engineering mechanisms.

Boltzmann had mixed reactions to Hertz's *Mechanics*, as did most readers. He admired its structure, the simplicity of its premises and the care with which they had been thought through, but still considered it at best a program for the future. He deliberately chose to follow the traditional order of exposition in his own *Lectures on the Principles of Mechanics*[33], preferring forces to the more awkward and artificial

30. L. Boltzmann, *op. cit.* note 26, pp. 1–10, 70–72, 104–161, 198–227.

31. L. Boltzmann: (a) "Über die Eigenschaften monozyklischer und anderer damit verwandter Systeme", Crelles Journal, **98** (1884), p. 68. *Wiss. Abh.* III, p. 122.

(b) "Über einige Fälle wo die lebendige Kraft nicht integrierender Nenner des Differentials der zugeführten Energie ist", Wien. Ber. **92** (1885), p. 853, *Wiss. Abh.* III, p. 153.

(c) "Neuer Beweis eines von Helmholtz aufgestellten Theorems betreffend die Eigenschaften monozyklischer Systeme", Gött. Nachr. (1886), p. 209. *Wiss. Abh.* III, p. 176.

32. L. Boltzmann, *Vorlesungen über Maxwells Theorie der Elektricität und des Lichtes* (Leipzig: J. A. Barth, I 1891, II 1893).

33. L. Boltzmann, *Vorlesungen über die Principe der Mechanik* (Leipzig: J. A. Barth, I 1897, II 1904). See I, pp. 1–6. Also see L. Boltzmann, *op. cit.* note 26, pp. 253–270.

looking concealed masses. He made a similar choice in his book on the kinetic theory of gases[34], where the use of Hertz's methods would only have required "more new arbitrary assumptions".

In 1900 Boltzmann had occasion to deal with Hertz's mechanics in a very specific context. Hertz had always worked with a system of an arbitrary but finite number of discrete masses – a system of particles – although he evidently anticipated that there would be no difficulty in extending his methods to a continuous distribution of mass, such as a fluid. Such an extension was attempted by R. Reiff[35], but Boltzmann was quick to point out that it was fundamentally in error[36]. Reiff had mistakenly tried to derive the pressure forces in a fluid from just the conservation of mass, and had obtained a result only because of a compensating error in his mathematics. In addition to pointing out Reiff's error and showing how to correct it, Boltzmann took the opportunity to make several remarks.

The first was very general, but negative. He commented that a proper extension of Hertz's mechanics to fluids might possibly be developed into a detailed picture of the entire physical world. Boltzmann meant that one would then be in a position to construct a definite representation of the Hertzian concealed motions by thinking of them as vortex rings in an ideal fluid, just as Kelvin had imagined the atom as a vortex. But what would be the advantage in such a construction? It would necessarily involve at least as many arbitrary and artificial assumptions as the old picture of discrete atomic particles interacting by forces exerted at a distance. Boltzmann thought that the value as well as the beauty of Hertz's approach were to be found in its generality; these would necessarily be sacrificed if one had to introduce arbitrary assumptions in trying to understand phenomena by using concealed motions.

Boltzmann's second remark was more specific, and positive. He agreed with a suggestion published the preceding year by A. Brill about the proper line to follow in extending Hertz's mechanics to

---

34. L. Boltzmann, *Lectures on Gas Theory*, transl. S. G. Brush (Berkeley and Los Angeles: University of California Press, 1964), p. 26. (Original edition published by Barth in 1896, 1898).
35. R. Reiff, "Die Druckkräfte der Hydrodynamik und die Hertz'sche Mechanik", Ann. Phys. (4) **1** (1900), p. 226.
36. L. Boltzmann, "Die Druckkräfte der Hydrodynamik und die Hertz'sche Mechanik", Ann. Phys. (4), **1**, (1900), p. 673. Wiss. Abh. III, p. 665.

fluids[37]. Instead of taking the Eulerian approach to hydrodynamics, (as Reiff had tried to do), where one works out the equations of motion of the fluid by considering what happens at a point fixed in space as the fluid flows by, one ought to adopt the Lagrangian method. Here one's attention is fixed on an element of the fluid as it moves through space, and this would be the natural approach to follow in generalizing a particle theory. Boltzmann was sure that the equations of motion of an incompressible fluid, and also of a rigid body submerged in such a fluid, could be derived from Hertz's principles in this way.

This suggestion of the proper method for generalizing Hertz's mechanics to fluids, made by Brill and endorsed by Boltzmann, was not immediately taken up. Boltzmann had an opportunity to repeat the suggestion, however, in his seminar on Hertz's mechanics at the University of Vienna in 1903, and this time it reached the right person – Paul Ehrenfest.

6. Ehrenfest decided to follow up Boltzmann's seminar remark and study the problem. He had already been introduced to Hertz's mechanics through several lectures by Lorentz which he attended during his visit to Leyden the previous spring.[38] Lorentz had become interested enough in Hertz's work to write a paper on it himself,[39] so that Ehrenfest had good precedents to justify his studies.

Ehrenfest's problem turned out to be a good deal richer than might have been expected. It was not that there were any great difficulties in extending Hertz's methods to the continuous fluid, but rather that treating the motion of a rigid body in such a fluid required a good deal of careful analysis of some nice points in classical dynamics. In order to see what was involved let us begin with Hertz's treatment of a system of a finite number of particles.

Suppose that we have a system of n particles. The cartesian coordinates of the system are denoted by x_i, where the index i runs from 1 to $3n$, and where, for example, x_1, x_2, x_3 are the x, y, z coordinates of the first particle. The masses of the particles are given by m_i, $(i = 1, \dots 3n)$, where $m_1 = m_2 = m_3$, $m_4 = m_5 = m_6$, etc. The total mass m of the system is then $\frac{1}{3}\Sigma_{i=1}^{3n} m_i$. Hertz defines an infinitesimal displace-

37. A. Brill, Mitteilungen d. naturw. Vereins in Württemberg (1899).

38. P. Ehrenfest, Diary for 1903.

39. H. A. Lorentz, "Some Considerations on the Principles of Dynamics, in Connexion with Hertz's 'Prinzipien der Mechanik'", Proc. Acad. Amst. **3** (1902), p. 713. *Collected Papers* (The Hague: M. Nijhoff, 1934–39) IV, p. 36.

ment of the system in $3n$-dimensional space by the differentials dx_i, and the length ds of the system's displacement by the equation

$$m(ds)^2 = \sum_{i=1}^{3n} m_i(dx_i)^2 . \qquad (4.1)$$

The curvature c of the path followed by the system in $3n$ space is given by the equation

$$mc^2 = \sum_{i=1}^{3n} m_i \left(\frac{d^2x_i}{ds^2}\right)^2 . \qquad (4.2)$$

For a free system, one subject to no external forces (and Hertz considers only such systems), his single postulate is this: the actual path is the one of least curvature (the straightest path), and it is traversed at a uniform rate. Expressed symbolically, this postulate has the form

$$\delta(mc^2) = 0. \qquad (4.3)$$

Because of the uniform rate, the derivatives with respect to path length s are just proportional to time derivatives, and the principle of least curvature is equivalent to Gauss's principle of least constraint for a system under no external forces. The variations in (4.3) are considered by Hertz as variations in the accelerations $\ddot{x}_i$ for fixed values of all positions x_i and velocities $\dot{x}_i$.

If all the particles moved independently of one another, (4.3) would simply lead to Newton's first law of motion: all accelerations would vanish on the actual path. The principle of least curvature becomes interesting only when there are interactions among the particles, which, for Hertz, means constraints on the values of the x_i, since he also does not use internal forces explicitly. Although the constraints need not be holonomic (integrable), let us assume that they are. They can be satisfied identically by introducing an appropriate set of independent generalized coordinates p_u where u runs from 1 to r. Here r is the requisite number of generalized coordinates, the number of degrees of freedom of the system, and is equal to $3n$ less the number of constraint equations. Any configuration of the system compatible with the constraints is specified by giving the values of $p_1, \dots, p_r$, and we have a set of equations of the type

$$x_i = x_i(p_1, \dots, p_r) \qquad (i = 1, \dots, 3n). \qquad (4.4)$$

The principle of least curvature can now be applied by expressing the accelerations $\ddot{x}_i$, which enter into the curvature, in terms of the generalized velocities $\dot{p}_u$ and generalized accelerations $\ddot{p}_u$. When this is done the equations of motion derived from Hertz's principle are, not unexpectedly, Lagrange's equations for a system having no potential energy:

$$\frac{\mathrm{d}}{\mathrm{d}t}\left(\frac{\partial T}{\partial \dot{p}_u}\right) - \frac{\partial T}{\partial p_u} = 0 \qquad (u = 1, \ldots, r). \tag{4.5}$$

Here T is the kinetic energy of the system:

$$T = \tfrac{1}{2}\sum_{i=1}^{3n} m_i \dot{x}_i^2 = \tfrac{1}{2}\sum_{u=1}^{r}\sum_{v=1}^{r} \dot{p}_u \dot{p}_v \sum_{i=1}^{3n} m_i \alpha_{iu}\alpha_{iv} \tag{4.6}$$

where

$$\alpha_{iu} = \frac{\partial x_i}{\partial p_u}. \tag{4.7}$$

Now let us turn to Ehrenfest's problem. In the Lagrangian method for hydrodynamics one follows the particles of the fluid as they move. The particle at (x, y, z) at time t is identified by the position (a, b, c) it occupied at a fixed time t_0. Instead of a discrete sum over the coordinates, of the type $\Sigma_{i=1}^{3n} f(x_i)$, one now has an integral extended over the region of space in which the fluid is to be found at t_0,

$$\int\int\int \mathrm{d}a\, \mathrm{d}b\, \mathrm{d}c\, \{f(x) + f(y) + f(z)\}.$$

The mass m_i is replaced by $\rho\, \mathrm{d}a\, \mathrm{d}b\, \mathrm{d}c$, where ρ is the local density, and so on.

Ehrenfest first showed how the pressure could be introduced in a Hertzian treatment. Since forces could not be used as such, the pressure had to appear as the undetermined multiplier of a constraint, and that constraint was the equation of incompressibility. In Lagrangian form imcompressibility is expressed by the condition[40]

$$\Delta \equiv \begin{vmatrix} \frac{\partial x}{\partial a} & \frac{\partial y}{\partial a} & \frac{\partial z}{\partial a} \\ \frac{\partial x}{\partial b} & \frac{\partial y}{\partial b} & \frac{\partial z}{\partial b} \\ \frac{\partial x}{\partial c} & \frac{\partial y}{\partial c} & \frac{\partial z}{\partial c} \end{vmatrix} = 1. \tag{4.8}$$

40. H. Lamb, *Hydrodynamics* (Cambridge: Cambridge University Press, 4th Edn. 1916) pp. 13–14.

Variations of the curvature of the path of the fluid must be made allowing for this constraint, using a multiplier λ. Since it is the accelerations at different points of space that are varied, the basic equation has the form

$$\delta \iiint \tfrac{1}{2}\rho(\ddot{x}^2+\ddot{y}^2+\ddot{z}^2)\, \mathrm{d}a\, \mathrm{d}b\, \mathrm{d}c - \iiint \lambda\,\delta\left(\frac{\mathrm{d}^2\Delta}{\mathrm{d}t^2}\right) \mathrm{d}a\, \mathrm{d}b\, \mathrm{d}c = 0. \tag{4.9}$$

Working from this equation Ehrenfest showed that λ had to satisfy the defining equations for the pressure. That is to say, for a freely moving "ball" of liquid, one has first the condition that ρ times the acceleration at any point in the interior is equal to the negative gradient of λ, and also the condition that λ must vanish on the free surface.

Much more interesting than the free fluid alone was the problem of the rigid body moving in a fluid, particularly under certain additional restrictions. Suppose one has a system of rigid bodies and incompressible fluid such that: a) the fluid is bounded exclusively by the rigid bodies, i.e. no free surfaces; b) the region occupied by the fluid is simply connected; and c) the fluid motion is irrotational. Under these circumstances there exists a velocity potential for the fluid, and this potential Φ is completely determined by the motion of the rigid bodies. If $p_1 \ldots p_r$ are a set of generalized coordinates for the rigid bodies, the velocity potential has the form[41],

$$\Phi = \sum_{u=1}^{r} \phi_u \dot{p}_u . \tag{4.10}$$

The ϕ_u are functions of the p's and x, y, z, and each ϕ_u satisfies Laplace's equation inside the fluid. The fluid velocity at any point is the gradient of Φ, so that one can write

$$\dot{x} = \frac{\partial \Phi}{\partial x} = \sum_{u=1}^{r} \frac{\partial \phi_u}{\partial x}\, \dot{p}_u \tag{4.11}$$

with similar equations for $\dot{y}$ and $\dot{z}$. If we define coefficients $\alpha_u, \beta_u, \gamma_u$ by the equations

$$\alpha_u \equiv \frac{\partial \phi_u}{\partial x}, \qquad \beta_u \equiv \frac{\partial \phi_u}{\partial y}, \qquad \gamma_u \equiv \frac{\partial \phi_u}{\partial z}, \tag{4.12}$$

then (4.11) can be written in the form

$$\begin{aligned} \dot{x} &= \sum_{u=1}^{r} \alpha_u \dot{p}_u \\ \dot{y} &= \sum_{u=1}^{r} \beta_u \dot{p}_u \\ \dot{z} &= \sum_{u=1}^{r} \gamma_u \dot{p}_u \end{aligned} \tag{4.13}$$

41. *Ibid.*, p. 152.

These equations show that the fluid velocity at any point is determined by the set of generalized velocities $\dot{p}_1, \ldots \dot{p}_r$ of the rigid bodies. The coefficients $\alpha_u, \beta_u, \gamma_u$ are themselves functions of the p's and the position x, y, z. With the help of these equations Ehrenfest could now express the "curvature" for the combined system of fluid and rigid bodies in terms of only the generalized velocities, $\dot{p}_u$, and the generalized accelerations, $\ddot{p}_u$. There were no further constraints to be allowed for; the use of the velocity potential (4.10) automatically took care of the various hydrodynamic conditions. When he worked out the variation in the curvature and determined the equations of motion Ehrenfest found that they could be written in the form

$$\frac{\mathrm{d}}{\mathrm{d}t}\left(\frac{\partial \mathscr{T}}{\partial \dot{p}_u}\right) - \frac{\partial \mathscr{T}}{\partial p_u} + \sum_{v=1}^{r} \sum_{w=1}^{r} \dot{p}_v \dot{p}_w \iiint \mathrm{d}a\, \mathrm{d}b\, \mathrm{d}c\, \rho \left\{ \alpha_v \left(\frac{\partial \alpha_w}{\partial p_u} - \frac{\partial \alpha_u}{\partial p_w} \right) + \beta_v \left(\frac{\partial \beta_w}{\partial p_u} - \frac{\partial \beta_u}{\partial p_w} \right) + \gamma_v \left(\frac{\partial \gamma_w}{\partial p_u} - \frac{\partial \gamma_u}{\partial p_w} \right) \right\} = 0, \tag{4.14}$$

where $\mathscr{T}$ is the total kinetic energy of fluid and rigid bodies. Thus, the equations of motion deduced from Hertz's principle of least curvature were not simply of the Lagrangian form, but had additional terms of a rather complex character. This was a most remarkable result: the equations of motion for this problem were "well known", had already appeared in various textbooks, and *did* have the Lagrangian form.[42]

Ehrenfest's real problem, then, was not to derive the equations of motion by Hertz's methods, but to understand why these methods gave a result that differed from that obtained by more standard routes. And since the simpler equations already studied by others were almost surely the correct ones, Ehrenfest also had to figure out how his extra terms could be made to vanish.

7. The supplementary terms have an interesting structure involving combinations of derivatives like $(\partial\alpha_w/\partial p_u) - (\partial\alpha_u/\partial p_w)$. If we look at (4.13) we see that the vanishing of such differences of derivatives would imply the integrability of these equations. In other words, if all of these cross derivatives were equal, there would be equations determining x, y and z in terms of the generalized coordinates of the rigid bodies. The location and orientation of the rigid bodies would fix the

42. *Ibid.*, pp. 175–181.

positions of all points in the fluid. There is no reason to believe that this should occur.

To put it another way, the equations (4.13) look very much like the differential form of the equations that express the cartesian coordinates of a system in terms of generalized coordinates, equations of the type (4.4). For, the differential form of (4.4) could be written

$$\mathrm{d}x_i = \sum_{u=1}^{r} \alpha_{iu}\mathrm{d}p_u \tag{4.15}$$

where α_{iu} was already defined in (4.7). But the existence of the integral relations (4.4) would require that $(\partial\alpha_{iw}/\partial p_u) - (\partial\alpha_{iu}/\partial p_w)$ vanishes. Ehrenfest's extra terms can therefore be attributed to the fact that the quantities $p_1 \ldots p_r$ do not have all the properties required of a set of generalized coordinates: their differentials determine the differentials $\mathrm{d}x$, $\mathrm{d}y$, $\mathrm{d}z$, but the equations are not generally integrable.

Boltzmann had actually studied the exact analogue of this situation for a discrete system only a year or so earlier.[43] Boltzmann considered a system of particles described by generalized coordinates whose values did not fully specify the cartesian coordinates, a system where the basic relations had the form (4.15) but were not integrable. He called such a set $\{p_u\}$ non-holonomic coordinates. Boltzmann showed that, if one started from d'Alembert's principle and proceeded as one usually does in deriving Lagrange's equations, the equations of motion had certain extra terms in them:

$$\frac{\mathrm{d}}{\mathrm{d}t}\left(\frac{\partial L}{\partial \dot{p}_u}\right) - \frac{\partial L}{\partial p_u} + \sum_{v=1}^{r}\sum_{w=1}^{r} \dot{p}_v\dot{p}_w \sum_{i=1}^{3n} m_i\alpha_{iv}\left(\frac{\partial\alpha_{iw}}{\partial p_u} - \frac{\partial\alpha_{iu}}{\partial p_w}\right) = 0. \tag{4.16}$$

In these equations L is the Lagrangian of the system, which reduces to just the kinetic energy T if there are no forces. Clearly (4.16) is the precise discrete equivalent of Ehrenfest's equation (4.14). The extra terms in these equations naturally vanish if the coordinates p_u are holonomic, i.e. if equations (4.15) are integrable.

(In the present context it is especially interesting that Boltzmann came upon the problem of non-holonomic coordinates in connection

43. L. Boltzmann, "Über die Form der Lagrangeschen Gleichungen für nichtholonome, generalisierte Koordinaten", Wien. Ber. **111** (1902), p. 1603. *Wiss. Abh.* III, p. 682. Also see L. Boltzmann, *op. cit.* note 33, II, pp. 104–115, E. T. Whittaker, *Analytical Dynamics* (Cambridge: Cambridge University Press, 4th Edn. 1937), pp. 41–44, and G. Hamel, Zs. f. Math. u. Phys. **15** (1904), p. 1.

with his study of Helmholtz's monocyclic systems.[31a] He noticed that, for one of the mechanical models he investigated, Lagrange's equations were obviously not the correct equations of motion. Further analysis showed him that the coordinates he was using were non-holonomic. Under these circumstances he felt it was "well worth the trouble to ask which equations appear instead of Lagrange's equations for non-holonomic coordinates", and wrote a paper in which the question was answered and the general equations derived.)

The extra terms in the equations of motion, which both Boltzmann and Ehrenfest had found, had their origin in the use of non-holonomic coordinates, and could be derived by using Hertz's principle or d'Alembert's principle. The next question Ehrenfest had to answer was why these terms had not appeared in earlier discussions of the motion of rigid bodies through a liquid. Earlier derivations of the equations had started from an integral variational principle rather than a differential one, and the relations between these had to be examined carefully.

Here too, Ehrenfest had turned up a question in his hydrodynamic problem that had just been discussed for ordinary discrete mechanical systems. The question was whether or not one could use Hamilton's principle, for example, when dealing with a system subject to non-holonomic conditions. Hertz himself had raised the question in his book and gave as his opinion that one could not. Subsequent analysis had shown that Hertz was wrong on this point. Hamilton's principle could be applied to non-holonomic systems, but only if the variational process were handled in one particular fashion.[44]

The point is this: if the constraints on a system are holonomic, then requiring that the varied path be compatible with the constraints is equivalent to requiring that the variations themselves satisfy the constraints. This equivalence no longer holds for non-holonomic constraints. Just because the constraints are not integrable, it no longer follows that variations compatible with the constraints will produce a varied path that is compatible with the constraints. Now d'Alembert's principle requires that the variations themselves be compatible with the constraints, and with this interpretation of allowed variations Hamilton's principle does apply to a system subject to non-holonomic constraints.

Ehrenfest took this result as his point of departure and considered

---

44. O. Hölder, Gött. Nachr. (1896), p. 122. E. T. Whittaker, *op. cit.*, p. 249.

such a dynamical system as described by non-holonomic coordinates (in the sense of (4.15) and the subsequent discussion). He showed that Hamilton's principle, with variations compatible with the non-holonomic constraints, then leads to Boltzmann's generalization of the Lagrangian equations, (4.16). He applied the same reasoning to the problem of the rigid body in an incompressible fluid, restricted as above, and showed that, if the proper variational procedure is used, his extra terms are indeed present in the equations of motion that result from Hamilton's principle. They were absent in earlier discussions only because these treatments had used an improper variational procedure – one in which the varied path, rather than the variations themselves, satisfied the non-holonomic conditions.

There was still one gap that had to be filled. Even if his own derivation from Hertz's principle of least curvature was consistent with a proper derivation from Hamilton's principle, why was it that the extra terms in the equations of motion had never shown up before? Ehrenfest had to prove that these extra terms do vanish, although the reason for their disappearance would have to be some special feature of the situation. The individual terms $(\partial\alpha_w/\partial p_u) - (\partial\alpha_u/\partial p_w)$ were not zero, as already discussed, but what about the whole expression in (4.14), which involved sums and integrals over these terms? Ehrenfest was able to complete his discussion by proving that his whole expression did indeed vanish as a consequence of the special form of the velocity potential, (4.10), for the vortex-free motion of an incompressible liquid in a simply connected region whose surfaces are fully specified by the rigid bodies themselves. But all these special features of the situation seemed to be necessary to reduce the equations of motion to their usual forms.

8. Few doctoral dissertations are major contributions to science, and Paul Ehrenfest's was not one of this rare group. His thesis problem had only technical interest, in the sense that one can hardly imagine any solution of it providing new insight into the natural world. It was a careful and thorough analysis of an essentially restricted problem, even though this problem had its place squarely within the principal tradition of theoretical physics—the mechanical explanation of nature. Ehrenfest's thesis made no advance toward Hertz's goal of understanding phenomena in terms of the hidden motions of hidden masses, but he really had not tried to do so. He did clarify the relationships between the differential and the integral variational principles for non-

holonomic systems; that, rather than any progress in Hertzian mechanics as such, was the contribution of his thesis.

It is always interesting to look at an author's earliest works for indications of what is yet to come, to see if the firm characteristics of his mature style have already begun to appear in his youthful efforts. Ehrenfest's thesis certainly exhibits his talent for clear exposition. Both the overall structure of the work and the details of the arguments are constructed in such a way as to keep the reader fully aware of the problem and to bring out the essential features that make the arguments work. The overall form is still the conventional one – a doctoral thesis is hardly the place to make innovations in form – and Ehrenfest has not yet adopted the highly schematic style of some of his later papers. There is, however, the characteristic stress on what assumptions have to be made, and where these assumptions are actually brought to bear in the argument.

Also characteristic of Ehrenfest is the fact that his thesis is not a calculation. He does not solve a problem, in the usual sense of applying a well formulated theory to a new situation. There are plenty of equations, of course, but the goal of the investigation is not a single formula. In this respect the thesis is typical of Ehrenfest's work, but, in this as in other respects, he was a very atypical theoretical physicist.

CHAPTER 5

Göttingen and Petersburg

The Ehrenfests returned to Göttingen in September 1906 with mixed emotions. They had left the Vienna which Paul would never again call his home, and come back to the pleasant and stimulating university atmosphere in which they had met almost four years earlier. Ehrenfest had no academic position at the university, but the small incomes he and his wife had inherited made it possible for them to live modestly and to pursue their own studies. Göttingen offered once again the intellectual excitement of its mathematicians – Klein and Hilbert, Minkowski and Carathéodory. This time there was also the pleasure of renewing and strengthening an old friendship: Gustav Herglotz, Paul's high school and university friend, was in Göttingen where he had already been *Privatdozent* in mathematics for two years. Ehrenfest always found Herglotz to be one of those mathematicians, lamentably few in number, who could really communicate with physicists, able both to understand physical problems and to explain relevant mathematical ideas in a way that made sense to a physicist. Herglotz became a close friend of all the Ehrenfests, a frequent visitor at their home, and an honorary uncle to young Tanya.

But the delightful aspects of this return to Göttingen could not cushion the shock of the news that Ludwig Boltzmann had committed suicide on the 6th of September.[1] Ehrenfest's life had been tied to

1. E. Broda, *op. cit.*, p. 28, and R. Dugas, *op. cit.*, p. 138. (See note 9, Chapter 3.)

Boltzmann's since he entered the university, closely tied during the last few years, in which he worked on his dissertation under Boltzmann's direction and then continued to participate in Boltzmann's seminar even after receiving his degree. Boltzmann, more than anyone else, by his teaching and by his example, helped set the direction of Ehrenfest's scientific interests and helped form his intellectual style. In the inaugural lecture he had given on his return to Vienna from Leipzig in 1902, Boltzmann had offered his audience not only his lectures on theoretical physics, but his whole way of thinking and perceiving, his whole self, and asked in return nothing less from them.[2] The suicide of such a teacher must have had a heavy impact on one of his immediate students, particularly on one so sensitive as Paul Ehrenfest.

Boltzmann was sixty two. He had suffered for years from periods of serious depression, and from the perhaps not unrelated burden of severe asthma. During his later years he was plagued by "the worst sickness that can strike a professor – the fear of lecturing".[3] He could not overcome the anxiety that his wits and his memory would suddenly leave him in the midst of a lecture. This anxiety had kept him from enjoying the Leipzig professorship he had looked forward to with so much pleasure in 1900, and had helped to drive him back to Vienna only two years after he had left.[4] The combination of his recurrent depressions and fears became too much for him to bear, and he took his own life while on summer vacation at Duino, near Trieste.

Ehrenfest had, of course, known Boltzmann only in his last years. Boltzmann then felt, with some justice, that he was scientifically isolated, since his views had been under vigorous attack for some years. He actually felt himself to be fighting a rearguard action in defense of a traditional atomic and mechanical view of nature against the onslaughts of what seemed to him to be the victorious barbarian hordes of the energeticists. "I might say that I am the only one left of those who embraced the old theories with all their hearts; at least the only one left to fight for them with all my strength."[5] In 1898 he had written

---

2. L. Boltzmann, *Populäre Schriften*, p. 337. (See note 10, Chapter 3.)

3. W. Ostwald, *Grosse Männer* (Leipzig: Akademische Verlagsgesellschaft, 1909) p. 405.

4. See the series of letters exchanged between Boltzmann and Ostwald in the years 1898–1900, in *Aus dem wissenschaftlichen Briefwechsel Wilhelm Ostwalds* I, ed. H.-G. Körber (Berlin: Akademie-Verlag, 1961) pp. 22–30.

5. L. Boltzmann, *op. cit.*, p. 205.

in the preface to the second volume of his book on the theory of gases[6] that he was "convinced that these attacks [on the kinetic theory] are merely based on a misunderstanding, and that the role of the theory of gases in science has not yet been played out". He admitted that he was "conscious of being only an individual struggling weakly against the stream of time", but he felt that "it still remains in my power to contribute in such a way that, when the theory of gases is again revived, not too much will have to be rediscovered". This was indeed a cry from the depths, but Boltzmann was not broken by his critics, any more than Keats was by his. Boltzmann continued to enter the lists to battle for his ideas in the same spirit that his heroes,[7] Beethoven and Schiller, had celebrated in the "Ode to Joy".

Even though one rejects the oversimplified view that Boltzmann's suicide was caused by the attacks on his scientific ideas, it is still possible to say that Boltzmann's suicide is inseparable from the sacrifices he made in forging his monumental contributions to science. So it seemed, at least, to a perceptive contemporary, Boltzmann's Leipzig colleague, good friend and scientific antagonist, the arch-energeticist Wilhelm Ostwald,[8] who began his obituary with the words: "This falls under the laws to which almost all servants of the strict goddess Science are subject: their lives end in grief, and this is all the more likely, the more completely they have dedicated their lives to Her." Ostwald found this law to hold all too often, even in cases when an outsider saw a life that seemed to be an uninterrupted series of successes. The irresistible fascination of scientific research excluded any consideration for one's health or strength, and so "every important discovery must be paid for by a human life, for whether the discoverer is completely destroyed or survives as a cripple makes a difference only to the extent that the latter fate is worse." Ostwald saw Boltzmann as such a crippled veteran, a victim of the intense struggle that constitutes scientific research.

How much these remarks meant to Paul Ehrenfest then we do not know, but he clipped Ostwald's obituary notice of Boltzmann from the newspaper in which it first appeared and saved it through the years.

Ehrenfest himself was hardly known or distinguished enough in 1906 to have been asked to write an obituary of his former teacher for any of

6. L. Boltzmann, *Lectures on Gas Theory*, transl. S. G. Brush (Berkeley and Los Angeles, University of California Press, 1964), p. 215.

7. L. Boltzmann, *Populäre Schriften*, pp. v–vi.

8. W. Ostwald, *op. cit.*, p. 401.

the leading journals. But there was a relatively new journal published by a group of students' scientific societies, the *Mathematisch-Naturwissenschaftliche Blätter*, for which Ehrenfest had written before, and for which he now wrote a notice of Boltzmann. This brief article,[9] in which Ehrenfest gave a masterful summary of the main themes and characteristic features of Boltzmann's work, also tells us a great deal about its author. The organization of the article, the choice of subject matter to be emphasized and to be ignored, are all very much Ehrenfest's own.

He devoted the major part of his obituary to Boltzmann's principal interest – the kinetic theory of gases. Ehrenfest gave this a "genetic presentation", to use his own phrase, following the ideas as they developed and searching out the questions that had really provoked the investigations. Ehrenfest began with Boltzmann's arrival on the scientific scene just after the kinetic theory had taken a great step forward with the work of Rudolf Clausius and James Clerk Maxwell. The choices for a new worker in the field seemed to be either to improve and extend experiments of a type already performed, or to refine and elaborate the approximate but complicated calculations of the theory. Boltzmann, however, had raised a new question and had thereby opened his own field of investigation. Boltzmann had asked whether and how the second law of thermodynamics could be understood as a consequence of the principles of mechanics. Ehrenfest followed the evolution of Boltzmann's thinking from his first attempt to treat the second law as purely a theorem in mechanics, (in 1866 when he was only 22), through a decade of deeper and deeper probing into the problem. He pointed out that Boltzmann had to reformulate the question when mechanics alone proved insufficient for understanding the origin of thermodynamic behavior, and to ask for those assumptions that were needed, in addition to the axioms of mechanics, in order to obtain the entropy law. Ehrenfest described how Boltzmann was led to the general energy distribution law, to the theorem of the equipartition of energy, and above all, to the essential role of statistical ideas in the theory. The second law of thermodynamics was a law of *statistical* mechanics, not of mechanics alone, and its key concept, the entropy, was directly related to the probability of finding the system in a given state. Ehrenfest distinguished the several different ways in which

9. P. Ehrenfest, "Ludwig Boltzmann†", Mathematisch-Naturwissenschaftliche Blätter, **3**, No. 12 (1906), *Papers*, p. 131. Also see Chapter 6.

Boltzmann had used statistical ideas and emphasized the underlying unity of a development that had, in effect, created a whole new branch of science – statistical mechanics.

Now anyone writing on Boltzmann's work would have had to emphasize his contributions to the kinetic theory and associated problems, but no one else was in a position to disentangle the main line of Boltzmann's thinking as Ehrenfest had done. One can, for example, read Lorentz's longer and more detailed analysis of Boltzmann's work (in the obituary that Lorentz wrote some months later[10]) and not have any real idea of the evolving character of Boltzmann's ideas. Ehrenfest appreciated the way in which Boltzmann built such a vast structure of thought around a central problem: he keenly regretted the lack of a corresponding center for his own work. Also interesting is the slight attention that Ehrenfest gave to Boltzmann's elaborate calculations, of viscosity coefficients, for example. This, too, was something that Ehrenfest himself could not do, although calculating as such never had a real appeal for him.

Ehrenfest pointed out that despite the apparent diversity of problems that Boltzmann had treated in his many papers, there was a fundamental unity in his work. That unity was maintained by Boltzmann's constant striving for mechanical interpretations of all of physics. This was obviously true of his work on the mechanical theory of heat, but it was just as true of his work on electromagnetism. Boltzmann, one of the first on the continent to appreciate and develop Maxwell's electromagnetic ideas, placed even more emphasis than Maxwell had on the possibility of using mechanical models to understand the conceptual structure of Maxwell's theory. Ehrenfest emphasized Boltzmann's remarkable ability to visualize and to grasp the complex motions of the mechanical systems he imagined, and the satisfaction he derived from this. Boltzmann made very effective use of examples that were easily understood but that still exhibited all the characteristic features of some major problem, and he gave these examples "an exhaustive and loving treatment" that made his detailed arguments "unusually lively". Ehrenfest also thought that Boltzmann's way of working through these examples "provided insight into all the weary labor which this immense imagination had to perform before it achieved

---

10. H. A. Lorentz, "Ludwig Boltzmann", Verh. d. Deutsch. Phys. Ges. **9** (1907), p. 206. Reprinted in H. A. Lorentz, *Collected Papers*, Vol. IX (The Hague: Nijhoff, 1939), p. 359.

its far-reaching results". Ehrenfest himself became a master of the art of inventing and using simple conceptual models to elucidate subtle points of a theory.

Most remarkable is the absence of any personal comment on Boltzmann. One could perhaps attribute this to a certain diffidence on the part of a young man writing about his famous mentor, but that would hardly seem to fit Ehrenfest's personality. There would be no way of telling from the obituary notice alone that Ehrenfest had actually known Boltzmann personally. At one point Ehrenfest does refer to Boltzmann's "intimate contacts" with his teachers Josef Stefan and Josef Loschmidt, and suggests that these close relationships contributed in an essential way to Boltzmann's rapid scientific development and early familiarity with the kinetic theory. He refers to two passages in Boltzmann's own obituary of Stefan. In the first[11] Boltzmann described how Professor Stefan had given him Maxwell's papers to read when he was still a student, and had also given him an English grammar with them since Boltzmann did not know a word of English yet. Ehrenfest also referred to Boltzmann's description[12] of the atmosphere surrounding Stefan in Vienna, the humor and joy with which Stefan made discussions on the most difficult subjects into enjoyable sport, and the contrast between this atmosphere and the stiffness surrounding Helmholtz at Berlin. One wonders whether Ehrenfest was unconsciously expressing his own sorrow for the passing of this spirit, or perhaps his regret that he had never experienced such a close relationship with Boltzmann. In any case, it is significant that in later years Ehrenfest, who loved to "praise famous men", rarely referred to Boltzmann in his correspondence, and never expressed about him anything like the veneration and love that he felt so strongly, and expressed so profusely, for Lorentz. Whatever the nature of his personal relationship to Boltzmann may have been, it was one that Ehrenfest could not, or would not, put into words.

October 1906 was a busy month for Ehrenfest. He finished his Boltzmann obituary, wrote a rejoinder to James Jeans[13] (as part of a controversy over Wien's displacement law that the two were carrying on in the pages of the *Physikalische Zeitschrift*), and wrote up a first

11. L. Boltzmann, *Populäre Schriften*, p. 96.
12. *Ibid.*, p. 102.
13. P. Ehrenfest, "Bemerkung zu einer neuen Ableitung des Wienschen Verschiebungsgesetzes", Phys. Z. **7** (1906), p. 850. *Papers*, p. 125.

Fig. 5 Tatyana and Paul Ehrenfest with Miss Föhringer and K. Baumgardt, c. 1910.

Fig. 6 Paul Ehrenfest, c. 1910.

version of some work that he and his wife had been doing the previous summer.[14] Towards the end of the month he was invited by Professor Felix Klein[15] to lecture to the Mathematical Colloquium and Paul chose this new joint work of the Ehrenfests as his subject. It was an ideal topic for a lecture.

The Ehrenfests had been analyzing some of the objections raised against Boltzmann's statistical mechanical explanation of the second law of thermodynamics.[16] Boltzmann himself had already answered his critics, but he had not fully succeeded in making the rather subtle points at issue sufficiently clear. In the process of discussing the logic of the various objections with his wife, Paul Ehrenfest had hit upon an extremely simple conceptual model with the help of which he could make the essential features of Boltzmann's argument stand out clearly and simply. Ehrenfest was illustrating in his own work one of the features of Boltzmann's scientific style which he had just emphasized in his obituary article. The model (the Ehrenfest urn model, or, more earthily, the dog-flea model) was easily understood; the lecture was a great success.

Just how great a success it had been became evident a few weeks later when Ehrenfest received a note from Felix Klein[17] inviting him to write, in collaboration with his wife if he so desired, a review article on the principles of statistical mechanics for the *Encyklopädie der mathematischen Wissenschaften*, the great Mathematical Encyclopedia of which Felix Klein was one of the guiding spirits. The first idea for such an encyclopedia had been conceived a dozen years earlier by Franz Meyer together with Klein and Heinrich Weber.[18] It was published under the auspices of the Academies at Göttingen, Leipzig, Munich, and Vienna; the actual writing was done by outstanding mathematicians and scientists from many countries. It all added up to a major and somewhat ponderous undertaking. The purpose of this large scale enterprise was to make available the

---

14. P. and T. Ehrenfest, "Über eine Aufgabe aus der Wahrscheinlichkeitsrechnung, die mit der kinetischen Deutung der Entropievermehrung zusammenhängt", Mathematisch-Naturwissenschaftliche Blätter, **3**, No. 11 and 12 (1906). *Papers*, p. 128.

15. F. Klein to P. Ehrenfest, 25 October 1906.

16. See Chapter 6.

17. F. Klein to P. Ehrenfest, 16 November 1906.

18. *Encyklopädie der mathematischen Wissenschaften mit Einschluss ihrer Anwendungen*, Vol. I, Part 1, ed. W. F. Meyer (Leipzig: Teubner, 1898–1904). Introduction by W. von Dyck, p. v.

principal results and methods developed in the mathematical sciences during the nineteenth century, and, by including the various applications of mathematics to physics and other sciences, to present a picture of the position of mathematics in the culture of the day. The plans called for some six large volumes, three devoted to the principal branches of pure mathematics, algebra, geometry, and analysis, two more on applications of mathematics, and a final volume on historical, philosophical, and didactic questions. The articles were to be addressed to the mathematically educated, so that the instructions[19] to the authors of articles in applied mathematics in particular, called for an emphasis on the mathematical side of the theories discussed; these articles were to inform the mathematician about the questions posed to him by the sciences, and also to inform the scientist about the answers that mathematics had provided to such questions.

Felix Klein had assumed the editorship of the fourth volume, on mechanics, and his former student and collaborator Arnold Sommerfeld, (by 1906 professor of theoretical physics at Munich), edited the fifth volume, on mathematical physics. Ludwig Boltzmann had become a scientific adviser to the Encyclopedia project at its beginning and had prepared an article with J. Nabl on the kinetic theory of matter[20] for Sommerfeld's volume; this article had been published in 1905. Boltzmann had also agreed to write an article for Klein's volume on the basic concepts of statistical mechanics. In his essay, "A German Professor's Journey to Eldorado", Boltzmann devoted several pages[21] to the problems involved in organizing the Encyclopedia activities; (they are written in the vein suggested by the title he had chosen for an essay describing a trip to California, the first stage of which involved a stop at Leipzig for a meeting to discuss the Encyclopedia.) Boltzmann described the enormous energy and enthusiasm with which Felix Klein pursued his encyclopedic ideal, and paid his respects to the sure insight that enabled Klein to find just the right argument to convince each intended author that he should undertake the major task of writing the article Klein wanted from him. In Boltzmann's case Klein had threatened to give the job of writing the article on the foundations of statistical mechanics to E. Zermelo, Boltzmann's arch-critic. The threat worked. Boltzmann agreed to write the article.

19. *Ibid.*, p. xi, footnote.

20. L. Boltzmann and J. Nabl, "Kinetische Theorie der Materie", *Encykl. d. math. Wiss.*, Vol. V, Part 1, Article V 8 (Leipzig: Teubner, 1905).

21. L. Boltzmann, *Populäre Schriften*, pp. 404–408.

Boltzmann's death had left Felix Klein with the problem of finding a new author. Now Ehrenfest's colloquium lecture gave Klein the solution: Paul Ehrenfest, Boltzmann's student, not only understood even the most complex of his teacher's ideas, but he had an extraordinary talent for making ideas come alive. His mastery of the art of "genetic presentation" insured that he could handle the expository problems involved in writing such an article. It was not an easy job. As Boltzmann had written:[22] "Such an undertaking would not be so enormously difficult if it were only a question of citing the most outstanding results and of recording the most important of these, which are naturally also the best known, all without too much close criticism. But if one wants to drag forth out of obscurity everything that is really useful in all fields, to reject everything inessential, to aim at as complete an account of the literature as possible, and if one wants to present all this clearly, in a form that is convenient for the reader's use, then the difficulties seem to be almost terrifying for anyone who has looked into the mathematical literature even a little bit."

When Klein asked Ehrenfest to undertake the article on statistical mechanics, he immediately reassured him on two counts. First, only the fundamental questions of statistical mechanics would have to be treated, since Boltzmann's already published article in the fifth volume dealt with the applications. In addition, the unresolved difficulties of the subject need not be settled for the Encyclopedia article; all that was necessary was to clarify the questions, and how they had developed. Paul Ehrenfest accepted the invitation as a joint venture to be carried out with his wife.

It would be three years before the main text of the article was drafted, and almost two more before the work was finally completed and appeared in print.[23]

At the time of their marriage Paul Ehrenfest had promised his bride that they would move to Russia and settle there as soon as she wanted to and it looked at all possible to do so.[24] That time arrived in 1907, and in the fall of that year the Ehrenfest family moved from Göttingen to St. Petersburg. There was even less chance for Ehrenfest to obtain an

---

22. *Ibid.*, p. 405.

23. P. and T. Ehrenfest, "Begriffliche Grundlagen der statistischen Auffassung in der Mechanik", *Encykl. d. math. Wiss.*, Vol. IV, Part 2, Article IV 32. (Leipzig: Teubner, 1911).

24. "Red Book", entry for 21 December 1904.

academic appointment in Russia than there was in Germany or the Austro-Hungarian Empire, but since the chances everywhere seemed so unpromising other factors governed the decision to move. Russia was Tatyana Alexeyevna's homeland, the homeland to which she was attached in a way that her husband respected, appreciated, and probably envied, since his own emotions about Vienna and Viennese life were such a dark tangle. The move to Russia might provide him with a place he too could honestly call home; it should at least give his daughter the strength and independence that her mother's solid roots seemed to have nourished in her. Their small inherited incomes would make it possible for them to live in modest comfort, and he could hope that the academic situation in Russia would change for the better. Such hopes could at least be entertained in the wake of the reforms brought about by the Revolution of 1905. These included the reopening of the universities in the fall of 1906 and the restoration of their autonomy, after they had been shut down by the Czar's government early in 1905.[25]

As soon as possible after arriving in Petersburg, Ehrenfest sought out and made contact with the younger generation of Russian physicists and mathematicians.[26] The Russians were delighted with the lively new arrival on the Petersburg scientific scene and with the fresh vigor that he brought with him. Ehrenfest met the young mathematicians, such as J. D. Tamarkin and A. A. Friedmann, as well as the formidable O. D. Chwolson, Professor of Physics at the University of St. Petersburg and the more sympathetic mathematician V. A. Stekloff, who had recently been appointed Professor of Analytical Mechanics at the University. His most important meeting during that first winter in Petersburg was with Abram Fedorovich Joffe.[27] Abram Fedorovich was a Ukrainian Jew, a few months younger than Ehrenfest, who had studied engineering at the Technological Institute in Petersburg, and then spent four years in Munich studying physics and working in the laboratory of the discoverer of X-rays, Wilhelm Röntgen. He had

25. See J. Walkin, *The Rise of Democracy in Pre-Revolutionary Russia* (New York: Praeger, 1962), Chapter 6.

26. Personal communication from Mrs. T. Ehrenfest-Afanassjewa.

27. A. F. Joffe, "Supplement to 'Reminiscences of Professor Paul Ehrenfest' by G. E. Uhlenbeck", Usp. Fiz. Nauk **62** (1957), p. 371. Also see I. K. Kikoin and M. S. Sominskii, "Abram Fedorovich Joffe (on his eightieth birthday)", Usp. Fiz. Nauk **72** (1960), p. 307. Translation in Soviet Physics Uspekhi **3** (1961), p. 798. Also see The New York Times, October 15, 1960, p. 23.

returned to Russia in 1906, a year after obtaining his doctorate in physics, summa cum laude, with Röntgen as his thesis director. Joffe was full of plans for continuing his research into the plastic properties of crystals using X-rays, and into the photoelectric effect. His plans were met with something less than enthusiasm in Petersburg. When he outlined them to Professor I. I. Borgman, Joffe was asked who had already done this research, and where it was done. When he explained that it had not been done anywhere and represented a new line of investigation, Joffe was told: "You are an opinionated young man. We should not attempt to seek anything new; it would be very well if we were able to duplicate correctly in our country the measurements made abroad!"[28] Physics in Petersburg was sorely in need of the new spirit that Joffe and Ehrenfest brought to it.

Abram Fedorovich Joffe and Pavel Sigismondovich Ehrenfest, (as he was known in Russia), hit it off at once and soon became close friends. They saw each other much less often than they would have liked since their homes were far apart. Although they could not meet more than once every two weeks they kept in close contact through the mails as Ehrenfest sent his new friend a steady stream of almost daily letters reporting his newest ideas and calculations.[29] Abram Fedorovich was one of a small group of young Petersburg physicists that began to meet quite regularly, usually once every other week, most often at the Ehrenfests' home, for lively discussions of subjects of current interest in physics. The initial suggestion for the organization of this discussion group came from Tatyana Alexeyevna's aunt, the one in whose home she had been brought up. This aunt had no scientific education or interest, but she saw that the young couple would simply not feel fully alive without this kind of scientific circle.

This physics discussion club was actually an informal and unofficial colloquium, organized by Paul Ehrenfest along the same lines he would follow a few years later in setting up the physics colloquium at Leyden. He kept records[30] of the meetings held, the subjects discussed and the members of the group who attended each time. At the third meeting, for example, on November 6, 1908, Ehrenfest himself spoke on Planck's energy quanta, and he noted that there was a vigorous

28. Quoted by I. E. Tamm in "Yakov Il'ich Frenkel", Usp. Fiz. Nauk **76** (1962), p. 397. English translation in Soviet Physics Uspekhi **4** (1962), p. 175.

29. A. F. Joffe, *op. cit.*, and personal communication from Mrs. T. Ehrenfest-Afanassjewa.

30. A small notebook on the Petersburg discussions is part of the Ehrenfest Collection.

discussion, with Joffe contributing a spirited defense of the concept of light quanta, Einstein's atoms of energy. The meetings usually began at nine and often lasted until twelve or one. There were some ten to fifteen participants at the sessions, the attendance varying from month to month. In the spring of 1909 Ehrenfest noted that he and his wife had attended the fifteen meetings held that academic year, while all the other members had missed one or more of the sessions; Joffe, for example, attended only eleven.

During the five years they lived in Russia the Ehrenfests spent their summers at Kanuka, a tiny Estonian village not far from Narva on the Gulf of Finland, some 90 or 100 miles west of Petersburg. This gave them a change of scene and a new set of surroundings in which to continue their regular pursuits. For no matter where they were, there was no real break in their activity.[31] The constant drive to understand, to clarify, and to discuss was too strong to be turned off for summer vacation, nor would they have wanted to turn it off. That first summer at Kanuka they had plenty to occupy them, in addition to summer visitors like the mathematician Tamarkin. For one thing they still had much to do on the Encyclopedia article promised to Felix Klein. Klein had already written a reassuring letter[32] to Ehrenfest in January 1908, commenting that he too had the feeling of having undertaken more than he could possibly accomplish in the time available, and comforting Ehrenfest with the news that he was far from the only one who was way behind schedule in his writing. At that time Klein had advised Paul and Tatyana to proceed at their own pace without trying to overwork to meet a deadline. By the end of July Klein had occasion to write[33] again to reassure the Ehrenfests; this time he was reminding them that they were not expected to solve all of the thorny problems that surrounded the foundations of statistical mechanics, but only to report on the present status of the problems and controversies, far from final though it might be. By early September Klein was delighted to report[34] that he found the Ehrenfests' copy to be unusually fine and that it was already at the printer, even though the footnotes (and there were to be many!) had not yet been turned in to him.

Paul Ehrenfest was busy with another project that summer, a self-

31. Notes on the summers of 1908–1912 in the "Red Book".
32. F. Klein to P. Ehrenfest, 20 January 1908.
33. F. Klein to P. Ehrenfest, 22 July 1908.
34. F. Klein to P. Ehrenfest, 7 September 1908.

imposed burden: he had decided to prepare for the examination for the degree of *Magister* in Physics. The University of St. Petersburg had just resumed the granting of the degree *Magister* after a lapse of twenty five years or so, and Ehrenfest was one of the first to try to jump the hurdles. It would seem that the University now required its own degree of *Magister* of anyone who aspired to a faculty post, even though this degree only corresponded roughly to a Master's degree rather than the research doctorate, and even though the applicant might already hold a doctorate from a university like Vienna or Munich. It was for this reason, presumably, that Ehrenfest and then Joffe submitted themselves to the ordeal. And it was indeed an ordeal since the rather old-fashioned Petersburg faculty demanded an excessive amount of detailed knowledge, including, in particular, a fluent knowledge of a great many formulas from the theory of special functions. Ehrenfest detested this sort of work, referring to it as *Chineserei*, which can be translated approximately as "Chinese puzzles". (His young daughter, who often heard him complain about the amount of this *Chineserei* that was done at the University, fully expected to find the academic halls crowded with Orientals when she was taken there for a visit.)

Ehrenfest succeeded in the examinations, but no position at the University was offered to him. (He was not only a foreigner, and a Jew, but was even officially certified as being without religion on his Austrian papers.) When Abram Fedorovich was preparing for the exams, Ehrenfest worked with him. Joffe was not terribly interested in the mathematics, but he had a phenomenal memory and an excellent teacher. He, too, did very well and impressed Professor Stekloff with the mathematical erudition he had acquired from his friend. Ehrenfest always considered it a great tribute to his teaching that one of the examiners declared that Joffe had done "even better than Ehrenfest".

It may have been something of a game for young men as brilliant as Ehrenfest and Joffe to overcome the obstacles set by the Petersburg faculty in the *Magister* examinations, but these obstacles kept other able men from doing productive work for years, to no useful end. Little wonder that Joffe described the system as "infuriating" and that Ehrenfest, too, referred to his own "rage" against it. When the opportunity arose to strike a blow against this outmoded set of examinations, Ehrenfest seized it. There was a meeting of physicists in Moscow at which the young Petersburg physicist D. S. Roschdestvenski was to give a paper reporting on a particularly elegant piece of work that he had done. Roschdestvenski could not go to Moscow, however, since he was in the throes of preparing for the *Magister* examinations, but Ehrenfest, who was going, agreed to deliver the paper for him. Ehrenfest presented the work in his usual brilliant and forceful style, but neither at the beginning nor during his talk did he mention that the work was Roschdestvenski's. The Petersburg professors in the audience were fuming over what looked like flagrant plagiarism. But as Ehrenfest finished his exposition he announced that this work was done, not by himself, but by Roschdestvenski who could not be here because he was busy in Petersburg cudgeling his brains to learn the following things necessary for the *Magister* examinations. And Ehrenfest erased the board and proceeded to cover it again with a selection of the unsightly equations in question. The Moscow professors roared with laughter, and the Petersburg faculty was thoroughly discomfited. Ehrenfest took some credit for the subsequent reform of the examinations, and in any event he loved to tell the story of his *coup de théâtre*.[35]

Despite the fact that he did not hold a regular university position, Ehrenfest could never be accused of inactivity during his years in Russia. His notebooks bear witness to his constant scientific questioning, and the work he published during his Russian years shows the wide range of his active interests. He corrected an error of Lord Rayleigh's in the theory of wave motion, (no mean achievement), raised a fundamental question about the status of the concept of rigid body in the theory of relativity, and carried out a fundamental study of the Le Chatelier-Braun principle in thermodynamics. In addition to all these, Ehrenfest wrote a major paper on one of the central issues in physics at the time, the quantum theory, as well as completing the Encyclopedia article on statistical mechanics in collaboration with Tatyana Alexeyevna. These are listed here simply as an indication of the range of Ehrenfest's completed work during this period.[36]

Ehrenfest engaged in a variety of activities apart from his own research. He served as a member of the editorial board of the only research journal in physics that was published in prerevolutionary Russia, the *Journal of the Russian Physical and Chemical Society*.

35. See A. F. Joffe, *op. cit.* His version of the Roschdestvenski story is somewhat different from that given here, which is based on information from Mrs. T. Ehrenfest-Afanassjewa and Mrs. T. van Aardenne-Ehrenfest.
36. The scientific work mentioned is discussed in Chapters 6, 7 and 10.

This journal had been divided into separate sections for physics and chemistry in 1907, to keep the more highly developed and active Russian chemistry from completely crowding out the physics. V. K. Lebedinski was the new editor, and he was anxious to improve the journal. He began the publication of a second part of the physics section, entitled *Physics Problems*, devoted to abstracts of the foreign literature, review articles, bibliographies, and so on. He also started a concerted attempt to persuade Russian physicists to publish in the Russian journal, which many of them had not been doing. They were urged to do this even if they also published the same article abroad. Ehrenfest enjoyed his activity on the editorial board, particularly the opportunity it gave him to make contact with physicists from all over Russia, who were being encouraged to treat the Journal as a national journal and not the exclusive property of the Petersburg group.[37]

Ehrenfest could not have carried on this editorial work, given lectures, and generally adapted himself to the Russian scene as thoroughly as he did without having quickly become fluent in the Russian language. He never mastered all the grammatical niceties of Russian (or of Dutch or English, later on), but he could move about freely in it. Some years later his speech was described by a native speaker as being "a precise Russian", "although broken in the sense of endings, cases and genders".[38] In addition to all the obvious occasions for using Russian in his daily life, he made a fair amount of scientific use of it in the frequent talks he gave to his physics discussion club, to one or another of the official gatherings of physicists or mathematicians, and to the seminar for students which he also founded and which met weekly at his home.

During his five years in Russia, Paul Ehrenfest actually had one opportunity to do some regular, officially sponsored, teaching. In 1910 the Polytechnic Institute at Petersburg suffered from that most atypical of academic complaints – a budget surplus, and decided to expend it by inviting Ehrenfest to give a special course of lectures on the differential equations of mathematical physics.[39] "And what lectures they were!", as Abram Fedorovich Joffe wrote almost half a century later; "No presentation of this classical subject ever combined the mathematics and the physics into such an harmonious unity."[40]

---

37. P. Ehrenfest to H. A. Lorentz, 19 May 1912. Also see M. I. Radovskii, "V. K. Lebedinskii: Editor of Physics Journals", Usp. Fiz. Nauk **78** (1962), p. 345. English translation in Soviet Physics Uspekhi **5** (1963), p. 843.
38. J. I. Frenkel's remark, quoted in I. E. Tamm, *op. cit.*, p. 180.
39. P. Ehrenfest to H. A. Lorentz, 19 May 1912.

But despite the impact of his lectures there was no reappointment when the surplus funds had been spent. Ehrenfest had already confided to his family notebook: "I often have the feeling now that I shall never be able to establish myself in Russia."[41]

The Ehrenfests took enormous pleasure in the development of their daughter, little Tatyana or Tanitschka. Not surprisingly, it was her rapid intellectual development that seemed to delight the intellectual couple, but every aspect of the little girl's growth was encouraged by the warm and loving atmosphere that surrounded her. The proud parents wrote down accounts of many of the incidents of her early childhood, some that showed her precocious intelligence, and others that simply recorded the inimitable and irrecoverable joys of life with their first born child. Paul Ehrenfest spent much time with his daughter, and enjoyed her conversation as she grew more articulate and her interests developed. He took pleasure in telling her about life in other lands, in seeing her reactions to the behavior of a magnet, in explaining the meaning of maps (which had so strongly impressed him in his own childhood). Sometimes, probably often, his intellectual projects aimed too high for his daughter's level of development: the model of a hyperboloid that he made for her out of cardboard and string was treasured – but as such a nice place to keep her dolls.[42]

When their daughter was only four, the proud parents were already dreaming of her future intellectual development, hoping that she would "obtain a serious and profound training in the logical thinking of science and mathematics very early", and then she could follow her own interests, "attacking them with youthful vigor".[43]

A few years later Ehrenfest wrote out an elaborate sketch of the range of subjects which he and his wife hoped that little Tanya would eventually master.[44] The list suggests his own joy in learning and his desire to share it with those he loved. Ehrenfest began with mathematics – "all sides of this glorious science: the conceptual niceties, the relations between its most heterogeneous parts – grandiose, and almost uncanny relations – and the visualization of these relations, (especially in their geometrical interpretation!), the purely methodological aspects,

---

40. A. F. Joffe, *op. cit.*
41. The "Red Book", September 1909.
42. Personal communication from Mrs. T. van Aardenne-Ehrenfest.
43. The "Red Book", September 1909.
44. The "Red Book", 1912.

and not least its technical side: mastery of the ability to calculate, geometrical imagination and logical intuition." Ehrenfest listed "the mathematicians whom we particularly love", a list that began with Riemann and Klein and included Lie, Poincaré, Hilbert and Hadamard, among others. After mathematics came mechanics – "as the English love it", with Thomson and Tait, Maxwell and Routh as examples; then physics – "as done by Faraday, Maxwell, Boltzmann, Helmholtz, J. J. Thomson, Lorentz, Planck, Nernst and Einstein". The list continued through chemistry, "especially physical chemistry" and enough organic chemistry "to know her way around in it," crystallography, biology – "the wonderful results on fertilization... mutation, Darwin's theory,... Loeb's work,... physiology of sense organs (Helmholtz), of muscle contraction and the brain, immunotherapy and bacteriology". Ehrenfest's list was almost a catalogue of what was exciting in science in 1912: the psychology of James, Freud and Wundt; the mathematical logic of Schröder, Couturat and Russell; the mathematical economics of Pareto, Jevons, Walras, Marx, Böhm-Bawerk and Fisher, and so on and on.

The educational program for his daughter was not restricted to the sciences. Ehrenfest noted some of the historical themes that seemed especially interesting to him, and so might be expected to appeal to his daughter: the age of Pericles and the Jews in the Middle Ages, the great discoveries of the Spanish and Portuguese explorers and the constitutional development of England, the Crusades and the background of the French Revolution, and a variety of others. It was a long and diversified list, and it was accompanied by others on cultural history, on the history of science and on other subjects. Ehrenfest was letting his mind play over all the subjects he had found pleasure in studying, and probably also those he would have liked to study but had never found time for. It is just this personal character of the list that is striking: it is neither a systematic guide to "scientific culture" nor a rigorous training program for a future scientific athlete preparing to perform new intellectual feats.

In July 1910 Tatyana Ehrenfest gave birth to their second child, another daughter, at Kanuka. The baby was named Anna, but she was always called Galya or its diminutive, Galinka. A second child meant increased responsibilities and more complicated household arrangements, especially since Tatyana's mother or else her aunt was often with the family. There was a household servant and also a nursemaid to help look after the girls, so that Tatyana Alexeyevna had some time to devote to her own scientific work. She was able to take part in the current movement to reform the teaching of mathematics in Russia.[45] But it became more difficult to live on their small fixed incomes and the prospects of a position at Moscow or Petersburg seemed no brighter than they had ever been.

Early in 1911 the Russian outlook was particularly bleak. The new Minister of Education, M. Kasso, had just tightened his control over the universities, students were being expelled en masse at Moscow, and many of the best men of the faculty had either been dismissed or had resigned in protest.[46] Ehrenfest found it almost impossible to work, and noted that his desk was piled high with unfinished fragments. He sadly concluded: "We *must* move away".[47]

It was not an easy decision to make. Paul Ehrenfest had moved to Russia in 1907 with every hope of settling there for good. The thought of having to pull up the roots that had hardly been allowed to form was deeply disturbing. "I still have found no *home*", he wrote in the family diary in a note addressed to his children.[48] "If I could have been active as a teacher in Russia, then Russia could probably have become *home* to me in the innermost sense of the word. Perhaps I will always remain homeless. That is *very* bad. A person without a home can produce something deep and real only if he is most exceptionally talented." And then he added: "Your mother *has* a home. I am deeply concerned over how she will be able to live without this home. And what will your home be?!!"

"Being without a home means experiencing everything as only temporary," Ehrenfest noted. While this made one freer and more flexible, both physically and mentally, he was sure that this was not a healthy sort of freedom.

Many years later, and many thousands of miles away from Russia, Paul Ehrenfest came back to this theme once more,[49] writing that it

45. P. Ehrenfest to H. A. Lorentz, 19 May 1912.

46. P. J. Novgorotsev, "Institutions of Higher Education Before the War", in P. N. Ignatiev, D. M. Odinitz and P. J. Novgorotsev, *Russian Schools and Universities in the World War* (New Haven: Yale University Press, 1929). Also see P. P. Lazarev, "My Recollections of P. N. Lebedev", Usp. Fiz. Nauk **77** (1962), p. 571; English translation in Soviet Physics Uspekhi **5** (1963), p. 623. A report on the university situation in Russia is to be found in the Physikalische Zeitschrift **12** (1911), p. 224.

47. The "Red Book", 3 February 1911.

48. The "Red Book", 30 May 1912.

49. P. Ehrenfest to T. Ehrenfest-Afanassjewa, 15 November 1930.

was "completely impossible to feel any place at all as 'home'." This was so despite the fact that there were so many things he had "learned really to love in the course of the long time since [his] childhood". Among these beloved images were "the sky of the 'white nights' over the buildings and the Neva bridges in Petersburg, in conversation with Russian friends." "The pine woods around the Polytechnic Institute in Petersburg – in deep snow. Heavy hoar frost on the trees, faint morning sun, the dark hurrying forms of the students, wrapped up in thick clothing." "The Baltic coast near Kanuka and Sillamaggi – the spruces right behind the beach, on mossy ground with mushrooms – gloriously fragrant." But perhaps most vivid was his memory of the freely ranging talk of those Russian years: "Conversations over tea in a tightly packed room until late into the night, and then out with one or two friends through the endless winter streets of Petersburg in the piercing cold, bundled up to the eyes...."

CHAPTER 6

The Critic of Statistical Mechanics

1. The statistical mechanics that Ludwig Boltzmann worked on so long and so hard did not appear in the dissertation of his student, Paul Ehrenfest, but it became one of the major themes of all his subsequent work. Ehrenfest's writings on statistical mechanics were crucial for the development of his own career, and we must now examine them closely to see why they made such an impression on his contemporaries. What was it that Ehrenfest actually contributed to statistical mechanics?

His work on this subject clearly reflects Boltzmann's influence, but at the same time it is completely typical of Ehrenfest's own characteristic mode of doing theoretical physics. He did not proceed in the manner of most bright beginners by pushing his master's calculations a step or two further, or by applying his methods to some previously untreated physical problem. Nor did he have the creative genius to propose some fresh approach, introducing new concepts and abruptly changing the course of the development. Ehrenfest's unique ability as a theoretical physicist lay neither in calculation nor in creation; it lay in criticism, in "his unusually well developed faculty to grasp the essence of a theoretical notion, to strip a theory of its mathematical accouterments until the simple basic idea emerged with clarity", in the words of Albert Einstein.[1] His greatest efforts went into trying to

1. A. Einstein, "Paul Ehrenfest in Memoriam", in his *Out of my Later Years* (New York: Philosophical Library, 1950), p. 215.

clarify the issues of the continuing discussion which accompanies, and in part constitutes, the development of science.

In statistical mechanics this meant striving to grasp precisely what his teacher Boltzmann had done in the series of long memoirs that he had written over the years. Boltzmann's ideas had provoked intense discussion and sharp controversy. There was much confusion about what he had meant, and how much of it had or had not been solidly established. Criticism, "the endeavor to see the object as in itself it really is", was called for, and criticism is what Paul Ehrenfest undertook to provide, particularly in his review article for the *Encyklopädie der mathematischen Wissenschaften.* If we are to understand and assess his contributions, we must start with a sketch of Boltzmann's work and the situation to which it had led.

2. Boltzmann seized upon the problem of understanding the second law of thermodynamics from the standpoint of a molecular theory at the very outset of his career. He thought he had solved it in his first paper on the subject, in 1866, but this early attempt proved to be only the beginning of a long struggle. It took more than a decade of effort before Boltzmann had a solution that accounted for all aspects of the second law to his own satisfaction. And then, for the rest of his life, he had to defend his work against neglect and misunderstandings.

When Boltzmann set out to give a "purely analytical, completely general proof of the second law of thermodynamics", he was taking up a very live issue in physics. Thermodynamics as a science based on two fundamental laws dates from Rudolf Clausius' memoir of 1850.[2] It was in 1865, only months before Boltzmann's first paper appeared, that Clausius had introduced the concept of entropy, recast the second law into its now standard form, and provided, as he said in the title of his paper, "Several Forms of the Principal Equations of the Mechanical Theory of Heat Convenient for Applications."[3] The assertions of the second law, as Clausius left it, were these. First, there exists a function of the thermodynamic variables defining the state of the system, the entropy function S. Its differential $\mathrm{d}S$ is equal to the ratio of

---

2. R. Clausius, "Über die bewegende Kraft der Wärme, und die Gesetze, welche sich daraus für die Wärmelehre selbst ableiten lassen", Pogg. Ann. **79** (1850), pp. 376,500.
3. R. Clausius, "Über verschiedene für die Anwendung bequeme Formen der Hauptgleichungen der mechanischen Wärmetheorie", Pogg. Ann. **125** (1865), p. 353.

the heat $đQ$, added to the system in an infinitesimal *reversible* process, to the absolute temperature T of the system:

$$dS = \frac{đQ}{T}. \tag{6.1}$$

(The symbol $đ$ is used to stress the point that there is no function Q of which $đQ$ is the differential: $đQ$ is an inexact differential while dS is exact.) Second, in any *irreversible* process the entropy change always exceeds the value of $đQ/T$, so that for an isolated system, (one that exchanges no heat with its surroundings), the entropy must increase in an irreversible process. The problems for the molecular theory were then to demonstrate the existence of an entropy function satisfying (6.1), and to prove that this function can only increase in an irreversible adiabatic process.

Boltzmann's first paper[4] attempted to show that the second law, as formulated above, could be derived as a purely mechanical theorem. He was able to prove the existence of an entropy function, but only under very restrictive conditions on the kinds of motions that the molecules carry out. He had to assume, in effect, that the motion of the whole gas was simply periodic, which is certainly very far from the case, and his expression for the entropy function depended upon the integral of the molecular kinetic energy over one period of the motion.[5] Boltzmann's additional argument for the increase of entropy in irreversible processes cannot be considered valid.

Not long after this first try at supplying a molecular basis for the second law, Boltzmann read Maxwell's studies on the kinetic theory of gases, particularly his paper of 1867.[6] Maxwell's work showed him that the proper way to handle the complexities of a gas composed of an unbelievably large number of molecules was to use a statistical description. Maxwell had already introduced statistical methods, deriving the velocity distribution for an ideal gas and using it to treat a variety of problems of theoretical and experimental interest. In a

4. L. Boltzmann, "Über die mechanische Bedeutung des zweiten Hauptsatzes der Wärmetheorie", Wien. Ber. **53** (1866), p. 195. *Wiss. Abh.* I, p.9.

5. Despite its inadequacy for its intended purpose, Boltzmann's 1866 theorem did figure in an important way in later developments, particularly in Ehrenfest's adiabatic hypothesis. See Chapter 11.

6. J. C. Maxwell, "On the Dynamical Theory of Gases", Phil. Trans. Roy. Soc. **157** (1867), p. 49. *Sci. Papers* II, p. 26.

series of lengthy papers published in 1868 and 1871,[7] Boltzmann rederived and extended Maxwell's results on the distribution of molecular velocities and on the average energy of a molecule. (Maxwell's brevity and Boltzmann's expansiveness stand in sharp contrast, as observed by Maxwell himself in a letter to Tait. "By the study of Boltzmann I have been unable to understand him. He could not understand me on account of my shortness, and his length was and is an equal stumbling-block to me. Hence I am very much inclined to join the glorious company of supplanters and to put the whole business in about six lines."[8]) There were two principal results of Boltzmann's calculations – the general form of the distribution law and, as a consequence, the general form of the equipartition theorem on average energies. Apart from the intrinsic interest and importance of this work, it provided the concepts and methods Boltzmann needed for a new attempt at proving the second law of thermodynamics.[9]

Boltzmann had already found the distribution law for the configuration of an arbitrary body in thermal equilibrium with a gas. The probability of finding the atoms of the body in a given configuration, i.e. with their positions and velocities specified to lie in certain intervals, depended only on the energy of the configuration. The single parameter in the distribution law could be associated with the temperature with the help of the equipartition theorem and the gas law. In order to establish the existence of an entropy function Boltzmann had to identify the heat added to the system and the work done on it in any process. This had been one of the basic difficulties in his earlier treatment of the problem, but now he could manage it with his statistical approach. Let us look at the structure of the argument.

Consider an arbitrary system whose r atoms have coordinates x_1, y_1, z_1, ..., x_r, y_r, z_r and velocity components u_1, v_1, w_1, ..., u_r, v_r, w_r. Boltzmann's basic distribution law gives the probability dp that the

---

7. L. Boltzmann, (a) "Studien über das Gleichgewicht der lebendigen Kraft zwischen bewegten materiellen Punkten", Wien. Ber. **58** (1868) p. 517. *Wiss. Abh.* I, p. 49. (b) "Über das Wärmegleichgewicht zwischen mehratomigen Gasmolekülen", Wien. Ber. **63** (1871), p. 397. *Wiss. Abh.* I, p. 237. (c) "Einige allgemeine Sätze über Wärmegleichgewicht", Wien. Ber. **63** (1871), p. 679. *Wiss. Abh.* I, p. 259.
8. J. C. Maxwell to P. G. Tait, August 1873. Quoted in C. G. Knott, *Life and Scientific Work of Peter Guthrie Tait* (Cambridge: Cambridge University Press, 1911) p. 114.
9. L. Boltzmann, "Analytischer Beweis des zweiten Hauptsatzes der mechanischen Wärmetheorie aus den Sätzen über das Gleichgewicht der lebendigen Kraft", Wien. Ber. **63** (1871), p. 712. Wiss. Abh. I, p. 288.

atomic positions lie in the ranges x_1 to $x_1+\mathrm{d}x_1$, etc. and the velocities in the ranges u_1 to $u_1+\mathrm{d}u_1$, etc. The distribution law has the form,

$$\mathrm{d}p=\frac{\exp(-hE)\,\mathrm{d}x_1\ldots\mathrm{d}z_r\,\mathrm{d}u_1\ldots\mathrm{d}w_r}{\int\ldots\int\exp(-hE)\,\mathrm{d}x_1\ldots\mathrm{d}z_r\,\mathrm{d}u_1\ldots\mathrm{d}w_r},\tag{6.2}$$

where h is the basic parameter of the distribution and E is the total energy of the system,

$$E=\tfrac{1}{2}\sum_{i=1}^{r}m_i\,(u_i^2+v_i^2+w_i^2)+\chi(x_1,\ldots,z_r).\tag{6.3}$$

Here m_i is the mass of the ith atom and χ is the potential energy of the system. (Boltzmann also equated the probability $\mathrm{d}p$ to the fraction $\mathrm{d}t/\tau$, where τ is some long time interval and $\mathrm{d}t$ is that subinterval during which the atoms of the system are found in the specified ranges. Averages over the distribution are then also time averages.)

A direct calculation of the average kinetic energy using (6.2) shows that the term corresponding to each atomic velocity component, e.g. $\frac{1}{2}m_1u_1^2$, contributes $1/2h$, and that the average of the total kinetic energy is therefore $3r/2h$. Boltzmann actually identified the temperature T with the average kinetic energy of a single atom,

$$T=\langle\tfrac{1}{2}m_i(u_i^2+v_i^2+w_i^2)\rangle=\frac{3}{2h},\tag{6.4}$$

where the angular brackets denote an average over the distribution (6.2). In doing so Boltzmann omitted the proportionality constant now universally known as Boltzmann's constant k. In current notation one would write (6.4) as

$$\langle\tfrac{1}{2}m_i(u_i^2+v_i^2+w_i^2)\rangle=\tfrac{3}{2}kT.$$

In order to discuss the existence of the entropy one must examine changes in the average energy $\langle E\rangle$, where

$$\langle E\rangle=\frac{3r}{2h}+\frac{\int\chi\exp(-hE)\,\mathrm{d}\omega}{\int\exp(-hE)\,\mathrm{d}\omega}.\tag{6.5}$$

The symbol $\mathrm{d}\omega$ has been used as an abbreviation for the full set of the differentials in the multiple integral. Note that the structure of E

and the fact that χ is independent of velocities allow us to rewrite the second term of (6.5) as

$$\frac{\int \chi \exp(-h\chi)\,\mathrm{d}\sigma}{\int \exp(-h\chi)\,\mathrm{d}\sigma},$$

where $\mathrm{d}\sigma$ is $\mathrm{d}x_1 \ldots \mathrm{d}z_r$.

The average energy can be changed by changing the distribution function, (which means changing the temperature), and also by changing the form of the potential energy function χ. This latter function depends on the parameters of the system, such as external fields or, using an appropriate mathematical artifice, the volume, and can be changed in form by varying these parameters. If one writes for the variation in $\langle E\rangle$ the expression

$$\delta\langle E\rangle = -\frac{3r}{2h^2}\,\delta h + \delta\langle\chi\rangle, \tag{6.6}$$

the first term is the change in the average kinetic energy due to a temperature change, but the second term includes changes in the average potential energy produced by changes in both the distribution function and in the potential energy function itself. These latter changes, however, are what one means by external work. Thus we can write for the heat δQ added to the system, the difference between the energy change and the work done on the system,

$$\delta Q = \delta\langle E\rangle - \langle\delta\chi\rangle = -\frac{3r}{2h^2}\,\delta h + \delta\langle\chi\rangle - \langle\delta\chi\rangle, \tag{6.7}$$

or, more explicitly,

$$\delta Q = -\frac{3r}{2h^2}\,\delta h + \delta\left[\frac{\int \chi \exp(-h\chi)\,\mathrm{d}\sigma}{\int \exp(-h\chi)\,\mathrm{d}\sigma}\right] - \frac{\int \delta\chi \exp(-h\chi)\,\mathrm{d}\sigma}{\int \exp(-h\chi)\,\mathrm{d}\sigma}. \tag{6.8}$$

This form for δQ is not an exact differential, nor would one expect it to be, but if one divides by the temperature, that is to say by $3/2h$, it is a straightforward matter to show that an exact differential results, namely,

$$\frac{\delta Q}{T} = \delta\left[-r\ln h + \frac{2h}{3}\langle\chi\rangle + \tfrac{2}{3}\ln\left(\int \exp(-h\chi)\,\mathrm{d}\sigma\right) + \text{const.}\right]. \tag{6.9}$$

The function inside the square brackets is the entropy, S, of the system.

It is clearly a function of the state of the system, and it can be written more compactly in the form

$$S = \frac{\langle E \rangle}{T} + \tfrac{2}{3} \ln \left[\int \exp(-hE)\, d\omega \right] + \text{const.} \tag{6.10}$$

This equation specifies a definite procedure for calculating the entropy of any system in thermodynamic equilibrium. Boltzmann's argument not only demonstrated that an entropy function exists, but it also fixed a procedure for finding it.[10]

3. Providing a molecular theory of the second law required more than just demonstrating the existence of an entropy function related in the proper way to heat and temperature. Boltzmann still had not discussed the behavior of this entropy function in irreversible processes. The methods he had been using did not allow him to investigate the way in which a system evolves in time.

The following year, 1872, he published another memoir[11] in which the focus of attention was on just the question of how the velocity distribution changes with time. Boltzmann was not satisfied with previous derivations of the equilibrium velocity distribution, neither his own nor Maxwell's, because no proof had yet been given that the equilibrium distribution, (6.2), was unique. This was the gap he wanted to fill by proving that the Maxwell-Boltzmann distribution law is the only one that is stationary in time; this distribution law is not only a sufficient condition for equilibrium, it is also a necessary one.

Boltzmann undertook a detailed analysis of the collision processes by means of which the molecules in a gas can change their velocities, extending earlier work by Maxwell. In contrast to the methods of his 1871 paper, this analysis was restricted to gases, and, in fact, to gases at densities sufficiently low that one need consider only encounters between pairs of molecules. Limiting himself to the case that the gas is spatially homogeneous and has no external forces acting on it, Boltzmann determined the rate at which the velocity distribution changes due to binary collisions. Let $f(\mathbf{v}, t)\, d\mathbf{v}$ be the number of molecules per unit volume which, at time t, have vector velocities in

10. The essential features of Boltzmann's argument are still valid. See, for example, E. Schrödinger, *Statistical Thermodynamics* (Cambridge: Cambridge University Press, 1952), p. 11.

11. L. Boltzmann, "Weitere Studien über das Wärmegleichgewicht unter Gasmolekülen", Wien. Ber. **66** (1872), p. 275. *Wiss. Abh.* I, p. 316.

the range from $\boldsymbol{v}$ to $\boldsymbol{v}+\mathrm{d}\boldsymbol{v}$. Those collisions in which one of the colliding molecules has a velocity in this range before collision will decrease f, while those collisions in which one of the colliding molecules has a velocity in the specified range after collision will increase f. If $\boldsymbol{v}_1$, $\boldsymbol{v}_2$ represent the molecular velocities before a collision, and $\boldsymbol{v}_1'$, $\boldsymbol{v}_2'$ represent the corresponding velocities after collision, the rate of change of f can be put in the form

$$\frac{\partial f}{\partial t}=\int \mathrm{d}\boldsymbol{v}_2 \int \mathrm{d}\Omega\, \sigma(\Omega)\, |\boldsymbol{v}_1-\boldsymbol{v}_2|\, \{f(\boldsymbol{v}_1')f(\boldsymbol{v}_2')-f(\boldsymbol{v}_1)f(\boldsymbol{v}_2)\}. \quad (6.11)$$

Here $\sigma(\Omega)\mathrm{d}\Omega$ is the collision cross section for a collision in which the relative velocity afterwards is within the solid angle $\mathrm{d}\Omega$ at Ω compared to the relative velocity before. The integral is over all possible scattering angles and all possible velocities $\boldsymbol{v}_2$ of the collision partner. (The final velocity vectors $\boldsymbol{v}_1'$, $\boldsymbol{v}'_2$ can be expressed in terms of $\boldsymbol{v}_1$, $\boldsymbol{v}_2$, and the two angles in Ω using the conservation laws for energy and momentum.) Equation (6.11) is based on a plausible randomness assumption, the Stosszahlansatz, to which we shall have to return later on.

It is clear from (6.11) that $\partial f/\partial t$ will be zero if $f(\boldsymbol{v})$ is the Maxwell-Boltzmann distribution, where $f(\boldsymbol{v})$ is proportional to $\exp(-hmv^2/2)$. (See (6.2).) For, in that case $f(\boldsymbol{v}_1')f(\boldsymbol{v}_2')$ is equal to $f(\boldsymbol{v}_1)f(\boldsymbol{v}_2)$ because of energy conservation. In order to prove the converse, Boltzmann introduced the function that he later called H, defined by the equation

$$H(t)=\int \mathrm{d}\boldsymbol{v}\, f(\boldsymbol{v},t)\ln f(\boldsymbol{v},t). \quad (6.12)$$

Because the total number of molecules does not change with time, the rate of change of H is given by the equation

$$\frac{\mathrm{d}H}{\mathrm{d}t}=\int \mathrm{d}\boldsymbol{v}\frac{\partial f}{\partial t}\ln f, \quad (6.13)$$

which can be evaluated by using (6.11).

Under the additional assumption that the inverse collision (in which molecules with velocities $\boldsymbol{v}_1'$, $\boldsymbol{v}_2'$ before collision acquire velocities $\boldsymbol{v}_1$, $\boldsymbol{v}_2$), is always possible, $\mathrm{d}H/\mathrm{d}t$ can be put into a very useful form. (Inverse collisions may not exist if the molecules are not spherical, but even then the theorem can still be proved at the expense of some further complications in the argument.[12] The transformed expression for $\mathrm{d}H/\mathrm{d}t$ can be written as

$$\frac{\mathrm{d}H}{\mathrm{d}t} = \tfrac{1}{4}\int \mathrm{d}\boldsymbol{v}_1 \int \mathrm{d}\boldsymbol{v}_2 \int \mathrm{d}\Omega\, \sigma(\Omega)\, |\boldsymbol{v}_1 - \boldsymbol{v}_2|\, (f'_1 f'_2 - f_1 f_2) \ln\left(\frac{f_1 f_2}{f'_1 f'_2}\right). \qquad (6.14)$$

(The notation has been abbreviated by writing $f(\boldsymbol{v}'_1)$ as f'_1, etc.) Now $\sigma(\Omega)$ and $|\boldsymbol{v}_1 - \boldsymbol{v}_2|$ are intrinsically positive, and the function $(x-y)\ln(y/x)$ is always negative, except when $x = y$, when it is zero. It follows then that H can never increase,

$$\frac{\mathrm{d}H}{\mathrm{d}t} \leqq 0. \qquad (6.15)$$

Furthermore, because the integrand is of one sign only, $\mathrm{d}H/\mathrm{d}t$ (and therefore $\partial f/\partial t$), can vanish only when the integrand vanishes, i.e. when $f_1 f_2 = f'_1 f'_1$, a condition that can be shown to require the Maxwell distribution. Boltzmann's use of the auxiliary function H had allowed him to prove the uniqueness of the equilibrium velocity distribution. He immediately recognized that his arguments actually proved a great deal more.

Boltzmann's "H-theorem" showed that if the distribution of molecular velocities in a gas is initially something other than the Maxwell distribution, then, as a result of intermolecular collisions, the distribution will evolve into the stationary Maxwell distribution. In the course of this evolution the quantity H will always decrease, attaining its minimum value at equilibrium when f is Maxwellian. The quantity $(-H)$ not only behaves like the entropy, increasing in the irreversible approach to equilibrium, it is actually proportional to the entropy, as Boltzmann showed by calculating it with the help of (6.12), using the equilibrium distribution for f. In other words, the H-theorem appeared to complete the proof of the second law of thermodynamics on the basis of molecular theory, at least for the case of a gas of low density.

4. The first important criticism of Boltzmann's work came from one of his colleagues at Vienna, Josef Loschmidt, in 1876.[13] Loschmidt

---

12. H. A. Lorentz, "Über das Gleichgewicht der lebendigen Kraft unter Gasmolekülen", Wien. Ber. **95** (1887), p. 115.
L. Boltzmann, "Neuer Beweis zweier Sätze über das Wärmegleichgewicht unter mehratomigen Gasmolekülen", Wien. Ber. **95** (1887), p. 153. *Wiss. Abh.* III. p. 272.
13. J. Loschmidt, "Über den Zustand des Wärmegleichgewichtes eines Systems von Körpern mit Rücksicht auf die Schwerkraft", Wien. Ber. **73** (1876), pp. 135, 366; **75**

had already shown how staunch an atomist he was by making the first determination of the number of molecules to be found in one cubic centimeter of a gas under normal conditions. This number, still referred to as Loschmidt's number in the German literature, can be used to find the mass of a single atom, and is *the* critical number on the atomic scale. Loschmidt had no doubts about the need for a molecular theory, but he had some serious questions about what Boltzmann had done. A number of these concerned Boltzmann's results on the distribution of molecular velocities and positions for a gas in a gravitational field, the problem of the atmosphere. These were answered by Boltzmann without any great difficulty.[14]

One of Loschmidt's arguments cut deeper, however, and even seemed to rule out any possibility of providing a molecular foundation for the second law of thermodynamics. Boltzmann took this objection very seriously[15], describing it as "extremely ingenious", and "of great importance for the correct understanding of the second law". He answered it promptly and then used the new insight provided by this exchange with Loschmidt to deepen his own conception of what the second law meant.

Loschmidt argued that one could never derive the approach to equilibrium and the associated increase of entropy from the reversible laws of mechanics, and that those observed characteristics of the natural world must, therefore, depend only on the initial conditions that actually prevail. For, suppose one has an isolated system that proceeds from some initial state (i) at $t=0$ to another, more "mixed" state (f) at $t=\tau$ with a monotonic increase of entropy. Consider now an initial state ($-f$) at $t=0$ in which all the molecules have the positions they would have attained at $t=\tau$ in the former motion, and also the velocities they would have attained, but with the directions of these velocities reversed. Then, since the laws of mechanics are invariant

(1877), p. 67; **76** (1877), p. 209. Also see R. Dugas, *La théorie physique au sens de Boltzmann* (Neuchâtel: Éditions du Griffon, 1959), p. 158.

14. L. Boltzmann, (a) "Über die Aufstellung und Integration der Gleichungen, welche die Molekularbewegung in Gasen bestimmen", Wien. Ber. **74** (1876) p. 503. *Wiss. Abh.* II, p. 55. (b) "Weitere Bemerkungen über einige Probleme der mechanischen Wärmetheorie", Wien. Ber. **78** (1878), p. 7. *Wiss. Abh.* II, p. 250. Especially Section II.
For a discussion of Maxwell's work on this problem see S. G. Brush, "Foundations of Statistical Mechanics 1845–1915", Archive for History of Exact Sciences **4** (1967), p. 157.

15. L. Boltzmann, "Bemerkungen über einige Probleme der mechanischen Wärmetheorie", Wien. Ber. **75** (1877), p. 62. *Wiss. Abh.* II, p. 112. Especially Section II.

under time reversal, the motion that ensues will be just the original motion occurring in the opposite sequence in time. After a time τ the molecules will have attained the former initial state, with reversed velocities, $(-i)$. This latter process will thus be one that proceeds away from equilibrium, with a monotonic decrease in entropy. Since for every entropy increasing process one can construct an equally allowed entropy decreasing process by making use of time reversal this way, it follows that the observed occurrence of only entropy increasing processes is a consequence of the special initial conditions in the world and not of the laws governing molecular motions. This is, as Boltzmann wrote, a "very seductive" argument.

In his answer Boltzmann emphasized that the molecular proof of the second law was not really based on mechanics alone, but on mechanics together with the laws of probability. Loschmidt was correct in asserting that the reversed motion would produce a decrease in entropy, and also that this motion was just as consistent with the laws of mechanics as the original, entropy increasing, motion. But Loschmidt's argument took no account of one essential point: the probability that the initial state of the system is such as to produce an entropy increasing motion is overwhelmingly greater than the probability that it will produce an entropy decreasing motion. In this context Boltzmann meant by probability simply the number of ways one could choose the molecular initial conditions in a manner compatible with the few observed macroscopic variables. There are so many more disordered molecular configurations which lead to entropy increasing, equilibrium attaining, processes than there are specially ordered configurations which lead to entropy decreasing processes going away from equilibrium, that the latter can be safely ignored. Loschmidt's argument gave a way of determining initial conditions that would surely give rise to entropy decreasing processes, but that did not prove that there were not incomparably more initial conditions giving rise to normal, entropy increasing processes.

For Boltzmann, Loschmidt's objections to the *H*-theorem were valuable just because they drew attention to the essential role of probability considerations in any attempt to understand the second law. Pushing the argument a step further, Boltzmann suggested that one could even calculate the probabilities of the various molecular distributions from the ratios of the number of ways they could be achieved, and that this "would perhaps lead to an interesting method of determining the state of thermodynamic equilibrium".

5. This suggestion was fully developed in a memoir,[16] published later in the same year 1877, dealing with the relationship between the second law of thermodynamics and the theory of probability. Boltzmann looked upon this memoir as the culmination of the series of papers we have been discussing, all of which had involved this relationship in one way or another. He was now ready to claim that the second law of thermodynamics was, in its very essence, a law of probability: a system evolves from states of low intrinsic probability to states of higher probability, the equilibrium state being the state of maximum probability. That entropy increases in this irreversible evolution is no accident, since entropy is nothing but a suitable measure of this probability. These were Boltzmann's assertions; let us see how he argued for them.

He began with a highly oversimplified case, introduced to facilitate the understanding of his new ideas. This was a gas of N "molecules" imagined to be of such a nature that each molecule could have for its kinetic energy only one of the set of values $0, \epsilon, 2\epsilon, 3\epsilon, \ldots, p\epsilon$, where ϵ is some arbitrary energy unit and p is an integer.[17] The total energy of the gas is fixed and has the value $\lambda\epsilon$ where λ is some integer. The state of the gas would be completely specified by a listing of the energy of each individual molecule, e.g. molecule number one has energy 4ϵ, molecule number two has energy 7ϵ, and so on. Such a detailed specification Boltzmann called a complexion, and he made the fundamental assumption that each complexion was as likely to occur as any other. This is an assumption of the essential randomness of the process, whatever it may be in detail, by means of which the molecules exchange energy in collisions. (In a similar way, if one assumes that each throw of a die is a random event, then in five successive throws the probability of obtaining the sequence 6,6,6,6,6 is no less and no more than that of obtaining the specific sequence 2,3,1,1,5 or any other specific one of the 6^5 possible outcomes.)

---

16. L. Boltzmann, "Über die Beziehung zwischen dem zweiten Hauptsatze der mechanischen Wärmetheorie und der Wahrscheinlichkeitsrechnung respektive den Sätzen über das Wärmegleichgewicht", Wien. Ber. **76** (1877), p. 373. *Wiss. Abh.* II, p. 164.

17. This model has much in common with the set of oscillators in Max Planck's first paper on quanta. M. Planck, "Zur Theorie des Gesetzes der Energieverteilung im Normalspectrum", Verh. der Deutsch. Phys. Gesellschaft, **2** (1900), p. 237. The connection is not completely accidental. See R. Dugas, *op. cit.* p. 250 and M. J. Klein, "Max Planck and the Beginnings of the Quantum Theory", Archive for History of Exact Sciences, **1** (1962), p. 472.

This full specification corresponding to a complexion is not what one means when one asks for the molecular distribution in some state. A distribution is determined by the set of numbers $w_0, w_1, w_2, \ldots, w_p$, where w_k is the number of molecules having energy $k\epsilon$. It does not matter which particular molecules have these energies, but only how many molecules have them. There are, in general, many complexions compatible with such a given distribution $w_0, w_1, w_2, \ldots, w_p$. Let P be the number of complexions for the distribution, so that P is determined by the equation

$$P = \frac{N!}{w_0!\, w_1!\, w_2! \ldots w_p!}\,. \tag{6.16}$$

Boltzmann took P as a measure of the probability of the distribution $w_0, w_1, \ldots, w_p$; more precisely, the ratio of P to the sum of the P's, for all possible distributions compatible with the assigned total energy, was set equal to the probability of the distribution defined by the w's.

The most probable distribution is the one that has the largest value of P associated with it. To find it, one must determine that set of numbers $w_0, w_1, w_2, \ldots, w_p$ which maximize the expression of (6.16) and also satisfy the equations,

$$\sum_{r=0}^{p} w_r = N, \tag{6.17}$$

which expresses the fact that the number of molecules is fixed, and

$$\sum_{r=0}^{p} r\epsilon w_r = \lambda\epsilon, \tag{6.18}$$

which expresses the fact that the total energy is fixed. This problem can be solved to an adequate approximation, and the resulting most probable distribution is well represented by an exponential; w_r is proportional to $\exp(-\beta r\epsilon)$, where β is a parameter determined by the energy constraint.

Boltzmann showed that the same method can be followed without artificially restricting the possible molecular energies to a particular discrete set of values. The quantities w_r must now be replaced by the function $f(x, y, z, u, v, w)$, the molecular distribution function, where $f\,\mathrm{d}x\mathrm{d}y\mathrm{d}z\mathrm{d}u\mathrm{d}v\mathrm{d}w$ is the number of molecules to be found in the region x to $x+\mathrm{d}x$, etc., with velocity components in the range u to $u+\mathrm{d}u$, etc. His basic result was that the most probable distribution is just the

Maxwell-Boltzmann distribution, the already familiar description of the equilibrium state. Boltzmann's arguments provided a new way of looking at equilibrium as the most probable state, the one that can be realized in the largest number of ways.

(Boltzmann pointed out that one must be careful to select the domains of equal probability for the gas molecules as he did, that is, to use domains of the form $dxdydzdudvdw$. If one tried to use equal energy intervals, for example, as domains of equal probability, one would not obtain the proper form for the equilibrium distribution.)

In the process of determining the most probable distribution it was the function ln P, rather than P, which was actually maximized. This replacement of P by its logarithm was made for technical reasons, but ln P turned out to have a real significance of its own. In the discrete case of (6.16), ln P has the form

$$\ln P = N \ln N - \sum_{r} w_r \ln w_r, \tag{6.19}$$

where Stirling's approximation is used for the factorials (i.e. $\ln n! \cong n \ln n - n$), and the restriction of (6.17) is taken into account. In the actual case of the gas, the analogous expression, dropping the uninteresting first term, becomes

$$\ln P = -\int \ldots \int f \ln f \, \mathrm{d}x\mathrm{d}y\mathrm{d}z\mathrm{d}u\mathrm{d}v\mathrm{d}w, \tag{6.20}$$

where f is the molecular distribution function. Boltzmann evaluated ln P explicitly for the equilibrium distribution, where f has the Maxwell-Boltzmann form, and showed that it was essentially the same quantity as the entropy of the gas calculated directly by thermodynamic methods. (The two functions differed by a scale factor and also by an additive constant, considered as having no significance for the entropy, since thermodynamic arguments determine only entropy differences.)

In thermodynamics, the entropy of a system is only defined when the system is in equilibrium. It was under these circumstances that Boltzmann showed the entropy was equivalent to ln P. But this latter quantity is defined for all states, no matter how different f may be from its equilibrium form. Boltzmann saw that the entropy concept could be given a much wider range of applicability if one extended its definition by always identifying it with ln P. The second law of thermodynamics would then require that this generalized entropy must increase in any irreversible process in a closed system, regardless of whether or not the initial and final states are states of equilibrium. And because of the new definition of entropy, this statement would be equivalent to the simpler one: all systems evolve from less probable into more probable states.

6. Boltzmann thought he had disposed of the problem of providing a theoretical foundation for the second law of thermodynamics by the series of papers ending with the 1877 memoir we have just been considering. He then directed his attention to other matters, returning to this problem only occasionally, to add a footnote or two to his earlier expositions, or to comment on some other physicist's discussion of a related point. There was, however, relatively little discussion of his work to which he could respond.

One of the reasons for some of the hesitancy and even reluctance about accepting Boltzmann's ideas was the uncertain status of one of the principal consequences of the kinetic theory of gases, the equipartition theorem. This theorem, anticipated by Clausius, developed and extended by Maxwell and Boltzmann, required, as its name suggests, that the energy of a gas be evenly distributed, on the average, over all modes of motion of all molecules.[18] More specifically, it follows directly from the Maxwell-Boltzmann distribution law that the average of each quadratic term in the total energy of a gas must have the same value, a constant multiple of the absolute temperature of the gas. In modern notation, this common average value is written as $\frac{1}{2}kT$. (See above, (6.4) and the subsequent discussion.) If each molecule has n such quadratic contributions to the kinetic energy and there is no internal potential energy, the total energy of a gas of N molecules would be $\frac{1}{2}nNkT$.

Although the energy of the gas is not directly observable, the temperature derivative of this energy is the heat capacity of the gas, which can be measured; it is, in fact, the heat capacity at constant volume, when no work can be done and the heat added is equal to the change in the internal energy. Considering one gram molecular weight (one

---

18. Clausius did not state the full equipartition theorem. He required only a proportionality between the energy of translational motion and any internal kinetic energy of the molecules. See R. Clausius, "Über die Art der Bewegung, welche wir Wärme nennen", Pogg. Ann. **100** (1857), p. 353. For Maxwell's work see his "Illustrations of the Dynamical Theory of Gases", Phil. Mag. **19** (1860), p. 19; **20** (1860), p. 21. *Sci. Papers* I, p. 377. Boltzmann discussed the problem in the paper cited in note 7 (b). For a modern, general proof see, e.g. R. C. Tolman, *The Principles of Statistical Mechanics* (Oxford: Oxford University Press, 1938), p. 95.

mole) of the gas, for which Nk is just equal to the constant R in the gas law, the heat capacity at constant volume, c_V, can then be written

$$c_V = \tfrac{1}{2} nR. \qquad (6.21)$$

Measurement of c_V thus allowed one to determine n, the number of quadratic terms in the kinetic energy of a single molecule, usually the number of its degrees of freedom. The trouble came in trying to reconcile the experimentally determined values of n with any reasonable picture of molecular structure.

As Maxwell put it in a lecture in 1875, "And here we are brought face to face with the greatest difficulty which the molecular theory has yet encountered."[19] For, if the molecules were simply material points, n would have to be 3, corresponding to the three independent translations which are the only motions a point can carry out. If, on the other hand, the molecule is an extended structure, n should be at least 6, since there would be three possible independent rotations in addition to the translational motion. "But," as Maxwell added, "the spectroscope tells us that some molecules can execute a great many different kinds of vibrations. They must therefore be systems of a very considerable degree of complexity, having far more than six variables." The experimental values of n, determined from heat capacities, simply did not fit this picture. A whole series of common diatomic gases including oxygen, hydrogen, nitrogen and hydrogen chloride, had heat capacities corresponding to n equal to 5, too small for even a rigid structure and apparently irreconcilable with the internal motions suggested by spectroscopic evidence.

Boltzmann suggested a partial solution to this difficulty in 1876,[20] after measurements by A. Kundt and E. Warburg[21] showed that mercury vapor, a monatomic gas, had a heat capacity corresponding to n equal to 3. He proposed that the individual atom did behave like a point

---

19. J. C. Maxwell, "On the Dynamical Evidence of the Molecular Constitution of Bodies", Nature **11** (1875), pp. 357, 374. *Sci. Papers* II, p. 418. (Quotation from p. 433).
20. L. Boltzmann, "Über die Natur der Gasmoleküle", Wien. Ber. **74** (1876), p. 553. *Wiss. Abh.* II, p. 103.
21. A. Kundt and E. Warburg, "Über die spezifische Wärme des Quecksilbergases", Pogg. Ann. **157** (1876), p. 353. The heat capacity ratio of argon, measured right after its discovery, was one of the major pieces of evidence supporting the assertion that argon is monatomic. See the Note in Nature **51** (1895), p. 323 announcing the discovery of argon, the editorial discussion after the lectures to the Royal Society, *ibid.* p. 337, the abstract of the paper by Lord Rayleigh and William Ramsay, *ibid.* p. 352, and the report of the subsequent discussion, *ibid.* p. 356.

particle, as the mercury value would imply, or at least like a smooth, hard sphere whose rotations would not be affected by collisions and so would not contribute to the thermal energy of the gas. The anomalous value 5 for the n of diatomic molecules could be accounted for in a similar way if the molecule were a smooth solid of revolution, with its mass symmetrically distributed around the axis. Then rotations about this symmetry axis would be unaffected by collisions, while the two independent rotations about axes perpendicular to the symmetry axis would make their proper contribution to the thermal energy.

Boltzmann's solution did not really settle the question of the internal molecular motions demanded by the existence of spectral lines, and it did not win general acceptance. The equipartition problem continued to be felt as a very real and urgent one, particularly in England, throughout the last quarter of the nineteenth century. This is evident in a number of works, the most famous of which is Lord Kelvin's lecture in 1900 on "Nineteenth Century Clouds over the Dynamical Theory of Heat and Light".[22] The second of Kelvin's clouds was, in fact, "the Maxwell-Boltzmann doctrine regarding the partition of energy", a doctrine in which Kelvin put very little faith and which he tried hard, but unsuccessfully, to overthrow by means of counter-examples. The standard English exposition of some of the work of Maxwell and Boltzmann was H. W. Watson's little book, *A Treatise on the Kinetic Theory of Gases*. Watson ended his second edition in 1893 with a discussion of the unresolved equipartition problem, and his last sentence expressed the hope that it was "not necessarily a fatal objection to the theory."[23]

There was enough interest in these matters in England so that the British Association for the Advancement of Science appointed a committee consisting of Joseph Larmor and George H. Bryan to report to it on "The Present State of our Knowledge of Thermodynamics, specially with regard to the Second Law." The committee submitted its lengthy report, written by Bryan, in two parts, in 1891 and 1894.[24]

22. Lord Kelvin's lecture was delivered at the Royal Institution on April 27, 1900. It is printed as an Appendix in his *Baltimore Lectures on Molecular Dynamics and the Wave Theory of Light* (London: C. J. Clay and Sons, 1904), p. 486.

23. H. W. Watson, *A Treatise on the Kinetic Theory of Gases* (Oxford: Clarendon Press, 2nd Ed. 1893), p. 87.

24. (a) G. H. Bryan, "I-Researches Relating to the Connection of the Second Law with Dynamical Principles", B. A. A. S. Report **61** (1891), p. 85

(b) G. H. Bryan, "II-The Laws of Distribution of Energy and their Limitations", B. A. A. S. Report **64** (1894), p. 64.

Bryan's first report covered the various attempts to relate the second law to the principles of dynamics, and this included the early efforts to construct a purely mechanical theory and Helmholtz's monocycle analogies, as well as the statistical theories. The second part of the report, dealing with "The Laws of Distribution of Energy and their Limitations", was presented to the British Association at its Oxford meeting in August, 1894. Boltzmann himself was there and took part in the discussion of Bryan's report after its presentation. This "brilliant discussion, which had to be curtailed from want of time at Oxford", was continued in the columns of *Nature* during the year that followed.[25]

Two points were discussed again and again in the course of twenty or so Letters to the Editor of *Nature* by some half a dozen correspondents. The first was the difficulty in reconciling the existence of spectral lines with the equipartition theorem.[26] This issue could not be resolved by the discussion. No one knew how to answer the objection that the vibrations involved in producing spectral lines should, but do not, contribute to the heat capacity of the gas. Some expressed the hope that electromagnetic theory, or a more detailed understanding of the aether, would show a way out. Boltzmann himself was inclined to lay the blame on an absence of equilibrium between the gas and the aether. But all agreed that the problem was unsolved. The second point that provoked a good deal of discussion was the apparent paradox involved in maintaining both the reversibility of mechanics and the irreversibility implied by Boltzmann's H-theorem.[27] This discussion did help to clarify the issues, partly because it showed Boltzmann where his arguments had been misunderstood. His answer,[28] therefore, emphasized an aspect of the situation which had not been brought out so explicitly

---

25. Boltzmann referred to it in 1895 as "the unforgettable meeting of the British Association at Oxford." L. Boltzmann, *Lectures on Gas Theory*, transl. S. G. Brush (Berkeley and Los Angeles: University of California Press, 1964), p. 22. The discussion is to be found in Nature **51** (1894–95) and **52** (1895). The participants were Boltzmann, G. H. Bryan, S. H. Burbury, E. P. Culverwell, G. F. Fitzgerald, J. Larmor, A. Schuster, and H. W. Watson. The quotation in the text is from G. H. Bryan, "Prof. Boltzmann and the Kinetic Theory of Gases", Nature **51** (1894), p. 31.
26. See Nature **51** (1894–95). pp. 31 (G. H. Bryan), 78 (E. P. Culverwell), 127 (S. H. Burbury), 221 (G. F. Fitzgerald), 293 (A. Schuster), 453 (G. F. Fitzgerald).
27. See Nature **50** (1894) p. 617 (E. P. Culverwell), and Nature **51** (1894–95) pp. 78 (S. H. Burbury), 105 (H. W. Watson and E. P. Culverwell), 152 (J. Larmor), 175 (G. H. Bryan), 246 (E. P. Culverwell), 319 (G. H. Bryan), 581 (E. P. Culverwell).
28. L. Boltzmann, "On Certain Questions of the Theory of Gases", Nature **51** (1895), pp. 413, 581. *Wiss. Abh.* III, p. 535.

before. If one examined the value of H (the negative of the entropy) in a closed system over a long period of time, then during most of this time H would be found at or very near its minimum value. And, whenever H had a value appreciably larger than its minimum, one could be almost certain that it would decrease, whether one followed it forward or backward in time. In other words, appreciable departures from the equilibrium or minimum value of H were most likely to show up as spikes, or local maxima, in a curve of H as a function of time. Thus, in a reversed motion of the kind proposed by Loschmidt, H would indeed increase for a time, back to the initial value of the unreversed (normal) motion, but it would then almost surely decrease once more, back towards its equilibrium minimum. Boltzmann tried to suggest the geometry of such an H-curve by describing it as resembling a succession of inverted trees, mostly very low and with nearly horizontal branches, occasionally higher, with inclined branches, "the improbability of such a tree increasing enormously with its height".

7. Less than a year after his response to his British critics, Boltzmann had to face a new argument against his explanation of the second law of thermodynamics. This time the attack came from Berlin, where Ernst Zermelo, a student of Max Planck's who would later become known for his work on the axiomatics of set theory, had thought out another objection to Boltzmann's derivation of irreversibility.[29] Zermelo's starting point was a theorem proved by Henri Poincaré in 1890.[30] Poincaré's theorem asserted that, in any system of particles acting on each other by arbitrary forces which depend only on their positions in space, any configuration, specified by the coordinates and velocities of the particles, will recur (infinitely many times), if not exactly, then within arbitrarily specified limits. Poincaré's theorem holds for almost all initial configurations and requires only that the coordinates and velocities are bounded. Zermelo argued that this theorem automatically ruled out any possibility of proving irreversibility since initial states were sure to recur, to as close an approximation as one wanted. No mechanical system could evolve irreversibly into an equilibrium state. There was a way out, to be sure, but it was not one that would appeal to physicists. For, as Zermelo remarked, Poincaré's theorem did not

29. E. Zermelo, "Über einen Satz der Dynamik und die mechanische Wärmetheorie", Wied. Ann. **57** (1896), p. 485.

30. H. Poincaré, "Sur le problème des trois corps et les équations de la dynamique", Acta Math. **13** (1890), p. 67.

apply to a certain highly restricted set of initial configurations, and one might try to construct a theory of irreversibility by assuming that only these singular initial configurations actually occur. While such an approach might be conceivable, Zermelo clearly recognized how unacceptable a theory would be if the initial states it allowed only constituted a set of measure zero. His conclusion, therefore, was that physics would have to choose between the second law of thermodynamics and the mechanical interpretation of nature; no middle ground was possible.

Boltzmann's reply came promptly.[31] He began by repeating the point he had already made so often: the laws governing the behavior of gases do not claim to be purely mechanical laws; they are essentially statistical. From the molecular point of view, the second law of thermodynamics is merely a law of probabilities. "Now Mr. Zermelo's paper shows that my writings have not been understood," Boltzmann went on bitterly, "but I am pleased with it in spite of that, because it is the first evidence that these writings have received any attention at all in Germany." Boltzmann did not dispute the validity or the applicability of Poincaré's theorem, but he disagreed strongly with the conclusions Zermelo had drawn from it. Repeating the description of the H-curve he had given in answer to his critics in *Nature*, Boltzmann emphasized that the Maxwellian distribution should be considered as the typical distribution. Most distributions that are at all possible are very similar to the Maxwellian and hardly distinguishable from it. It is the essentially non-Maxwellian distributions that are singular and occur only rarely. This was just the opposite of Zermelo's suggestion that only singular initial conditions could lead to equilibrium. As for the Poincaré recurrence itself, Boltzmann simply pointed out that the time necessary for such a return to the initial state of a gas would be almost inconceivably long. The fact that such recurrences are not observed is, therefore, hardly to be marveled at. The apparent irreversibility of the natural world is not inconsistent with the existence of Poincaré recurrences, if their time scale is so inordinately long. The fact that they are predicted, but not observed within the times available, is no reason to reject statistical mechanics. Boltzmann accused Zermelo of being willing to do just that, and compared him to a dice player who claims a die is false because he has never seen it come up on one a thousand

31. L. Boltzmann, "Entgegnung auf die Wärmetheoretischen Betrachtungen des Hrn. E. Zermelo", Wied. Ann. **57** (1896), p. 773. *Wiss. Abh.* III, p. 567.

times in succession, even though it has a non-zero probability of doing so.

Zermelo did not yield to Boltzmann's counterattack,[32] but continued to maintain that there was no firm foundation for a theory of irreversibility. He was willing to accept the consequences of such a position, rejecting all kinetic theories of heat conduction, viscosity, etc. along with the *H*-theorem. Only the equilibrium theories would be left standing, if Zermelo's views had prevailed. Needless to add, Boltzmann was not convinced.

Boltzmann did, however, try to clarify his description of the *H*-curve as a result of Zermelo's criticism.[33] Zermelo had objected that, according to Boltzmann, this curve, representing *H* as a function of time, would have to consist almost entirely of maxima, which was hardly conceivable, and that Boltzmann's description was self-contradictory. In his second response to Zermelo, Boltzmann tried to make clear that it was only when *H* was decidedly greater than its minimum, equilibrium, value that the *H*-curve was most likely to be at a maximum, and that such appreciable departures from equilibrium were, in any case, rare events, and became very much rarer the greater the departure from equilibrium. He even included a sketch showing the kind of thing he had in mind, although he warned that the sketch was a very sketchy one, to be taken "sehr cum grano". Boltzmann must have known that further explanation was called for, because a year or so later he submitted another article, "On the So-called *H*-Curve".[34] He could not, of course, calculate the detailed behavior of *H* for a gas, and so he had instead invented a model which showed those features of the *H*-curve that seemed so paradoxical.

This model, an elaboration of one Boltzmann had previously suggested in the *Nature* discussion, can be described as follows. Suppose we have an urn containing equal numbers of identical black and white balls. A ball is drawn at random at regular intervals of time, and replaced in the urn after its color has been noted. This process is carried

32. E. Zermelo, "Über mechanische Erklärungen irreversibler Vorgänge. Eine Antwort auf Hrn. Boltzmann's 'Entgegnung'", Wied. Ann. **59** (1896), p. 793.

33. L. Boltzmann, "Zu Hrn. Zermelos Abhandlung 'Über die mechanische Erklärung irreversibler Vorgänge'", Wied. Ann. **60** (1897), p. 392. *Wiss. Abh.* III, p. 579.
Boltzmann never doubted the truth of Poincaré's theorem, though he did not like Zermelo's proof of it. He offered his own proof in the paper, "Über einen mechanischen Satz Poincaré's", Wien. Ber. **106** (1897), p. 12. *Wiss. Abh.* III, p. 587.

34. L. Boltzmann, "Über die sogenannte H-Curve", Math. Ann. **50** (1898), p. 325. *Wiss. Abh.* II, p. 629.

on through some extremely long series of drawings. Consider now the quantity a_k, where a_k is the number of white balls drawn in a series of n drawings beginning at the kth drawing. Construct a "curve", really a discrete set of points, where the x and y coordinates of the kth point are defined by the equations,

$$x_k = k/n, \qquad y_k = |1-(2a_k/n)|. \tag{6.22}$$

If n is large, successive points will be close together, since their abcissas will differ by $1/n$ and their ordinates by at most $2/n$. It is expected that y will, on the average, be small, near zero, since the expectation value of a_k is $n/2$; half the drawings, on the average, will produce white balls. In the course of a very long series of drawings, however, it will be possible to find series of n successive whites or n successive blacks, for both of which cases the ordinate will be 1. One can determine the "average rate" at which the curve will fall off from one of these peaks by simply noting that a change in k by the amount Δk (also a large integer, but one small compared to n), means dropping the first Δk drawings of the original sequence of n, and including instead the subsequent Δk drawings. If the original group consisted of all white or all black balls, then the average change in a_k will be $\Delta k/2$. Hence the "quasi-derivative" will be $\Delta y/\Delta x$, where Δx is $\Delta k/n$ and Δy is $(-)(2/n)(\Delta k/2)$, so that the "quasi-slope" is -1. This indicates the average rate at which the curve falls off from one of its highest peaks. Many other details of this model could be worked out, but perhaps this is enough to show its general nature. It did suggest how some of the peculiar properties of the H-curve might be realized, but the model did not really clarify the issues involved in the H-theorem itself.

8. Another model that did illuminate the situation, and brought out clearly what Boltzmann had been trying to say, was proposed a few years later, in 1906, by his former student, Paul Ehrenfest. Ehrenfest described his model in two papers[35] written jointly with his wife,

35. (a) P. and T. Ehrenfest, "Über eine Aufgabe aus der Wahrscheinlichkeitsrechnung, die mit der kinetischen Deutung der Entropievermehrung zusammenhängt", Math.-Naturw. Blätter **3** (1906). *Papers*, p. 128.
(b) P. and T. Ehrenfest, "Über zwei bekannte Einwände gegen das Boltzmannsche H-Theorem", Phys. Z. **8** (1907), p. 311. *Papers*, p. 146.
The Ehrenfests seem to have introduced the terms "Reversibility Objection", and "Recurrence Objection" for Loschmidt's and Zermelo's arguments, respectively, in this paper.

Tatyana, who had worked with him in exploring its various subtleties. Here is Ehrenfest's own description of the model, taken from the earlier paper which appeared in the student journal *Mathematisch-Naturwissenschaftliche Blätter*, and is in a much more colloquial style than the later, more formal publication in the *Physikalische Zeitschrift*.

"Suppose we have N balls, (100, for example). They are numbered consecutively from 1 through N so they can be individually distinguished. They are at some moment distributed between two urns so that urn A contains P_0 balls (90, for example), and urn B has $Q_0 = N - P_0$ balls (therefore 10). But it is not known which individual balls are located in A and which in B. We also have, in a bag, N lottery tickets numbered from 1 through N. Every ten seconds a ticket is drawn, its number is announced, it is returned to the bag and the tickets are thoroughly mixed. Then another is chosen, announced, etc. Every time a number is announced, the ball that bears this number jumps out of the urn in which it happens to be located and over into the other urn, and it stays there until its number is drawn once again. Notice: it is always more likely that the ball just summoned will be found in the fuller urn rather than in the emptier one. As long, therefore, as urn A is still much fuller than urn B, most of the subsequent lottery drawings will produce an emptying out of urn A into urn B, and only rarely will urn A receive a ball from urn B."

It is already clear that Ehrenfest's model captures, at least qualitatively, Boltzmann's basic insight: the trend toward equilibrium (in this case an approximate equalization of the populations of the two urns) occurs because there are so many more ways the system can go toward equilibrium than away from it. (It is also clear why Ehrenfest referred to his model, though not in print, as the "dog-flea" model.) But much more could be seen from an analysis of the model.

One can see the behavior of the model graphically, as Ehrenfest suggested, by plotting the quantity $|P_r - Q_r|$, the absolute value of the difference between the populations of the two urns after r drawings, as the ordinate of a point whose abcissa is r units (i.e. proportional to the time). One obtains a sequence of points, a "curve", whose successive heights always differ by exactly two units, positive or negative. Starting with a large value, 80 in Ehrenfest's example, the "curve" will show a decided tendency to fall, showing only occasional rises on its way down. As the two populations become equalized, upward steps become more likely, and when the two urns are almost equally populated one can expect a fluctuating behavior with now one and now the other slightly fuller. But that is not the end. "Will it run along near the axis for all future times? – No." If one imagines the lottery continued for an enormously long period of time, one can expect to find sequences of drawings that lead to appreciable and even large differences in population. There is, in fact, a nonzero probability, very small but calculable, that one will even have sequences that put all the balls in one urn. Thus an enormously prolonged lottery curve will indeed lie near the axis for an overwhelmingly large fraction of the time, but it will have peaks of all possible heights, the frequency of these peaks falling off extremely sharply as their height increases. All of this was, of course, just in line with Boltzmann's often repeated remarks.

But now Ehrenfest could raise the old paradox in an especially sharp form. "Every time that, in the course of the drawings, 80 more balls are in one urn than in the other, we are naturally ready to bet that the difference will decrease to 78 on the next drawing. For it is nine times more likely that the next ball called will be one of the 90 balls in the fuller pot than one of the 10 balls in the emptier pot. How is this situation expressed in the behavior of the drawing-curve? Does it go down nine times as often as it goes up from those points that lie at a height $y=80$? The following picture forces itself on us: if we draw a horizontal line at the height $y=80$, it will cut our curve in precisely all those points under consideration. If this line hits the curve 10001 times at a point which lies on the right (decreasing) side of a hump, then it must hit the curve at least 10000 times at a point lying on the left (ascending) side of a hump. For in order to leave a peak on the right side, it must always first have been reached from the left. But wouldn't this lead one to say: the curve will go up as often as down from any intersection with the straight line. Mustn't one then be on one's guard against betting that the difference will decrease from 80 to 78? – Of course the bet is completely justified, and the error in the geometrical discussion is very easily found."

This error, which gave rise to the problems that had bothered so many of Boltzmann's critics, lay in neglecting the fact that the great majority of the points at height 80, in the example, are on neither the rising nor the falling side of a peak. They are, as Boltzmann had always asserted, precisely at the maxima themselves. Ehrenfest showed how this could be proved for his model.

For convenience let us put $N=2R$, and let $R+m$ be the number of balls in urn A and $R-m$ be the number in urn B. If $Z(m)$ is the probability that, in the very long run, the balls are divided this way, then clearly $Z(m)$ is given by the equation

$$Z(m) = \frac{1}{2^{2R}} \cdot \frac{(2R)!}{(R+m)!(R-m)!}. \tag{6.23}$$

Again, with this same distribution, the probability that the next lottery ticket takes a ball from A to B is $(R+m)/2R$, and the probability that the ball goes from B to A is $(R-m)/2R$.

Now the points for which there are $(R+m)$ balls in urn A must lie in one of the following sequences:

a) $(m-1) \rightarrow (m) \rightarrow (m-1)$
b) $(m-1) \rightarrow (m) \rightarrow (m+1)$
c) $(m+1) \rightarrow (m) \rightarrow (m-1)$
d) $(m+1) \rightarrow (m) \rightarrow (m+1)$.

In these four situations the point representing the state (m) is: a) at a local maximum, b) on the rising side of a peak, c) on the falling side of a peak, and d) at a local minimum. With the help of the results in the previous paragraph, one can easily show that the relative probabilities of these four situations are

a) $(R+m)/(R-m)$
b) 1
c) 1
d) $(R-m)/(R+m)$.

In other words, if m is near R, that is, if the state (m) is one in which most of the balls are in one urn, then it is by far most likely that this value $2m$ of the difference in populations is a local maximum. The two cases b) and c) are equally likely, as expected, and the last case, of a local minimum, is extremely unlikely. On the other hand, if m is very small compared to R, so that the state (m) is one of roughly equal populations in the two urns, then all four possibilities are about equally likely.

Ehrenfest also pointed out that a step $(m) \rightarrow (m-1)$ was, in the long run, just as likely as a step $(m-1) \rightarrow (m)$, both being much more likely for large m than $(m) \rightarrow (m+1)$. He summarized his discussion in two remarks. "I. The lottery 'curve', as a rule, descends from each of its high points, and only rarely continues to increase. II. This assertion holds – and this sounds especially paradoxical at first – just as well if one traverses the 'curve' from left to right or from right to left."

In the more elaborate of the two papers discussing the model,[35b] the

Ehrenfests emphasized that neither Loschmidt's reversibility objection nor Zermelo's recurrence objection to Boltzmann's *H*-theorem was sufficient to demonstrate an internal inconsistency in Boltzmann's ideas. The model suggested how one could imagine the behavior asserted by Boltzmann and, at the same time, have a system showing both reversibility and recurrence.[36] The Ehrenfests also pointed out, however, that much in Boltzmann's theory still required analysis. "Our discussion, accordingly, does not concern itself with the enormously difficult question: are the purely mechanical assumptions and the various frequency hypotheses (probability assumptions) of Boltzmann's calculations free of contradictions?" And again, "We have here left completely untouched the question of the extent to which the proof of the *H*-theorem can be considered to be free of gaps; in particular, what meaning one should give to the hypothesis of a continuing 'molecular disorder'." These questions were mentioned, but any further study of them was put off to the review article for the *Mathematical Encyclopedia* which the Ehrenfests had undertaken to write. (It was Paul Ehrenfest's lecture on his urn model in Göttingen which had prompted Felix Klein to offer him this assignment, as already described in Chapter Five.)

9. The Ehrenfests were not asked to review the entire field of statistical mechanics in their article for the *Mathematical Encyclopedia*.[37] Boltzmann had already written an article on "The Kinetic Theory of Matter", in collaboration with J. Nabl,[38] which covered much of the ground but emphasized the mathematical development of the theory and, especially, its physical results. Boltzmann did discuss the underlying assumptions and the controversies that had surrounded his work, but not in any great detail. What was wanted now was an analysis

36. The Ehrenfest urn model has been explored in detail. See, for example, E. Schrödinger and K. W. F. Kohlrausch, "Die Ehrenfestsche Modell der H-Kurve", Phys. Z. **27** (1926), p. 306. A discussion of more recent work, including the detailed theory of the evolution of the model in time, can be found in M. Kac, *Probability and Related Topics in Physical Sciences* (New York: Interscience, 1959).

37. P. and T. Ehrenfest, "Begriffliche Grundlagen der statistischen Auffasşung in der Mechanik", *Encyklopädie der mathematischen Wissenschaften* (Leipzig: B. G. Teubner) Vol. IV, Part 32 (1911). Reprinted in *Papers*, p. 213. English translation by M. J. Moravcsik: *The Conceptual Foundations of the Statistical Approach in Mechanics* (Ithaca: Cornell University Press, 1959). This will be referred to as *Transl.*

38. L. Boltzmann and J. Nabl, "Kinetische Theorie der Materie", *Encyklopädie der mathematischen Wissenschaften* (Leipzig: B. G. Teubner) Vol. V Part 8 (1905).

and critique of the foundations of the subject. This was no easy assignment. The controversies were still unresolved. The reviewers did not not have to settle the issues, but they did have to take some definite point of view, if they were to give a coherent picture of the subject. They would also have to deal seriously with a new and major work on their subject, Josiah Willard Gibbs's *Elementary Principles in Statistical Mechanics*.[39] Gibbs had described his book as a "study of the fundamental notions and principles of that [statistical] department of mechanics to which thermodynamics is especially related," so it obviously belonged to the domain which the Ehrenfests were going to survey.

One of the first and most difficult problems was deciding how to organize the article. Paul Ehrenfest decided that a "genetic presentation" would be the best way to bring out the essential features of the subject. As he saw it, the controversies created by Boltzmann's *H*-theorem of 1872 had changed the whole level of discussion. The objections raised first by Loschmidt and then by others had forced Boltzmann not only to restate and clarify his position, but actually to change it in essential ways. Ehrenfest believed that Boltzmann's answers to his critics were basically satisfactory, but that Boltzmann had not always stated these answers clearly enough, and had not stressed the changes in his own point of view. Ehrenfest's "genetic presentation" would not really be a historical account, but it would emphasize the fundamental distinction between the earlier, more naive, works of Clausius, Maxwell and the youthful Boltzmann, and the later, more self-conscious, papers of Boltzmann in which the concept of probability played a large and explicit role. Ehrenfest thought that a careful analysis of the "still sketchy" state of the theory, as Boltzmann had left it, could highlight those spots where Boltzmann's arguments were weak or even entirely missing. Boltzmann had expressed many of his basic ideas in the course of replying to his various critics over the years. Ehrenfest would try to show that these ideas could be built into a logical structure, although many of the key theorems had not been explicitly stated by Boltzmann, and it was by no means evident how – or even if – they could be proved. Gibbs's work was to be considered in the last of the three major sections of the monograph. Gibbs had attempted to create a deductive system of

39. J. W. Gibbs, *Elementary Principles in Statistical Mechanics* (New York: C. Scribner's Sons, 1902). German translation by E. Zermelo: *Elementare Grundlagen der statistische Mechanik* (Leipzig: J. A. Barth, 1905).

statistical mechanics, but the Ehrenfests considered this attempt to be premature and inadequate in many ways.

There were other questions, closely related to the way in which the article was organized, which had to be settled at the outset. At what length, and in what detail, and at what pace should the ideas of statistical mechanics be discussed? Articles in German scientific encyclopedias could be book length, and some of them served as the standard treatises on their subjects. (Recent generations of physicists learned much of their theoretical physics from Wolfgang Pauli's "articles" on relativity and on quantum theory, for example.) But an encyclopedic treatment of the subject, setting it forth in full detail with the proofs of all theorems, and so on, did not correspond to Ehrenfest's way of working. He did not give his courses that way in later years, either. The encyclopedia article is written in his characteristic style, each section or subsection making a point as sharply as possible. There is no padding, no excessive verbiage. There is also a minimum of mathematical machinery: only a few key equations are given, and no calculations or derivations. This style works well for exposing a conceptual structure. It could hardly be used for discussing, say, the kinetic theory of viscosity, or any of a great variety of subjects where calculations are simply inescapable if one is to get at the material at all. But then Ehrenfest almost never dealt with such subjects.

(Perhaps I should already have remarked explicitly that Paul Ehrenfest was primarily responsible for this article which appeared under the names of both Paul and Tatyana. As Tatyana herself wrote almost half a century later: "The great task of collecting the literature and of organizing the *Encyclopedia* article was done by Paul Ehrenfest. My contribution consisted only in discussing with him all the problems involved, and I feel that I succeeded in clarifying some concepts that were often incorrectly used."[40] One should not underestimate the importance of those discussions. Paul Ehrenfest needed them if he was to do physics at all. His wife's ability to pick out weak arguments, internal contradictions and all other offenses against logic proved to be a valuable complement to his own physical insight and enthusiasm. The text itself shows the characteristics of his own literary style.)

The principal line of argument in *The Conceptual Foundations of the Statistical Approach in Mechanics*, to give the monograph its full title,

---

40. See *Transl.* p. ix. Also see the footnote at the beginning of the original German text, *Papers*, p. 213.

concerns the status of Boltzmann's H-theorem. This theorem is at the heart of the subject, since it gives the basis in statistical mechanics for the second law of thermodynamics and the fundamental concepts of equilibrium and irreversibility. How did the Ehrenfests clarify the situation? What did their article add to the hundreds of pages Boltzmann and others had already written on these problems?

First of all, as I have already suggested, they made a fundamental distinction between Boltzmann's original (1872) proof of the H-theorem ("The Older Formulation") and his later investigations and remarks ("The Modern Formulation"). The older work is essentially characterized by its use of the distribution function for a single molecule, the function $f(v, t)$ of (6.11) to (6.14) above. Boltzmann's original proof had as its key assumption the *Stosszahlansatz*, the assumption about the number of collisions per unit time between molecules of specified velocities. Boltzmann had assumed, (just as Clausius and Maxwell had done earlier), that this was equal to the product of the volume swept out by one type of molecule, per unit time, and the number of molecules of the other type per unit volume. This latter number was itself assumed to be the same throughout the gas. Once this assumption is granted, Boltzmann's argument goes through and H always decreases to its minimum, equilibrium, value.

In order to bring out the significance of this assumption, Ehrenfest invented another of his models, "simplified to the very utmost".[41] In this model the molecules are all confined to one plane, and move only in directions parallel or antiparallel to the fixed x and y axes, always with one fixed speed. The molecules do not collide with each other, but do collide with stationary scattering centers. These collisions are elastic and always produce a change in direction of the colliding molecule of plus or minus 90°. The scattering centers are randomly distributed over the plane. (Ehrenfest referred to them as molecules of another type, square in shape, with their diagonals oriented along the x and y axes.) The molecular distribution function, f, for this gas now consists, at any time, of the set of four numbers $f_i (i = 1, 2, 3, 4)$, where f_1 is the number of molecules moving in the $+x$ direction, f_2 the number in the $+y$ direction, and so on. The equilibrium situation corresponds to all f_i being equal, with $\frac{1}{4}$ of the molecules moving in each of the four possible directions. Since all molecules move at the same speed and the

41. In the "Interlude" between Sections 5 and 6. This model is usually referred to as the "wind-tree" model, the fixed scatterers being thought of as the trees that divert the wind of molecules.

scatterers are randomly distributed, the *Stosszahlansatz* says that the number of collisions per second, for each of the geometrically allowed transitions $i \rightarrow j$, is a constant (always the same constant), times the distribution function f_i for the initial state. One can then prove very simply that, no matter what the initial distribution may be, the system will approach the uniform equilibrium state. Furthermore, all initial inequalities in the distribution converge to zero monotonically, and H always decreases.

It is typical of the *Encyclopedia* article that Ehrenfest did not write down even the simplest equations of the actual H-theorem (as they were given earlier in this chapter, for example).[42] Nevertheless, his brief discussion brought into prominence the evident contradiction between the original bald statement of the theorem and both the reversibility of the equations of motion and Poincaré's recurrence theorem. It also made clear the essential role played by the *Stosszahlansatz*. This was particularly easy to demonstrate for his model. Ehrenfest showed that one obtained an absurd result if one assumed that this simple form for the collision rate was valid for both the actual and the time-reversed motions.[43]

The second stage of the discussion, "The Modern Formulation", in which Boltzmann argued that the apparent contradictions of the reversibility and recurrence objections were not real, was characterized by a change in the concepts used. Boltzmann had written over and over again that the H-theorem had to be taken as a theorem about probabilities. But what, precisely, did Boltzmann mean by probability? Sometimes he had used the term probability of a state to mean the fraction of a long time interval during which the system was in that state. At other times, however, Boltzmann used the same term to mean the fraction of the number of identical systems, in one or another ensemble, which were in that state at some particular time. Ehrenfest tried to get rid of the resulting ambiguities and logical gaps by giving up the word probability entirely, referring always to frequencies in the particular ensemble under consideration, and being particularly careful about previously unnoticed shifts from one ensemble to another.[44]

The sub-titles of his chapters on the "older" and the "modern"

---

42. He did not even write down the set of equations that describe the simple "wind-tree" model.

43. This is shown in note 62, *Papers*, p. 233. Footnotes were unfortunately renumbered in the English translation. See *Transl.* note 65, p. 85.

44. *Papers*, pp. 221, 251, 259. *Trans.*, pp. 3, 31, 40.

formulations of the theory show what Ehrenfest considered to be the greatest conceptual change from one to the other. The older theory was a "kinetic-statistical theory of the molecule." Its principal tool was the molecular distribution function, f. The modern theory was, however, a "kinetic-statistical theory of the gas" as a whole. If one takes the gas to consist of N identical molecules, each molecule having r degrees of freedom, (where r depends on the structure of the particular gas molecules being studied), then the mechanical system under consideration is the whole gas, with its Nr degrees of freedom. Its phase, or state of motion, is specified by the generalized coordinates $q_i^{(j)}$ $(i=1,\ldots,r;$ $j=1,\ldots,N)$, and the corresponding canonical momenta $p_i^{(j)}$. One can represent the state of the gas by a point in a space of $2Nr$ dimensions, whose coordinates are the $q_i^{(j)}$ and $p_i^{(j)}$. Ehrenfest called this phase space the Γ space, and stressed the difference between it and the μ space, the molecular phase space, which has only the $2r$ dimensions needed to specify the state of motion of a molecule. The relationships between these two ways of describing the gas – a single point in Γ space, or N labeled points (one for each molecule) in μ space – had to be of some significance. Ehrenfest found these relationships to be crucial to an understanding of Boltzmann's ideas.

Suppose one imagines μ space, the phase space of the molecule, divided up, once and for all, into small but finite cells, each having the same volume ω. At any time t each cell ω_k contains a certain number, a_k, of the N points that represent the phases (coordinates and momenta) of the individual molecules. The set of numbers $\{a_k\}$ determines the distribution Z of the molecules at this time t, and is essentially equivalent to the molecular distribution function f of the earlier discussion. Now it is evident that the distribution Z is uniquely determined when the phase point in Γ space of the whole gas is given. But the converse is not true. A distribution Z corresponds to a domain in Γ space, because the molecular phase points are only placed in finite cells when Z is given, and also because Z tells the number of molecules in each cell, but does not tell which molecules are in which cell. Taking these two facts into account one can obtain the volume $[Z]$ in Γ space corresponding to the distribution Z from the equation

$$[Z]=\frac{N!}{a_1!\,a_2!\ldots a_k!\ldots}\omega^N. \tag{6.24}$$

Boltzmann, in his memoir of 1877, had implicitly taken this quantity $[Z]$ as a measure of the "probability" of the distribution Z. He had shown that the equilibrium distribution (the Maxwell-Boltzmann distribution) was the one that corresponded to a larger Γ volume, [*Z*], than any other distribution for which the energy had the same value. This is the sense in which it was the most probable distribution.[45] Boltzmann had also argued, although, as Ehrenfest pointed out, he had not proved, that the value of [*Z*] corresponding to any distribution appreciably different from the equilibrium one would be enormously less. In this sense the equilibrium distribution was overwhelmingly the most probable. But Boltzmann had also appealed to these results to justify his discussions of the behavior of a system in the course of time, shifting the meaning of probability from the quantity measured by phase volume [*Z*] to probability in the sense of frequency in time.

Ehrenfest called attention to the fundamental gap in the argument at this point. The gap could be filled, if one could prove what he referred to as "the dubious ergodic hypothesis". This is the assumption that, in the course of time, a system will pass through every Γ point compatible with its given energy.[46] What was really needed was a proof that the fraction of time spent by the system in some region of phase space was proportional to the volume of that region of phase space. For then the two senses of probability would be interchangeable, and Boltzmann's combinatorial or "statistical" arguments of 1877 could be

---

45. Ehrenfest emphasized the importance of Boltzmann's 1877 memoir which had received very little attention from most writers on the subject. See, for example, Watson's book (*op. cit.* note 23), which did not mention it at all. It was also ignored by S. H. Burbury in his book, *A Treatise on the Kinetic Theory of Gases* (Cambridge: Cambridge University Press, 1899). It was discussed, however, by Bryan (*op. cit.* note 24b, p. 91). Ehrenfest stressed the difference between Boltzmann's "statistical method" of the 1877 memoir and the "kinetic method" of his 1872 paper. See the thesis of Ehrenfest's student, George Uhlenbeck, *Over statistische methoden in de theorie der quanta* ('s-Gravenhage: Martinus Nijhoff, 1927), pp. 7, 12.

46. In section 10a, which is devoted to "Ergodic Mechanical Systems", Ehrenfest pointed out that *no* system had ever been shown to satisfy the ergodic hypothesis. Not only that, no system had ever been shown to satisfy the weaker hypothesis that its phase point passes arbitrarily near each point of the energy surface in Γ space. He called this the "quasi-ergodic" hypothesis, and pointed out that it could not substitute for the ergodic hypothesis in the development of statistical mechanics. See *Papers*, p. 240, *Transl.*, pp. 21, 89. As a result of Ehrenfest's sharp formulation of the existence problem for ergodic systems, two mathematicians took up the question. The following year both proved that ergodic systems are impossible. M. Plancherel, "Beweis der Unmöglichkeit ergodischer mechanischer Systeme", Ann. Phys. **42** (1913), p. 1061. A. Rosenthal, "Beweis der Unmöglichkeit ergodischer Gassysteme", Ann. Phys. **42** (1913), p. 796. See S. G. Brush (*op. cit.* note 14) for further discussion and references.

reinterpreted in a "kinetic" sense to explain the evolution of a system in time. One would then be justified in saying that "the Maxwell-Boltzmann distribution will predominate overwhelmingly in time over all other appreciably different state distributions."

In order to go any further one simply had to accept this last statement, recognizing that it badly needed a proof. Ehrenfest went on to explain how Boltzmann's reinterpretation of the H-theorem was to be understood. The quantity H itself would now be expressed in terms of the distribution Z, that is, the set of occupation numbers $\{a_k\}$:

$$H(Z) = \sum_k a_k \ln a_k. \tag{6.25}$$

Since H changes with time through the changes in the occupation numbers, its changes must come in discrete steps. A plot of H as a function of time would necessarily be a step function. What Boltzmann discussed as an "H-curve", however, was really a discrete set of points chosen from this step function at regular intervals of time Δt, these intervals chosen as short compared to any experimental time scale but long compared to the mean time between molecular collisions. As soon as one recognized that this was what Boltzmann really meant by an "H-curve", the apparently impossible geometrical properties that he had asserted about the "curve" were seen to be perfectly possible for the discrete set of points. This, of course, is what the Ehrenfests had shown so directly in their papers on the urn model.

If the system is described by some particular distribution Z_A at an initial time t_A, its subsequent behavior is not uniquely determined. Each distribution Z_A corresponds to a whole domain of points in Γ space, and it is the individual Γ point that lies on a unique trajectory in Γ space. (This is only a geometrical way of saying that if one gives the initial coordinates and momenta of all particles, then the motion of the gas is uniquely determined.) This means that there is a whole "bundle of H-curves in the (t, H) plane which radiate out of a point $t = t_A$, $H = H(Z_A)$." Boltzmann's "statistical" version of the H-theorem was really a series of assertions about the properties of this bundle of H-curves corresponding to a given initial distribution.

For any *individual* H-curve, (really a discrete set of points as described above), Ehrenfest put Boltzmann's assertions in this form.

"(a) The individual H-curve *almost always decreases immediately* from each point H_1 which lies above the minimum H_0.

(b) This statement is equally valid regardless of whether one traverses the curve from left to right (positive time sequence) or from right to left (inverted time sequence).

(c) Otherwise the H-curve almost always runs very near the minimum H_0."[47]

But in addition to these statements about the individual H-curves, Boltzmann had really been making a series of assertions about the bundle of curves as a set. Ehrenfest pointed out that Boltzmann had never said these things very explicitly, which was one principal source of the confusion and controversy about his ideas. "This is again a synthesis of a collection of remarks," wrote Ehrenfest, "which are dispersed over all the quoted works of Boltzmann. They show clearly how many purely intuitive statements are hidden behind the usual probabilistic terminology."[48] Ehrenfest put these additional assertions in the following form.

(a) "The totality of H values to which the bundle leads at later times $t_A + n\Delta t$ converges with an *enormously small dispersion around a certain value* $\mathcal{H}_n$. Let us collect these values $\mathcal{H}_1$, $\mathcal{H}_2$,... into a discrete set of points and call this briefly the *concentration curve* of the bundle. Then we assert in addition:

(b) The *concentration curve* of the bundle *monotonically* decreases from the initially high value $H(Z_A)$, converges to the minimum H_0, and *never* again departs from it.

(c) From t_A on, over a *long* time, an overwhelming majority of the curves will run very near to the concentration curve.

(d) Those motions, on the other hand, whose H-curves run *without interruption* very near to the concentration curve even if the motion is followed over an *unlimited* time interval, if they exist at all, form a set with a measure of lower order."[49]

Ehrenfest's concern was to make clear what Boltzmann had meant, and to show that there were no evident contradictions in his assertions, despite what his critics had said. But Ehrenfest was equally concerned to show that these assertions of Boltzmann's were as yet *unproved*, however plausible they might now seem. Still another assertion in this category was that relating the new and the old formulations of the H-theorem.

---

47. *Papers*, p. 253. *Transl.*, p. 33. I have revised the translation and restored Ehrenfest's italics. (He always used italics profusely, and the English translation has dropped almost all his emphases.)

48. *Papers*, p. 254, note 135. *Transl.*, p. 94, note 140.

49. *Papers*, p. 254. *Transl.*, p. 34.

In the old formulation, the *Stosszahlansatz* and the resulting equation for the time evolution of the distribution function, (6.11), determine a definite distribution Z_n coming from the initial distribution Z_A at the time $t_A + n\Delta t$. (This alone shows that the older approach is oversimplified, since the initial distribution does not really determine the later distribution uniquely, as we have seen.) The corresponding values of H, which decrease monotonically according to the old proof of 1872, form another H-curve, "the curve of the H-theorem", as Ehrenfest called it. Boltzmann's old and new formulations are then related by the unproved assertion: "The curve of the H-theorem is identical with the concentration curve of the bundle of H-curves."

To conclude this discussion Ehrenfest pointed out two more unproved assertions that underlie any use of statistical mechanics. One assumes that "the actually observed sequence of states of a gas from time t_A on is identical with that created by the overwhelming majority of the motions discussed above,"[50] i.e. the motions that result from Γ points corresponding to the initial distribution Z_A. But observations do not really determine the initial distribution Z_A at all; they determine only pressures, densities, temperatures, and so on, what Ehrenfest called the "visible state" of the gas. And so, the final assertion: "Of all the various distributions Z that correspond to a certain visible state, there is one distinguished distribution $\mathscr{Z}$, such that an enormously larger domain in Γ space belongs to this $\mathscr{Z}$ and those very close to it, than to all other Z's that correspond to this visible state."[50]

10. The last major section of the Ehrenfests' *Encyclopedia* article dealt with Gibbs's book on statistical mechanics.[39] Gibbs's ideas did not fit neatly into the pattern of their discussion, and one can see why from the only reference to Gibbs in the introduction to their article.[51] Ehrenfest had just pointed out that a new level of understanding of the role of probabilistic concepts in the theory had been reached as a result of the "struggle over the H-theorem". He went on to remark that Gibbs's attempt at a systematic treatment of the subject "covers only a small fraction of the ideas which have been brought to light in the course of this development, and by no means the most important ones". Ehrenfest was clearly rejecting any claim that Gibbs had really

---

50. *Papers*, p. 255. *Transl.*, p. 36.
51. *Papers*, p. 221. *Transl.*, p. 3.

systematized the foundations of the subject when he wrote, in the sentence immediately following: "For this reason the choice of an essentially *genetic* presentation in this report seemed to be obligatory."

Ehrenfest saw the problem of irreversibility as the central conceptual problem of statistical mechanics. His discussion had indeed been built around "the struggle over the H-theorem", and he was quite correct in saying that Gibbs had largely ignored almost all the issues over which this battle had raged. But Gibbs took a very different view of the structure of the subject from Ehrenfest, or Boltzmann, and he was not trying to solve the same problem.

Willard Gibbs wrote his *Elementary Principles in Statistical Mechanics* at the end of his life, and it was his only published work on this subject. His approach to statistical mechanics was that of the grand master of thermodynamics, which he had been ever since 1873 when his first papers appeared. In these early papers, and especially in his great memoir "On the Equilibrium of Heterogeneous Substances", published a few years later,[52] Gibbs developed an essentially new approach to thermodynamics, deepening its foundations and generalizing its range enormously. The point to be stressed for the present discussion is that Gibbs put his emphasis on the nature of equilibrium. A thermodynamic system at equilibrium is characterized for Gibbs by its "fundamental equation," which expresses its internal energy U as a function of its volume V and entropy S, if it consists of a single pure substance, or of volume, entropy, and the quantities of the various chemical substances of which it is composed, in the more general case. This fundamental equation contains all the thermodynamics of the system, since one can deduce from it the coefficients describing what the system will do in all imaginable thermodynamic processes. The laws of thermodynamics insure the existence of this fundamental equation (in the simplest case),

$$U = U(V, S) \tag{6.26}$$

and its differential form

$$\mathrm{d}U = T\mathrm{d}S - P\mathrm{d}V \tag{6.27}$$

---

52. Gibbs's principal papers on thermodynamics appeared from 1873 to 1878 in the Transactions of the Connecticut Academy of Arts and Sciences. They are collected in *The Scientific Papers of J. Willard Gibbs*, ed. H. A. Bumstead and R. G. Van Name (New York: Longmans, Green and Company, 1906) Volume I.

where P is the pressure and T is the absolute temperature. The laws of thermodynamics also serve to characterize the equilibrium state by the condition that U must be a minimum at equilibrium compared to all departures from equilibrium at fixed V and S. The fundamental equation and its minimum property at equilibrium are the basis of Gibbs's great thermodynamic structure.

Gibbs's book on statistical mechanics was written, as its subtitle made clear, "With Especial Reference to the Rational Foundation of Thermodynamics". In view of Gibbs's own approach to thermodynamics, this meant that statistical mechanics had to supply, in particular, an appropriate foundation for understanding the fundamental thermodynamic equation of a system and the way in which it characterizes equilibrium. *This* was the focus of attention in Gibbs's thermodynamics, and it had to occupy the same position in his statistical mechanics. In other words, Chapter XIV of Gibbs's book, "Discussion of Thermodynamic Analogies", marks the climax of his development.

Gibbs had begun his book by explaining how statistical considerations would be introduced into mechanics. One had a physical system, a gas or a solid or some more complicated structure, which could be described by a set of generalized coordinates and momenta satisfying the laws of mechanics. One then defined an ensemble of such systems by imagining a great number of independent systems, all alike in structure, but differing in phase, i.e. in the values of their coordinates and momenta. The state of each system is represented by one point in phase space (Ehrenfest's Γ space), and the ensemble as a whole is specified by its density function ρ. This function depends on coordinates, momenta and time; $\rho \mathrm{d}\Omega$ is the fraction of the total number of systems in the ensemble whose phase points lie within the volume element $\mathrm{d}\Omega$ of Γ space at the specified location at time t. As the system evolves in time, its phase point will move along a trajectory and the phase density function will vary. The equations of motion allow one to prove that, if one moves with the phase point, the value of ρ is unchanged: $\mathrm{d}\rho/\mathrm{d}t = 0$. This is Liouville's theorem, and it is perfectly general. A stationary ensemble, however, is one for which the function ρ does not change with time at a fixed location in Γ space, one for which $\partial\rho/\partial t$ is zero. An ensemble that is to represent a system in thermodynamic equilibrium must be a stationary ensemble in this sense. Practically all of Gibbs's book is devoted to the study of such stationary ensembles.

The two stationary ensembles that Gibbs studied most thoroughly were the ones he called canonical and microcanonical. The canonical ensemble is defined by the phase density function ρ given by the equation,

$$\rho = \exp\{(\psi - E)/\theta\}, \tag{6.28}$$

where E is the energy of the system as a function of its coordinates and momenta, and ψ and θ are constants (i.e. independent of position in phase space). The microcanonical ensemble is one in which all systems have the same energy, considered as the limit of the case in which the density ρ is a constant throughout the region of phase space bounded by the surfaces on which the energy has the constant values E and $E + \Delta E$, respectively.

Gibbs concentrated on the canonical ensemble, showing that its modulus θ had all the essential properties of the temperature, and that this ensemble was the appropriate way of representing a system whose temperature is established by contact with a heat reservoir. He showed that the average energy $\bar{E}$ of a system in the canonical ensemble, and its average pressure $\bar{P}$ could be calculated, as well as the small fluctuations about these averages. The averages satisfied the equation

$$\mathrm{d}\bar{E} = -\theta \mathrm{d}\eta - \bar{P}\mathrm{d}V, \tag{6.29}$$

where the one new quantity η is given by the equation

$$\eta = \int \rho \ln \rho \, \mathrm{d}\Omega, \tag{6.30}$$

the integral being extended over all phase space. Gibbs showed that $(-\eta)$, which can be seen to have the formal properties of the entropy by comparison of (6.29) with (6.27), behaves like the entropy in various kinds of processes.

In his Chapter XIV on thermodynamic analogies, Gibbs tied together the various theorems he had derived earlier and completed his basic line of argument, showing "the mathematical operations by which the fundamental thermodynamic equation, which in general is an equation of few variables, is derived from the fundamental mechanical equation, [for the energy as a function of coordinates and momenta], which in the case of the bodies of nature is one of an enormous number of variables."[53] Gibbs's final result for the relation between mechanical and thermodynamical quantities has the form

$$\exp(-\psi/\theta) = \int \exp(-E/\theta)\mathrm{d}\Omega, \tag{6.31}$$

53. Gibbs, *op. cit.* note 39, p. 165.

where the integral is over all phase space, E being the energy as a function of coordinates and momenta, and ψ being equal to $(\bar{E}+T\eta)$ or $(U-TS)$. This last equation, (6.31), is the basic equation of the statistical mechanics of equilibrium, since it connects the structure of the system, as defined by its mechanical energy, with the thermodynamic properties of the system, which can all be derived from the function ψ.[54] (ψ as a function of T and V is a fundamental thermodynamic equation in Gibbs's sense.) This is really the culmination of Gibbs's development of statistical mechanics.

This, however, is not the way that Ehrenfest looked at Gibbs's work. Since he considered the nature of irreversibility rather than the nature of equilibrium to be the fundamental issue, Ehrenfest took Gibbs's Chapter XII, "On the Motion of Systems and Ensembles of Systems Through Long Periods of Time," to be the crux of his argument. Thus, Ehrenfest described all the preceding material in Gibbs as "preparatory", and referred to the treatment of "nonstationary ensembles" as "the main object of Gibbs's book."[55] In his twelfth chapter, Gibbs discussed the question as to "whether an ensemble of isolated systems has any tendency in the course of time toward a state of statistical equilibrium." His answer to this question was a somewhat guarded "yes," but he did not really claim to have a proper proof of this result. It is significant that in this chapter, and only in this chapter, no equations appear, and also that his later references to this result are always qualified.[56] Gibbs compared the approach of an ensemble to statistical equilibrium with the mixing of ink in water produced by

54. This is closely related to the result obtained by Boltzmann in his 1871 memoir. Gibbs's equation (6.31) is much more convenient to handle than Boltzmann's, (6.10).

55. *Papers*, p. 270. *Transl.*, p. 52.

56. Thus Gibbs says, after discussing the analogy of mixing: "We might perhaps fairly infer from such considerations as have been adduced that an approach to a limiting condition of statistical equilibrium is the general rule . . ." (*op. cit.* p. 148). He adds, "But the subject is of such importance that it seems desirable to give it farther consideration." He does not call what follows a proof, but only a "farther consideration".

In the following chapter (XIII) when he has occasion to call on this material, Gibbs is equally guarded. "Now we have seen, in the last chapter, that when the distribution-in-phase is not one of statistical equilibrium, an ensemble of systems may, and in general will, after a longer or shorter time, come to a state which may be regarded, if very small differences of phase are neglected, as one of statistical equilibrium, and in which consequently the average value of the index (η) is less that at first." (*op. cit.* p. 154). And a little later on, "The considerations adduced in the last chapter show that it is safe to write [an inequality stating that the entropy has increased with time]." (*op. cit.* p. 159).

Gibbs knew when he had proved something rigorously and when he had not.

stirring (in the absence of diffusion.) Each filament of ink retains its color, but these filaments eventually wend their way through all volume elements, producing what appears to be a uniform mixture. In a similar way the true local phase density ρ remains constant (just like the color of the ink itself), but the average of ρ over any small but finite volume element in phase space approaches a limiting value.

Ehrenfest had already discussed this mixing argument from Gibbs's twelfth chapter in a paper written jointly with his wife, early in 1906.[57] He pointed out there that Gibbs had not consistently maintained the distinction between ρ, the "fine grained density", and its average value over a small but finite volume element, the "coarse grained density". (It is the latter which should appear in (6.30), for example, if the entropy is to be a quantity which can vary with time.) The Ehrenfests emphasized that because Gibbs had not kept this distinction in his notation and equations, (even though he had clearly introduced the concepts), his proof that the ensemble approaches equilibrium with increasing entropy was fallacious. And as a consequence, Gibbs's theory of irreversible phenomena lacked a foundation. The fact that the Ehrenfests' first published comments on Gibbs consisted of a clarification of one of his concepts and the demonstration of a gap in his reasoning certainly did not turn them into supporters of Gibbs's ideas and methods.

In the *Encyclopedia* article the Ehrenfests discussed the principal developments of Gibbs's book, making it clear, however, that they found Boltzmann's methods superior, even in discussing thermodynamic equilibrium, because they gave more physical insight into what was happening within the system.[58] The Ehrenfests did not put any weight on the great generality of Gibbs's approach, (which he himself considered, properly, to be foreshadowed in Boltzmann's own work,

57. P. and T. Ehrenfest, "Bemerkung zur Theorie der Entropiezunahme in der 'Statistischen Mechanik' von W. Gibbs", Wien Ber. **115** (1906), p. 89 *Papers*, p. 107.

58. Ehrenfest's attitude toward Gibbs's book may not be totally unexpected in a student of Boltzmann's. Near the end of a lecture on statistical mechanics in 1904, Boltzmann gave credit to Gibbs for having "brought this science into systematic form, set it forth in a major book, and given it a characteristic name,[statistical mechanics]". (See L. Boltzmann. *Populäre Schriften*, (Leipzig: J. A. Barth, 1905). p. 360.) But that is all Boltzmann ever wrote about Gibbs's contributions to the subject. It must however, be said that Boltzmann knew and admired Gibbs's work in thermodynamics. On one occasion he even prided himself on having cited it as early as 1883, "therefore long before the appearance of the meritorious German translation." (*Wiss. Abh.* III, p. 638. The earlier reference was to Wiss. Abh. III, p. 93, and the remark about the German transla-

particularly his 1871 memoirs.[59]) The great power of the canonical ensemble as a method of calculation had no special appeal for Ehrenfest, since he rarely carried out such calculations. He did mention that the use of the canonical ensemble, in place of the microcanonical ensemble which directly describes a system of fixed energy, "seems to be an analytical trick reminiscent of Dirichlet's 'discontinuity factor'."[60] He also added that L. S. Ornstein's work showed "that operations with canonical ensembles occasionally furnish a more convenient computational scheme for the treatment of complicated problems of equilibrium . . . than does Boltzmann's procedure."[61] But he did not emphasize that the canonical ensemble is precisely the appropriate one for describing a system whose temperature is fixed, but whose energy can vary, i.e. a system in contact with a heat reservoir. Ehrenfest did at least mention this aspect of the canonical ensemble, however, and also the closely related point (shown by Boltzmann as well as by Gibbs and also Albert Einstein) that a small part of a system of fixed energy is properly described by a canonical ensemble.

As their 1906 paper indicated, the Ehrenfests were not happy with Gibbs's treatment of nonstationary ensembles. They considered the methods of the H-theorem more direct, and thought that Gibbs's work gave no assurance that equilibrium would be reached and no insight into the time that might be required to attain it. At one point they remarked that this time would be of the order of the enormous Poincaré

---

tion by Wilhelm Ostwald refers both to Ostwald's satisfaction over having introduced Gibbs to the German speaking world and to the Boltzmann-Ostwald controversy over energetics.) Boltzmann also used Gibbs's now famous remark, "The impossibility of an uncompensated decrease of entropy seems to be reduced to improbability", as the motto to the second volume of his *Lectures on Gas Theory*. This put Gibbs in a class with Goethe and Schiller, from whom Boltzmann had earlier drawn mottos. Nevertheless, Boltzmann must have felt some parental possessiveness towards statistical mechanics.

59. Gibbs, *op. cit.*, p. viii.

60. *Papers*, p. 269. *Transl.*, p. 50.

61. *Papers*, p. 281. *Transl.*, p. 64. It must be pointed out that Ehrenfest's highly critical attitude towards Gibbs's work is not representative of the views of his contemporaries. See, for example, J. Hadamard, "La mécanique statistique," Bull. Am. Math. Soc., **12** (1906), p. 194. Most interesting are Einstein's comments: see A. Einstein, "Bemerkungen zu den P. Hertzschen Arbeiten: 'Über die mechanischen Grundlagen der Thermodynamik'", Ann. d. Phys. **34** (1911), p. 175. Also see H. A. Lorentz's discussion, taken from his lectures, "Über den zweiten Hauptsatz der Thermodynamik und dessen Beziehung zu den Molekulartheorien", in H. A. Lorentz, *Abhandlungen über theoretische Physik* (Leipzig: B. G. Teubner, 1907), p. 277.

recurrence periods, a remark which would be hard to justify.[62]

One feature of Gibbs's book which has often been commented on is the cautious way in which Gibbs limited his aims. "We avoid the gravest difficulties," he wrote, "when, giving up the attempt to frame hypotheses concerning the constitution of material bodies, we pursue statistical inquiries as a branch of rational mechanics." Gibbs was concerned about the apparent impossibility of extending the theory to include radiation and "the electrical manifestations which accompany the union of atoms." Even apart from such matters, there were the problems generated by the failure of the equipartition theorem to account properly for the specific heats of diatomic gases.

> "Difficulties of this kind have deterred the author from attempting to explain the mysteries of nature, and have forced him to be contented with the more modest aim of deducing some of the more obvious propositions relating to the statistical branch of mechanics. Here there can be no mistake in regard to the agreement of the hypotheses with the facts of nature, for nothing is assumed in that respect. The only error into which one can fall, is the want of agreement between the premises and the conclusions, and this, with care, one may hope, in the main, to avoid."[63]

This passage is often referred to, sometimes as an example of Gibbs's wisdom and foresight in developing his statistical mechanics in such a way that it would survive the quantum revolution with only minor changes,[64] and occasionally as an example of Gibbs's "timid and ambiguous attitude."[65] Ehrenfest cited it too, but immediately commented in a footnote: "Gibbs's discussion shows that later on he hardly keeps to this program of axiomatization any more than, say, Maxwell and Boltzmann."[66] Ehrenfest knew very well that Boltzmann had also made such disclaimers on various occasions, refusing to assert the reality of atoms and insisting only on their indispensability as theoretical tools. He too had written of scientific theories as metaphors

---

62. *Papers*, p. 271. *Transl.*, p. 53. G. E. Uhlenbeck has already expressed his doubts on this point in the Ehrenfests' article. See his Higgins Lectures at Princeton (1954), p.21.

63. Gibbs, *op. cit.*, pp. ix–x.

64. Tolman, *op. cit.*, (note 18), p. 14.

65. L. Rosenfeld, "On the Foundations of Statistical Thermodynamics", Acta Physica Polonica **14** (1955), p. 4.

66. *Papers*, p. 264 note 172. *Transl.*, p. 97 note 175.

and would hardly have been bothered by Gibbs's discussion of "thermodynamic analogies". Despite their cautionary statements, both men believed that statistical mechanics was, in some sense, a "real" theory and no mere analogy.

Gibbs did write quite explicitly about "Systems Composed of Molecules", but only in the last chapter of his book, since "it seemed desirable to separate sharply the purely thermodynamic laws from those special modifications which belong rather to the theory of the properties of matter."[67] (This kind of separation had precedents going back to Clausius' writings almost half a century earlier.) In this last chapter of his book Gibbs discussed the statistical equilibrium of ensembles of systems composed of many particles. The critical questions were the role of the indistinguishability of particles of a given kind, and the procedure for treating a system when the number of its particles could vary. Gibbs showed how to handle these with the help of a new ensemble, the grand canonical ensemble, appropriate to the description of a system which can exchange particles with its environment. He defined this ensemble in such a way as to allow for the identity of the particles, and showed how the use of "generic phases", ("*not* altered by the exchange of places between similar particles"), took care of a paradox he had pointed out in his thermodynamic studies many years earlier.[68]

This aspect of Gibbs's work was not appreciated by Ehrenfest, who, in this respect, was typical of Gibbs's contemporaries. The only reference to Gibbs's last chapter comes in a footnote in which Ehrenfest commented on the difference in the ways in which Boltzmann and Gibbs handled the problem of dissociation equilibrium. Boltzmann had had to take explicit account of the constraints fixing the total numbers of atoms of the various kinds involved, but "Gibbs with his formal trick manages to evade these auxiliary conditions also." He had introduced "a peculiar ensemble of gas models in which the numbers of atoms vary in the different members of the ensemble, running through all the integers from 0 to ∞."[69] Ehrenfest did not remark at all on Gibbs's discussion of the indistinguishability of identical particles, although he was well aware of the importance of the question. He even remarked that "the superiority of Boltzmann's theory regarding the increase of

67. Gibbs., *op. cit.*, p. xii and Chapter XV.
68. Gibbs, *op. cit.*, p. 206. Also Gibbs, *Scientific Papers*, p. 166.
69. *Papers*, p. 276 note 189. *Transl.*, p. 99 note 194.

entropy is based . . . on the proper consideration of the permutability of the molecules."[70] The question of particle indistinguishability was one to which Ehrenfest would return some years later.

11. It would be misleading to limit the description of the Ehrenfests' *Encyclopedia* article to its principal sections on the work of Boltzmann and Gibbs. Statistical mechanics had not come to a halt with the publication of Gibbs's *Elementary Principles* in 1902. Quite the contrary: the first decade of the twentieth century saw statistical mechanics acquire an unprecedented importance. In the concluding section of the main body of the article, written in December 1909, Ehrenfest commented on this situation. The article itself had drawn attention to the "large number of loosely formulated and perhaps even inconsistent statements" at the core of the subject, giving it a "serious logical incompleteness" without parallel in the other branches of mechanics. Despite this, Ehrenfest wrote, "the last few years have seen a sudden and wide dissemination of Boltzmann's ideas . . . . One cannot, however, point to a corresponding progress in the conceptual clarification of Boltzmann's system to which one could ascribe this sudden change."[71] The new interest in statistical mechanics, and atomism in general, came, not from a deepening of the theoretical foundations, but from successful new applications and dramatic new experiments. Ehrenfest could refer to a whole series of these: the electron theory of metals, the theory of Brownian motion developed by both Einstein and Smoluchowski, and its experimental confirmation by Jean Perrin. Perhaps the most remarkable confirmation of the old "atomic hypothesis" had been the counting of individual charged particles by E. Regener and by Ernest Rutherford and Hans Geiger. These were the successes that "have had the effect of reviving and deepening the concept that all bodies can be pictured as aggregates of a finite number of very small and identical elementary components, and that correspondingly every process in a physical or chemical problem . . . is a complex of an enormously large number of individual processes."

Another striking, but much more controversial, recent application of statistical mechanics had been its use in the theory of black body radiation. Ehrenfest had followed this early work on the quantum theory and devoted much effort to trying to understand its significance

70. *Papers*, p. 280. *Transl.*, p. 63.
71. *Papers*, p. 284. *Transl.*, p. 67.

and its implications. In the supplement that he added to the *Encyclopedia* article in September 1911 just before its publication, Ehrenfest made a number of comments on the very different contributions that Planck and Einstein had made in this field. Chapter 10, on Ehrenfest's work on quanta, will deal with these matters in detail.

Ehrenfest's ability to formulate challenging questions is clearly evident in the pages of the *Encyclopedia* article. A particularly interesting question appears in one of the many long footnotes; (these footnotes actually contain some of the most important new ideas in the article.) "What are the statistical, and also the corresponding thermodynamic features," Ehrenfest asked, "that distinguish the stationary irreversible processes from the nonstationary ones?"[72] This question applied to heat conduction, thermoelectricity, or any of the other irreversible flow processes. If the two ends of a copper rod, for example, are suddenly placed in contact with a steam bath and an ice bath respectively, heat will flow through the rod and the temperature of the rod will vary with position and time. Eventually a steady state of heat flow will be established in which the temperature at each point of the rod no longer changes with time. The rod is not in thermal equilibrium, but the irreversible process of heat conduction has reached a steady state. Ehrenfest's question applies to such states. It was an old question to him. His first research notebook contains an entry, going back to the beginning of 1903, asking if there is some sort of principle of minimum production of heat. Closely related questions appear in several later notebooks. Then in May 1910 Ehrenfest suggested that one could attack the problem with the help of an urn model, a generalization of the one he had used to clarify the reversibility and recurrence objections to the H-theorem. This approach apparently led nowhere.[73] In the *Encyclopedia* footnote he formulated the problem this way. "Consider an irreversible process with fixed external constraints which goes over spontaneously from a nonstationary to a stationary state. Can the distribution of states which is finally attained be described as in some sense the 'relatively most probable distribution', and can it be characterized by the minimum of some function which can be regarded as a generalization of the H-function? (Just as in the case of equilibrium $\mathrm{d}H/\mathrm{d}t$ is equal to zero here, too. But it is not immediately possible to characterize the stationary state

72. *Papers*, p. 292 note 231. *Transl.*, p. 103 note 239.
73. Ehrenfest research notebooks I (100) and XI (2604).

by a relative minimum of H.)" Ehrenfest's problem still awaits a general solution.[74]

The final section of the *Encyclopedia* article raises a very general and also very difficult question. "How can a 'probability hypothesis' be verified experimentally?" One can compare the outcome of one experiment with the theoretical prediction of the "most probable" occurrence. One can also compare the outcome of n_1 repetitions of the experiment with the theoretical prediction for the "most probable" ensemble of n_1 occurrences. This may be considered a second order test. But one can also consider n_2 repetitions of the second order event and compare this third order event with its expected outcome, and so on. Now the difficulty is this. "Let us assume that we observe the outcome of the experiment E_k [the kth order experiment], which from the viewpoint of the kth order hypothesis H_k is particularly unsatisfactory. This outcome, from the point of view of hypothesis H_{k+1} is not only admissible, but hypothesis H_{k+1} actually requires that if we repeat the experiment E_k n_k times, i.e. perform the experiment E_{k+1}, this particular outcome should indeed occur with a very definite frequency."[75] In other words, how does one decide when the probability hypothesis has been verified, or falsified, by experiment? This point was raised by Tatyana Ehrenfest, and she wrote a paper in 1911 explaining the difficulty and illustrating it with a simple example involving the drawing of a ball from an urn containing balls of two colors.[76] The problem comes up in many fields. When, for example, the original form of the H-theorem had to be given up, the *Stosszahlansatz*, originally viewed as giving *the* number of collisions per unit time, had to be reinterpreted. One really needed an assumption of higher

74. Some work along this line was done by P. S. Epstein. See his article, "On the Resistance Experienced by Spheres in their Motion through Gases", Phys. Rev. **23** (1924), p. 724, and references given there. The question reappeared in 1930 in the thesis of Ehrenfest's student, A. J. Rutgers, *Bijdrage tot de Theorie der Thermo-electriciteit in Kristallen*, (Amsterdam: H. J. Paris, 1930), p. 87. The most important contribution to the problem, and perhaps the closest one can come to an answer to Ehrenfest's question, is to be found in the work of Onsager and Machlup. See L. Onsager and S. Machlup, Phys. Rev. **91** (1953), pp. 1505, 1512. Other work on the subject can be found in *Transport Processes in Statistical Mechanics*, ed. I. Prigogine (New York: Interscience Publishers, 1958), pp. 311, 324, 327.

75. *Papers*, p. 298. *Transl.*, p. 78.

76. (a) T. Ehrenfest, "Die Anwendung der Wahrscheinlichkeitsrechnung auf gesetzmässige Erscheinungen", Jour. d. russ. phys. Ges. **43** (1911), p. 256. (b) T. Ehrenfest, "On the Use of the Notion 'Probability' in Physics", Am. J. Phys. **26** (1958), p. 388.

order, giving the frequency with which other values of the collision number would arise, making the value originally assumed in the *Stosszahlansatz* only the most probable value.

A closely related point is the status to be assigned to events predicted by the theory as extremely improbable.[77] There is, for example, a ridiculously small, but nonzero and calculable probability that all the molecules of a gas would, at a certain moment, be found in the left half of the container. Max Planck argued at one time that such improbable events must be explicitly excluded by a special hypothesis, "for a universe in which things can happen such as the return of heat into a hot body, or the separation of two interdiffused gases, would not be the same as our universe."[78] The exclusion is not so simply accomplished, however, since the same theory that predicts these enormously improbable fluctuations also predicts the smaller observed fluctuations that constitute the Brownian motion, for example. Ehrenfest did not have an answer for the question he had posed.[79] But he concluded the article with these rather prophetic words: "Every investigation into the structure of a physical theory now leads inevitably to the question of the nature of 'probability hypotheses'."[80]

---

77. *Papers*, p. 298. *Transl.*, p. 79.
78. M. Planck, 'Die Einheit des physikalischen Weltbildes", Phys. Z. **10** (1909), p. 62. Engl. transl. by R. Jones and D. H. Williams in M. Planck, *A Survey of Physical Theory*, Reprinted (New York: Dover, 1960), p. 16.
79. For a recent discussion that bears on these problems see I. Levi, *Gambling with Truth. An Essay on Induction and the Aims of Science* (New York: Alfred A. Knopf, 1967).
80. *Papers*, p. 300. *Transl.*, p. 79.

CHAPTER 7

A Theoretical Miscellany

1. Theoretical physicists vary widely in the range of subject matter they study and in the sheer quantity of the work they accomplish. In this respect they differ in no essential way from those who devote themselves to any of the other vocations that demand creative powers. Lord Rayleigh's four hundred fifty scientific papers fill six large thick volumes and deal with every aspect of mathematical physics, as well as a good many experimental problems. The thirty or so papers by Josiah Willard Gibbs treat only three or four principal subjects. H. A. Lorentz wrote at length and in depth on almost every problem that could concern a theorist during the half century of his activity. Niels Bohr, whose career was just as long, pursued only a few themes, but he was prepared to follow them into "the abyss where truth resides."

Paul Ehrenfest often lamented the absence of a central guiding principle in his work. He would have liked to be able to follow one main line, as his teacher, Boltzmann, had done in developing statistical mechanics, or as his contemporary, Einstein, was doing in his attempt to create a new foundation for physics. But Ehrenfest knew only too well that he lacked the extraordinary combination of insight, intellectual power and unrelenting drive needed to do physics at that level. His work does show a single minded concern with questions of fundamental principle. Ehrenfest might not be able to solve the deepest problems of his time, but he usually knew what they were, and he did what he could to sharpen the issues and ask the right questions.

When Ehrenfest complained that he had no central principle guiding his work, he meant he was not building a structure that he, and others, could enlarge and extend. With the exception of his work in the early quantum theory leading up to the adiabatic principle, this was generally true. Nevertheless, his principal interests were quite evidently in statistical mechanics and the quantum theory, a theory which seemed at first to be only a branch of statistical mechanics. In addition to his work on these subjects, however, Ehrenfest thought and wrote about a variety of other problems. Almost all of these papers are marked by the same concern with questions of principle. There is no busy work there, no papers written just to keep the theoretical tools sharp. Ehrenfest was *not* one of those for whom "science is their own special sport to which they look for vivid experience and the satisfaction of ambition."[1]

Some of these other, more miscellaneous, works of Ehrenfest's, written in Göttingen and St. Petersburg, are discussed in this chapter. These articles deal with a variety of themes, from optics to thermodynamics, and include his minor but always interesting papers related to the special theory of relativity. Not every paper is discussed, and, in the nature of the situation, the background against which these papers were written is sketched much more briefly than elsewhere in this book.

2. In 1908 Ehrenfest raised an interesting question about the velocity of light. The question was not particularly current at the time, and it very likely occurred to him in the course of his omnivorous reading. Ehrenfest's analysis led him to propose that the master of wave theory, Lord Rayleigh, had made an error in his own proper domain.

The velocity of a harmonic wave is usually described as its wave velocity, or phase velocity, v. This quantity is related to the wave length, λ, and frequency, ν, of the wave by the equation

$$v = \lambda\nu. \tag{7.1}$$

When a wave, let us say a light wave, passes through a medium, its velocity v may well depend on the wave length; in other words, the medium may be a dispersive one, in the same sense that a prism can disperse the light of different colors passing through it.

1. A. Einstein, *The World as I See It*, (New York: Covici Friede, 1934), p. 19.

If one wants to measure the velocity of a harmonic wave, however, it becomes necessary, as Rayleigh described it, "to impress some peculiarity, it may be of intensity, or of wave length, or of polarisation, upon a part of an otherwise continuous train of waves, and determine the velocity at which this *peculiarity* travels."[2] This velocity is called the group velocity, u, and it will be different from v if the medium is dispersive. Rayleigh had already shown that u and v are related by the equation,

$$u = v - \lambda \frac{dv}{d\lambda}. \tag{7.2}$$

The question, which velocity does an experiment actually determine, was discussed by Rayleigh in 1881.[2] New measurements of the velocity of light had seemed to indicate that this velocity depended on the wave length even when the light propagated in vacuum. This startling claim had to be analyzed very closely. Rayleigh went down the list of ways in which the velocity of light could be measured, classifying the methods into those that determine group velocity and those that determine wave velocity. The usual terrestrial methods, which employ Fizeau's toothed wheels or Foucault's rotating mirror, determine the group velocity, as does the astronomical method based on the eclipses of Jupiter's moons. There was another astronomical method, however, based on the observation of the angle of aberration. This angle, α, the apparent angular displacement of a star due to the earth's motion, is proportional to the ratio of the earth's velocity to the velocity of light, but which velocity of light? Rayleigh argued that the method of aberration "does not depend upon observing the propagation of a peculiarity impressed upon a train of waves, and therefore has no relation to u." Since the method of aberration determines the wave velocity, v, and any of the other methods determines the group velocity, u, a comparison of the results obtained by these different methods could decide the presence or absence of dispersion in vacuum, or rather, in the luminiferous aether itself.

Rayleigh's statement that the aberration method and only the aberration method determined the wave velocity rather than the group velocity, was repeated in many of the standard texts on optics early in this century. It was this statement that attracted Ehrenfest's notice.[3]

2. J. W. Strutt, Baron Rayleigh, "On the Velocity of Light," Nature **24** (1881), p. 382; **25** (1881), p. 52. *Scientific Papers* Vol. I (Cambridge: University Press, 1899), p. 537.

3. P. Ehrenfest, "Misst der Aberrationswinkel im Fall einer Dispersion des Äthers die Wellengeschwindigkeit?", Ann. Phys. **33** (1910), p. 1571. *Papers*, p. 161.

He constructed a very simple argument showing that Rayleigh was wrong, and that the aberration method, like all others, could not determine the wave velocity.

Ehrenfest replaced the telescope on the earth by a more schematic arrangement consisting of two parallel plates, each having a circular hole cut in it, both plates moving at the same constant velocity in some direction lying in their planes. Light is normally incident on the upper plane. One now asks for the angle through which the aperture in the lower plate must be displaced relative to that in the upper plate so that the rays of light that pass through one will also pass through the other. It is clear immediately that what counts is the velocity of propagation of those sections of the light rays cut off by the upper aperture. In other words, there is an "impressed peculiarity" here too. The argument could also be made without using the "dubious concept of light rays," as Ehrenfest pointed out. The upper aperture punches out a "circular disk" of waves from the incident plane wave fronts, and it is the velocity of this "circular disk" that determines how much offset is needed for the lower aperture to be in the right place at the right time to allow it to pass through. In the presence of dispersion this propagation velocity would certainly not be the wave velocity. The aberration method could be considered as identical in principle with Fizeau's toothed wheel, as Ehrenfest's argument indicated.

Ehrenfest wrote to Lord Rayleigh, pointing out the error in his old paper, and the answer[4] came back from Terling Place in Essex to Petersburg: "I think you are right. I think I must withdraw the opinion you quote." As Rayleigh mentioned, he "ought to have found the contradiction" himself since he had used an argument very similar to Ehrenfest's in a paper on aberration written in 1892.[5] "I think that when I wrote the erroneous passage," he went on, "I had not given much attention to aberration. It is terribly easy to make mistakes!"

In the fall of 1908 Ehrenfest gave a lecture on this work to the Physics Colloquium at Petersburg, but he did not publish it until two years later. The reason for his delay is not clear; perhaps he was waiting to see if Rayleigh would correct his work himself. When he did write it up, Ehrenfest used a question for his title, as he did every so

4. Lord Rayleigh to P. Ehrenfest, 20 October 1908.

5. J. W. Strutt, Baron Rayleigh, "Aberration," Nature **45** (1892), p. 499. *Scientific Papers* Vol. III (1902), p. 542.

often throughout his life: "Does the aberration angle measure the wave velocity when there is dispersion in the aether?"

The paper actually goes on to raise another and more difficult question: is it really the *group* velocity which is always measured when the velocity of light is determined? The question was not whether it might be the wave velocity, but rather whether it might be something different from both wave and group velocities. Ehrenfest thought there was no satisfactory derivation of the group velocity for the general case of a wave packet constructed as a Fourier integral. Michelson's measurements of the velocity of light in carbon disulfide did agree with the group velocity to be expected from data on the index of refraction, but this did not settle the question for all cases. What bothered Ehrenfest was the existence of situations in which the group velocity exceeded the velocity of the wave front. This could occur for certain more or less artificial systems, but it would also occur in any medium for wave lengths in the immediate neighbourhood of an absorption line. The usefulness of the group velocity had its limits, and further analysis of the velocity concept for wave motion in dispersive media was definitely called for. It would be supplied some years later in classic studies by Sommerfeld[6] and Léon Brillouin[7] which answered the question in great detail.

In 1911 Rayleigh wrote a short paper[8] acknowledging the error he had made thirty years earlier and giving Ehrenfest full credit for pointing it out. He also extended Ehrenfest's discussion of aberration in a quantitative way, demonstrating that the aberration angle actually did determine the group velocity, but he did not say anything about those situations where the group velocity does not adequately describe the propagation of light.

3. Ehrenfest's dissertation on Hertzian mechanics, (see Chapter 4), came at the very end of the long tradition that made the mechanical interpretation of all phenomena the goal of physics. By the time he wrote that dissertation in 1904, the mechanical ideal was no longer undisputed. It had been attacked by some who would replace it by a

6. A. Sommerfeld, "Über die Fortpflanzung des Lichtes in dispergierenden Medien," Ann. Phys. **44** (1914), p. 179. Sommerfeld refers briefly to Ehrenfest's paper, but his work was actually done earlier, and for other reasons.

7. L. Brillouin, "Über die Fortpflanzung des Lichtes in dispergierenden Medien," Ann. Phys. **44** (1914), p. 203.

8. J. W. Strutt, Baron Rayleigh, "Aberration in a Dispersive Medium," Phil. Mag. **22** (1911), p. 130. *Scientific Papers* Vol. VI (1920), p. 41.

more modest goal of phenomenological description and by others who would make mechanics merely a branch of a generalized thermodynamics. Probably the most significant attack on the mechanical world view, at least among physicists of the younger generation at the turn of the century, was one that had its origins within physics itself, in the successful theory of electromagnetism, and seemed to offer a serious and up to the minute alternative to mechanism.

Maxwell had shown that the equations of the electromagnetic field could be looked on as the equations of a mechanical system: "What I propose now to do is to examine the consequences of the assumption that the phenomena of the electric current are those of a moving system, the motion being communicated from one part of the system to another by forces, the nature and laws of which we do not yet even attempt to define, because we can eliminate these forces from the equations of motion by the method given by Lagrange for any connected system."[9] Maxwell's success still left open the question of the detailed structure of the mechanical system which constituted the electromagnetic aether. It was a question that fascinated Boltzmann and that had much to do with Hertz's attempt to create a mechanics which would reduce to the kinematics of masses, both visible and concealed, in order to grasp the structure of the aether.

By 1900, however, others were ready to interpret Maxwell's result in the reverse sense. "The notion of a mechanism in space empty of ponderable matter seems very artificial. But even those who do not accept it must recognize the importance of the discovery that the motions of ponderable bodies and the phenomena of electrodynamics are subject to the same laws. One need not, indeed, interpret this close relationship between mechanics and electrodynamics as being unconditionally in favor of the mechanical view of nature. One can, with equal justification, see it as an indication that the motions of bodies obey the laws of electrodynamics. Modern physics is, as a matter of fact, inclining more and more to this interpretation; we would like to derive mechanics from the equations of the electromagnetic field, and, in general, to give all of physics an electromagnetic foundation."[10] These words by Max Abraham were written in 1904 and express the

9. J. C. Maxwell, *A Treatise on Electricity and Magnetism* 3rd ed. 1891. Reprinted (New York: Dover Publications, 1954), Vol. II, p. 198.

10. M. Abraham, *Theorie der Elektrizität*, I *Einführung in die Maxwellsche Theorie der Elektrizität* (by A. Föppl, revised by M. Abraham), (Leipzig: B. G. Teubner, 1904), p. 273.

program of those who wanted to replace the mechanical world view with an electromagnetic world view.

Abraham stated this position in more detail the following year, in the context of his book dealing with the latest developments in the electromagnetic theory of radiation and the theory of electrons. All mass was to be interpreted as electromagnetic mass, the inertial effect of the electron's own magnetic field. All forces would be electromagnetic in origin. "The world would then consist exclusively of positive and negative electrons, and of the electromagnetic fields generated in space by them, and all natural phenomena would be considered as convection radiation of the electrons or as wave radiation sent out by them." Abraham admitted freely that "this electromagnetic world view is only a program up to now," but he hoped that efforts along this line "would be crowned with further successes."[11]

The theory of electrons, which was the heart of this program, held the central position in theoretical physics in those years. J. J. Thomson's experimental proof that the cathode rays consisted of "corpuscles," particles whose ratio of electric charge to mass was of the order of a thousand times greater than the corresponding ratio for hydrogen ions, had provided new and weighty evidence for the theory of electrons. Thomson's corpuscles were identified with the electrically charged particles of the theory already under development by many physicists and particularly by H. A. Lorentz. Lorentz showed how a great variety of electrical and optical phenomena could be understood in terms of this theory, from the newly discovered Zeeman effect to the Fresnel drag coefficient, which determined the velocity of light in a moving transparent medium.[12] It is not possible even to sketch the theory here, much less to show how it achieved its successes. But in the early 1900's it was the problems of the theory rather than its successes that concerned working physicists, as always.

One of these problems was the structure to be attributed to the electron. A variety of assumptions was possible here. Abraham had made the one that seemed simplest: the electron was to be treated as a rigid sphere.[13] He did not investigate the nature of the other forces needed

11. M. Abraham, *Theorie der Elektrizität*, II *Elektromagnetische Theorie der Strahlung*, (Leipzig: B. G. Teubner, 1905), p. 147.

12. For a unified presentation see H. A. Lorentz, *The Theory of Electrons* (Leipzig: B. G. Teubner, 1909; 2nd ed. 1916).

13. M. Abraham, *op. cit.* note 11, pp. 136–201. H. A. Lorentz, *op. cit.* note 12, pp. 35–43, 217–218.

to keep the electron in this configuration, but simply treated it as a kinematic constraint, in the sense of Hertz's mechanics. With this assumed form Abraham was able to calculate the effective mass exhibited by the electron as a function of its velocity. (Since the electromagnetic field created by the electron changes with its velocity, so does the inertial effect of this field. The effective mass of the electron, i.e. the ratio of an external force on it to the acceleration it produces, also depends on the direction of the force relative to the electron's motion, so that one had to deal with both longitudinal and transverse masses.)

Although Abraham's assumption seemed the simplest one could make, Lorentz chose another, on rather different grounds. Lorentz's electron theory had postulated an aether always and everywhere at rest, an aether that constituted an absolute frame of reference. On the basis of this assumption he had been able to explain a wide range of phenomena, but not Michelson's negative result of 1881, confirmed by Michelson and Morley in 1887. The absence of any observable effects of the earth's motion through the aether was a great puzzle, and it was to solve this puzzle that Lorentz introduced his contraction hypothesis: when a body moves through the aether with velocity v its dimension in the direction of motion is reduced by the factor $(1-\beta^2)^{1/2}$, where β is the ratio of v to c, the velocity of light; the dimensions perpendicular to the direction of motion are unchanged. Lorentz took this hypothesis as applying to the electron, so that its spherical shape when at rest was changed into an ellipsoidal shape when it moved. He, too, could then calculate the effective masses (both transverse and longitudinal) of the electron, as functions of its velocity.[14]

One further assumption was tried by both A. H. Bucherer and Paul Langevin. They took the electron to be a sphere at rest and an ellipsoid when in motion, with the ratio of longitudinal to transverse semi-axes of the ellipsoid equal to $(1-\beta^2)^{1/2}$, as in Lorentz's theory. The Bucherer-Langevin electron, however, was assumed to have the same volume when in motion as it had at rest. Consequently, the longitudinal and transverse contraction factors were not the same as those required by Lorentz to account for the Michelson-Morley result. The Bucherer-Langevin electron had a velocity-dependent effective mass of still another form.[15]

---

14. H. A. Lorentz, pp. 168–217. M. Abraham, pp. 201–208, 373–391.
15. H. A. Lorentz, pp. 219–220.

There was one more theory that had a prediction to make about the way in which the electron's mass should vary with its velocity. The prediction was not new – it coincided precisely with Lorentz's – but the theory was: it was Einstein's special theory of relativity of 1905. Einstein's theory was not a theory of electrons as such, but Einstein did deduce the effective mass values from it.[16] What is most relevant here is that one of the first discussions of Einstein's theory to be found in the literature, if not the very first, came in 1906, in a paper describing a careful experimental test of the variation of mass with velocity.[17]

W. Kaufmann had been studying the apparent mass of fast electrons, emitted as beta rays by radioactive nuclei, since 1901. His experimental results showed that the mass did increase with velocity and gave strong support to those who wanted to explain the mass as entirely electromagnetic. In 1906 he published the results of a new series of measurements, carried out with the express purpose of distinguishing between the various models of the electron and, especially, of testing Lorentz's hypothesis. Kaufmann stated his conclusion in a very definite form: "*The results of the measurements are not compatible with the fundamental assumption of Lorentz and Einstein.*" (The italics are Kaufmann's.) The observed variation of electronic mass with velocity was consistent with either Abraham's rigid electron or the Bucherer-Langevin electron. Kaufmann's results were not precise enough to distinguish between these two; they could only serve to reject the result obtained in very different ways by Lorentz and Einstein.

Lorentz took Kaufmann's work very seriously. He referred to it in his famous lectures on *The Theory of Electrons*, saying that the results were "decidedly unfavorable to the idea of a contraction," and that "it seems very likely that we shall have to relinquish this idea altogether."[18] It was a disastrous situation. The Lorentz contraction was the only known way to account for the Michelson-Morley result within a consistent electrodynamic theory. Einstein's relativity offered an alternative approach, to be sure, but it too contained the Lorentz contraction, as a theorem rather than a hypothesis. Both seemed to be explicitly rejected by Kaufmann's results. Lorentz had every reason to

---

16. A. Einstein, "Zur Elektrodynamik bewegter Körper," Ann. Phys. **17** (1905), p. 891.
17. W. Kaufmann, "Über die Konstitution des Elektrons," Ann. Phys. **19** (1906), p. 487.
18. H. A. Lorentz, *op. cit.*, p. 213.

be concerned. Nevertheless Einstein was not.[19] He took note of the fact that the differences between his predictions and the experiments were outside the limits of error, but he thought there might still be some source of systematic error of which Kaufmann was unaware. There would be time to come to a decision after more extensive experimental results were available. Einstein was right about this, as about so many other questions. Experiment eventually confirmed his confidence in his theory, though it took some years before the issue was fully resolved.[20]

At the time, however, the situation looked quite different. "Prof. Kaufmann has recently shown the extent to which his measurements on radium rays make it possible to decide between the various current theories on the structure of the electron. He demonstrates that Lorentz's deformable electron is ruled out by the measurements. A decision between Abraham's rigid electron and the Bucherer-Langevin deformable electron is not possible on the basis of the measurements already made." These are Ehrenfest's words, written in March 1906, summarizing the status of electronic models.[21] They appear at the beginning of his first short note on the theory of electrons, a note written to suggest another way in which one might try to distinguish between the two surviving models of the electron. Abraham's electron was necessarily stable against changes of shape since it was defined to be rigid. Ehrenfest argued that the Bucherer-Langevin electron would probably be unstable, that even at rest the spherical shape would have a higher potential energy than that of a deformed shape.

Perhaps the most interesting aspect of Ehrenfest's note is the absence of any mention of Einstein's theory. Although Kaufmann had referred to Einstein and had even tried to give some idea of what his theory was like, Ehrenfest spoke only of Lorentz's electron as being ruled out by the experimental results. This says something about the reception of the theory of relativity almost a year after the appearance of Einstein's original paper.

19. A. Einstein, "Über das Relativitätsprinzip und die aus demselben gezogenen Folgerungen," Jahrbuch der Radioactivität und Elektronik **4** (1907), p. 439. Einstein noted that both the Abraham and Bucherer theories led to mass vs. velocity curves in considerably better agreement with experiment than his own. He then remarked: "In my opinion, however, these theories are rather improbable because their basic assumptions, from which the mass of the moving electron is deduced, are not suggested by theoretical systems which encompass wider complexes of phenomena."

20. H. A. Lorentz, *op. cit.*, p. 339.

21. P. Ehrenfest, "Zur Stabilitätsfrage bei den Bucherer-Langevin-Elektronen," Phys. Z. **7** (1906), p. 302. *Papers*, p. 117.

Ehrenfest returned to the problem of the structure of the electron in another short note written a year later,[22] after he had done some studying in Einstein's papers. This time he picked up a point made by Abraham and raised a question about relativity based on it. Abraham had shown that a rigid ellipsoidal electron could not, in general, move in a straight line with constant velocity unless an external torque were supplied to balance out the torque due to its own field. Ehrenfest asked whether or not the same remark would apply to a deformable electron of arbitrary, non-ellipsoidal, shape: could such an electron move in uniform translation in an arbitrary direction in the absence of external forces? If it could not, one would need to introduce another hypothesis into the theory of relativity to exclude the possibility that such electrons exist, or else one would have a way of defining a state of absolute rest. If it could, one ought to be able to prove it from the "the Einsteinian system" with no additional assumptions.

This note of Ehrenfest's drew a response from Einstein,[23] who saw that Ehrenfest's expectations were based on a fundamental misconception of the very nature of the theory of relativity. Ehrenfest had referred to the "the Lorentz relativity-electrodynamics" as constituting "a closed system" in "the formulation published by Einstein," and so he thought his question about the deformable electron ought to have an answer which one could arrive at "purely deductively" within this system. This was not the way Einstein looked at his own work. "The principle of relativity or, more precisely, the principle of relativity together with the principle of the constancy of the velocity of light, is not to be interpreted as a 'closed system,' not really as a system at all, but rather merely as a heuristic principle which, considered by itself, contains only statements about rigid bodies, clocks and light signals. Anything beyond that that the theory of relativity supplies is in the connections it requires between laws that would otherwise appear to be independent of one another." If one wanted to discuss the motion of fast electrons, for example, one would have to use the known laws for slow electrons together with the relativistic transformations for both kinematical quantities and electromagnetic fields. "Thus we are by no means dealing with a 'system' here, a 'system' in which the individual laws would implicitly be contained and from which they

---

22. P. Ehrenfest, "Die Translation deformierbarer Elektronen und der Flächensatz," Ann. Phys. **23** (1907), p. 204. *Papers*, p. 144.

23. A. Einstein, "Bemerkungen zu der Notiz von Hrn. Paul Ehrenfest: 'Die Translation deformierbarer Elektronen und der Flächensatz'," Ann. Phys. **23** (1907), p. 206.

could be obtained just by deduction, but rather only with a principle that allows one to trace certain laws back to others, analogously to the second law of thermodynamics." In order to answer Ehrenfest's question one would have to have a relativistic dynamics of rigid bodies, but, as Einstein pointed out, no such theory existed yet.

4. In 1909 Max Born[24] proposed a theory of the rigid body that would be suitably invariant under Lorentz transformation. This was not just an attempt to generalize an idea which had been so useful in classical mechanics; it was intended to serve as the foundation for further developments in the theory of the electron. Born emphasized that "the practical value of the new definition of rigidity will have to be demonstrated in the dynamics of the electron." This new definition of rigidity was based on the requirement that, for every volume element of the body, the volume element must always be undeformed in its own rest frame. Born developed this idea at length in an article which appeared at the end of August 1909.

Ehrenfest had also been thinking about the problem of extending the idea of rigidity into relativistic mechanics. As Born had pointed out, the difficult thing was to treat general motions, and not just uniform translations. A few weeks after Born's long article came out, Ehrenfest sent off a one page note on the subject to the *Physikalische Zeitschrift*,[25] reporting his own reflections on this subject. He had constructed an argument showing that there must be an internal contradiction in the concept of a relativistic rigid body. Ehrenfest used a definition of a rigid body different from Born's, but equivalent to it: at any instant each element of such a body must exhibit to an observer at rest that Lorentz contraction appropriate to its velocity with respect to this observer. This led directly to a paradox as soon as one looked at a motion as simple as uniform rotation about a fixed axis.

Suppose one has a "relativistically rigid" cylinder of radius R and height H which is gradually set into uniform rotation about its axis. Let R' be its radius, as seen by an observer at rest, when the cylinder is rotating uniformly with angular velocity ω. Then R' must satisfy two contradictory requirements. Since every element of the periphery moves with velocity $R'\omega$ along its own direction it must suffer a Lorentz

24. M. Born, "Die Theorie des starren Elektrons in der Kinematik des Relativitätsprinzips," Ann. Phys. **30** (1909), p. 1.

25. P. Ehrenfest, "Gleichförmige Rotation starrer Körper und Relativitätstheorie," Phys. Z. **10** (1909), p. 918. *Papers*, p. 154.

contraction, so that $2\pi R' < 2\pi R$. But if one looks at any radial element, it is moving at right angles to its own length and so it must suffer no contraction at all, $R' = R$. Consequently, one cannot set such a "relativistically rigid" body into uniform rotation, since a deformation would have to occur as it goes from its state of rest to a state of uniform rotation.

Ehrenfest's paradox was the first of a series of critical analyses which showed that the concept of the rigid body had no place in the theory of relativity.[26] An attempt to defend the concept was made by W. von Ignatowsky,[27] who claimed that it was indeed possible to set such a "relativistically rigid" body into uniform rotation. He wrote that there were no contradictions and that everything could be explained by using the definition of simultaneous measurement. Ehrenfest answered Ignatowsky in crushing detail.[28] He cited the authorities including Born, Planck and Abraham who had accepted the arguments demonstrating the difficulty with "relativistic rigidity." He explained again that the problem was not in the state of uniform rotation once achieved, but in setting the body into uniform rotation from rest. Ehrenfest concluded by describing exactly how an observer at rest could measure the circumference and radius of the rotating cylinder with the help of markers and tracing paper, and how he would necessarily arrive at the paradoxical results previously described.

Ignatowsky admitted that Ehrenfest was right, including this admission at the end of an article on a relativistic theory of elasticity.[29] For some reason this seems to have provoked Ehrenfest into another devastating attack.[30] "Can a body that satisfies Born's definition of rigidity go from rest to a state of uniform rotation? Herglotz, Noether and I have shown that the answer is: no. Mr. Ignatowsky asserted that his calculations led to the result: yes. My protest has forced him to

---

26. See W. Pauli, *Relativitätstheorie* in *Encyclopädie der mathematischen Wissenschaften* Vol. 5, Part 2 (Leipzig: B. G. Teubner, 1921), p. 690. Reprinted in W. Pauli, *Collected Scientific Papers*, ed. R. Kronig and V. F. Weisskopf (New York: John Wiley, 1964), Vol. I, p. 152.

27. W. von Ignatowsky, "Der starre Körper und das Relativitätsprinzip", Ann. Phys. **33** (1910), p. 607.

28. P. Ehrenfest, "Zu Herrn v. Ignatowskys Behandlung der Bornschen Starrheitsdefinition," Phys. Z. **11** (1910), p. 1127. *Papers*, p. 156.

29. W. von Ignatowsky, "Zur Elastizitätstheorie vom Standpunkte des Relativitätsprinzips," Phys. Z. **12** (1911), p. 164.

30. P. Ehrenfest, "Zu Herrn v. Ignatowskys Behandlung der Bornschen Starrheitsdefinition, II", Phys. Z. **12** (1911), p. 412. *Papers*, p. 159.

admit: (a) that nothing of the sort follows from his calculations, and (b) that his assertion is itself false." Ehrenfest went on to point out with equal bluntness two more such errors of Ignatowsky's, and then added: "In his paper on 'The Rigid Body and the Principle of Relativity' Mr. Ignatowsky arrived at no new results except for the three just mentioned, which are incorrect. Consequently, one can easily understand that this author has abandoned the treatment of the all too precisely defined 'relativistically rigid' body and has now turned to the suggestively expandable problems of the 'relativistically elastic' body." (It must be added that in his later years Ehrenfest was not proud of the way he had disposed of Ignatowsky, and would have preferred to omit these two papers from any collection of his writings.)

Ehrenfest's note was also discussed by V. Varičak,[31] who tried to argue away the paradox by distinguishing between the "real" contraction assumed in the Lorentz theory and the "apparent, subjective" nature of the contraction deduced from Einstein's theory. This attempt to treat the theory of relativity as involving a contraction that was "so to speak, only a psychological and not a physical fact" could not be allowed to go without correction. Einstein wrote to Ehrenfest,[32] inviting him to answer it, but then prepared an answer himself.[33] Einstein explained that the contraction was completely "real" and subject to physical measurement by an observer not moving with the contracted body, "as Ehrenfest has made clear in a very pretty way." Of course it could not be detected by an observer moving along with the body, so for him it did not exist, but there were no questions of subjectivity or arbitrary conventions of measurement involved.

5. In 1911 Arnold Sommerfeld[34] declined to talk to a gathering of physicists about the special theory of relativity on the grounds that it was no longer one of the problems of current interest in physics. Despite the fact that the theory was only six years old, Sommerfeld considered it to be a "secured possession" of physics. There is a sense in which Sommerfeld's remarks were fully justified: Einstein's papers had established the essential features of the theory once and for all,

31. V. Varičak, "Zum Ehrenfestschen Paradoxon," Phys. Z. **12** (1911), p. 169.

32. A. Einstein to P. Ehrenfest, 12 April 1911.

33. A. Einstein, "Zum Ehrenfestschen Paradoxon. Bemerkung zu V. Varičaks Aufsatz," Phys. Z. **12** (1911), p. 509.

34. A. Sommerfeld, "Das Plancksche Wirkungsquantum und seine allgemeine Bedeutung für die Molekularphysik," Phys. Z. **12** (1911), p. 1057.

and H. Minkowski had shown how to give the theory its natural geometrical expression. Nevertheless, there were still physicists who were not fully convinced that Einstein's theory offered the only solution to the profound problem of securing an accommodation between mechanics and electrodynamics. These recalcitrant physicists were not all members of an older generation cut off from the new developments by their inflexible attitudes, although history might be tidier if they had been.

Walther Ritz devoted his efforts in the years 1908 and 1909, the last years of his tragically short life, to the attempt to create another kind of theory.[35] He wanted to make a theory that would satisfy the principle of relativity, so that there would be no absolute frame of reference, but he did not want a theory that led to such consequences as a velocity-dependent mass, relative simultaneity, a new law for addition of velocities, and the many other startling and seemingly paradoxical consequences of Einstein's theory. Ritz actually saw Einstein's theory as too conservative in one respect, too ready to preserve the Maxwell-Lorentz electrodynamics at the expense of the familiar concepts of mechanics. The crux of the matter for Ritz was Einstein's second postulate, his assumption that the velocity of light is independent of the motion of its source. Ritz considered that Einstein, by making this assumption, was unnecessarily preserving a vestige of the aether, since it was this universal medium, the aether, which had originally fixed the velocity of light in a natural way. Ritz was trying to develop an emission theory of light in which this condition would not hold. He was willing to accept the need to modify electrodynamics, and to replace the electromagnetic field with a retarded action at a distance. He would treat light as "projected" rather than "propagated" from its source through empty space.

Ritz died before he had time to work out the consequences of his ideas, but in November 1911 his friend, Paul Ehrenfest, wrote a paper[36] that brought out very clearly some of the points that had troubled Ritz. Ehrenfest pointed out that Einstein had dispensed with the hypothesis of an aether and had referred to the propagation of light as involving, not the state of a medium, but rather "independent entities, sent out by the light source, just as in the Newtonian emission theory of light."

---

35. W. Ritz, *Oeuvres* (Paris: Gauthier-Villars, 1911), pp. 317–492.
36. P. Ehrenfest, "Zur Frage nach der Entbehrlichkeit des Lichtäthers", Phys. Z. **13** (1912), p. 317. *Papers*, p. 303. See also Chapter 1.

On the other hand, Einstein also required that the velocity of light be independent of the velocity of its source, so that one had to stipulate that the kinematics of the emitted light certainly differed from what one would expect on the basis of the Newtonian emission theory. Ehrenfest then described Ritz's work by saying that he had undertaken to develop a theory that really could be described as an emission theory, preserving the usual kinematics and therefore making the velocity of the emitted light depend on the velocity of its source.

Ehrenfest discussed the sorts of experiments which might serve to distinguish between Ritz's theory and Einstein's. There were several possibilities, all more or less subtle. The necessary experiments had not yet been performed at the time Ehrenfest wrote, but subsequent work has effectively disposed of Ritz's emission theory.[37] Ehrenfest's closing remarks are still sound, and serve to emphasize an aspect of the theory of relativity which is too often ignored.

"Let us suppose," he wrote, "that someone soon manages to plan a crucial experiment which can actually be carried out and which will distinguish between assumptions B and D." (Assumption B was essentially Ritz's assumption that the velocity of light depends upon the velocity of its source, while assumption D was Einstein's assumption that it does not.) "What sort of situation would this lead to? The supporters of the aether hypothesis would have to wish that assumption D turns out to be correct. The supporters of the real emission theory, [Ritz's theory], would have to wish that assumption B is confirmed. The supporters of Einstein's theory of relativity would have to wish that this time the supporters of the aether hypothesis win out over the supporters of the real emission theory."

6. Thermodynamics was known as "a science with secure foundations, clear definitions, and distinct boundaries"[38] long before Ehrenfest started his career. It did not seem to be a likely subject for his kind of critical inquiry, and in fact Ehrenfest did not write very much about thermodynamics. But in 1909 he came upon just the sort of obscure and apparently paradoxical situation that always challenged him, and

37. See W. Pauli, *op. cit.* note 26, p. 549 in the *Encyclopädie*, or p. 11 in the *Collected Papers*. Also see J. G. Fox, "Evidence Against Emission Theories," Am. J. Phys. **33** (1965), p. 1.

38. J. C. Maxwell, "Tait's *Thermodynamics*," Nature **17** (1878), p. 257. See his *Scientific Papers*, Vol. II, p. 662.

he came upon it in the tightly organized discipline of thermodynamics.[39] The Le Chatelier-Braun principle had been in use for some twenty years as a qualitative guide to understanding the direction in which an equilibrium would be shifted when one of the parameters was changed. The principle was very loosely formulated, however, and had never been given a proper thermodynamic basis. Not only that, it had to be applied with the utmost tact and care or it would lead to a result of the wrong sign. This was the situation called to Ehrenfest's attention by V. R. Bursian, one of his friends among the Petersburg students. When Ehrenfest finished with it, the Le Chatelier-Braun principle was a theorem in thermodynamics.

To show the nature of the principle and the problem it gave rise to, Ehrenfest began with a very simple example. Suppose one has a definite quantity of an ideal gas which is compressed by increasing the pressure on it by an amount δP. This can be done by holding the temperature fixed, in which case there is a volume change $\delta_{\text{I}} V$, or it can be done adiabatically, leaving the temperature free to vary, in which case the volume change is $\delta_{\text{II}} V$. It is easy to show that the absolute values of the two volume changes satisfy the inequality,

$$|\delta_{\text{II}} V| < |\delta_{\text{I}} V|. \tag{7.3}$$

The volume, which is the parameter subjected to a direct influence, shows a greater resistance to change in the second case than in the first. This can be described by saying that when the second parameter, the temperature, is not fixed, it is "free to come to the aid of the first." This situation is generalized in the Le Chatelier-Braun principle.

Suppose a system in stable thermodynamic equilibrium is so constrained that only two parameters, ρ and σ, can vary. Let ρ be acted on directly by an external influence, and compare the change in ρ when σ is held fixed, $\delta_{\text{I}}\rho$, to the change in ρ when σ is left free to vary, $\delta_{\text{II}}\rho$. The assertion is that the absolute values obey the inequality

$$|\delta_{\text{II}}\rho| < |\delta_{\text{I}}\rho|. \tag{7.4}$$

Allowing the second parameter, σ, to remain free, increases the ability of ρ to resist the external influence trying to change it.

Despite the appealing simplicity of this principle, which suggests that situations of stable equilibrium can be analyzed by arguments that

39. P. Ehrenfest, "Das Prinzip von Le Chatelier-Braun und die Reziprozitätssätze der Thermodynamik," Z. Phys. Chemie **77** (1911), p. 227. *Papers*, p. 167. The paper was first published in Russian in 1909.

sound almost anthropomorphic, it is not true in general. One of the simple counterexamples cited by Ehrenfest is that of an elastic rectangular solid body, kept at constant temperature, whose dimensions are X_1, X_2, X_3. Suppose X_3 is fixed once and for all, and X_1 and X_2 are taken as the ρ and σ of the principle. One would then expect that the change in X_1 due to a force δF_1 along its own direction would be greater when X_2 is held fixed than when X_2 is left free to vary. In fact, however, it is easy to show by elementary elasticity theory that $|\delta_{II} X_1| > |\delta_I X_1|$ in this case. The parameter X_2 does not "come to the aid of X_1" when it is left free, but instead "reduces its ability to resist a change."

Ehrenfest went on to point out that even the first example, which seemed to be such an ideally simple illustration of the correct working of the principle, actually contained the same difficulty. The crux of the matter was the choice of the parameters ρ and σ. Suppose one had chosen ρ to be the volume, but had taken σ to be the entropy rather than the temperature. Then the two experimental situations would be exactly interchanged, so long as one agreed to interpret the entropy being left free to vary as meaning that the temperature was held fixed. In this case the ability of the volume to resist a change in the pressure would be less when the second parameter (entropy) was left free than when it was held fixed, exactly the opposite of what the Le Chatelier-Braun principle would require. And by an analogous change in the language of the description one could restate the example of the elastic solid in such a way as to make it exemplify rather than contradict the principle, (by taking σ to be the force in the X_2 direction, rather than X_2 itself.) There was another, equally vexing, ambiguity in the choice of the parameter ρ. If one took it to be pressure rather than volume in the case of the ideal gas one could once again reverse the inequality.

The Le Chatelier-Braun principle was obviously inadequate as stated. One needed a real physical insight into the particular experimental situation in order to choose the parameters in such a way as to make the principle work properly; but if one had that much insight into the situation one did not need the principle at all! (Ehrenfest took great pleasure in describing these ambiguities in the use of the principle to his friend Joffe, who had evidently defended it in their discussions. "This time," Ehrenfest wrote, "Tanya and I are certainly victorious over Uncle Experimentalist!"[40])

40. P. Ehrenfest to A. F. Joffe, undated but presumably spring or summer 1909.

Ehrenfest recognized that it would have to be possible to derive a suitably reformulated version of the principle from the second law of thermodynamics, since this law expresses the stability of equilibrium by the requirement that the entropy be an actual maximum. The first step in this reformulation was to systematize the choice of parameters, already shown to be crucial to the argument. Ehrenfest used the expression for $đW$, the infinitesimal work done on the system by its surroundings,

$$đW = \sum_{i=1}^{n} y_i \, \mathrm{d}x_i \tag{7.5}$$

The symbol $đ$ signifies that $đW$ is an inexact differential. The x_i are the extensive variables, quantity parameters, or generalized coordinates which specify the generalized "mechanical" state of the system. The y_i are the corresponding intensive variables, intensity parameters, or generalized forces. The distinction between these two classes of parameters had been strongly emphasized by Boltzmann's old antagonists, the Energeticists, and Ehrenfest referred to Georg Helm's book for further discussion.[41] The proper statement of the Le Chatelier-Braun principle depended upon whether the two relevant variables, ρ and σ, were of the same type, (both extensive or both intensive), or of opposite types, (one extensive and one intensive).

If an external influence acts to change ρ, and we denote by $\delta_{\mathrm{I}}\rho$ the change in ρ when σ is held fixed, and by $\delta_{\mathrm{II}}\rho$ the change in ρ when σ is left free to vary, then the absolute values of these two changes, $|\delta_{\mathrm{I}}\rho|$ and $|\delta_{\mathrm{II}}\rho|$, are related by the following inequalities. When ρ and σ are both extensive or both intensive, then

$$|\delta_{\mathrm{I}}\rho| < |\delta_{\mathrm{II}}\rho|. \qquad \text{(like variables)} \tag{7.6}$$

If, however, ρ and σ are a mixed pair, one extensive and one intensive, then the inequality is reversed,

$$|\delta_{\mathrm{I}}\rho| > |\delta_{\mathrm{II}}\rho|. \qquad \text{(unlike variables)} \tag{7.7}$$

In both cases, σ "left free to vary", is interpreted as meaning that the variable conjugate to σ is held fixed.

To see how the proof goes, consider the case in which both ρ and

41. G. Helm, *Die Energetik nach ihrer geschichtlichen Entwickelung* (Leipzig: Veit, 1898), pp. 266–276.

σ are extensive.[42] The internal energy U of the system satisfies the equation,

$$dU = TdS + đW = TdS + \sum_{i=1}^{n} y_i\, dx_i = \sum_{i=0}^{n} y_i\, dx_i, \qquad (7.8)$$

where the temperature T and the entropy S are also written as y_0 and x_0, respectively. Suppose that ρ is x_i and σ is x_j. The external influence attempting to change ρ directly is then δy_i, and the two changes satisfy the equations,

$$\delta y_i = \left(\frac{\partial y_i}{\partial x_i}\right)_{x_j} \delta_{\mathrm{I}} x_i \qquad (7.9)$$

$$\delta y_i = \left(\frac{\partial y_i}{\partial x_i}\right)_{y_j} \delta_{\mathrm{II}} x_i. \qquad (7.10)$$

To compare $\delta_{\mathrm{I}} x_i$ and $\delta_{\mathrm{II}} x_i$ one needs the relationship between the partial derivatives. The second of these can be transformed in a straightforward way into the form,

$$\left(\frac{\partial y_i}{\partial x_i}\right)_{y_j} = \left(\frac{\partial y_i}{\partial x_i}\right)_{x_j} - \left\{\left(\frac{\partial y_i}{\partial x_j}\right)_{x_i}\left(\frac{\partial y_j}{\partial x_i}\right)_{x_j} \Big/ \left(\frac{\partial y_j}{\partial x_j}\right)_{x_i}\right\}. \qquad (7.11)$$

Up to this point the condition of stable equilibrium has not been used. The second law requires that U be a minimum at a stable equilibrium, and when this is expressed with the help of (7.8), it can be shown that

$$\left(\frac{\partial y_i}{\partial x_i}\right)_{y_j} < \left(\frac{\partial y_i}{\partial x_i}\right)_{x_j} \qquad (7.12)$$

from which (7.6) then follows.[43]

For a mixed pair of variables the discussion has to be modified. Ehrenfest showed that one must use the minimum property not of the internal energy, U, but rather of a generalized Helmholtz function, which is just $A = U - x_i y_i$, in the case that y_i is ρ and x_j is σ. Where the former discussion used the relationship $y_i = (\partial U/\partial x_i)_{x_j}$, one now uses $x_i = -(\partial A/\partial y_i)_{x_j}$, and this critical minus sign ultimately reverses the inequality, leading to the result (7.7).

---

42. For more detailed discussions see P. S. Epstein, *Textbook of Thermodynamics* (New York: John Wiley, 1937), pp. 375–384, and L. D. Landau and E. M. Lifshitz, *Statistical Physics* (London: Pergamon Press, 1958), pp. 63–66.

43. It should be pointed out that Ehrenfest's proof was not quite correct. The proper conditions for U to be a minimum, considering only the two variables x_i and x_j, are that $(\partial y_i/\partial x_i) > 0$ and that $(\partial y_i/\partial x_i)(\partial y_j/\partial x_j) - (\partial y_i/\partial x_j)^2 > 0$. Ehrenfest used the first condition but replaced the second by $(\partial y_j/\partial x_j) > 0$ which is necessary but not sufficient. His argument for equation (22) of his paper (*Papers*, p. 180) is not cogent.

Ehrenfest was not quite satisfied with his analysis. Everything rested on the distinction between extensive and intensive variables, but this distinction needed further clarification, "perhaps along the lines that Carathéodory had recently followed for other basic concepts of thermodynamics." But even apart from that, Ehrenfest felt that his theorem, which gave a correct formulation of the Le Chatelier-Braun principle, lacked the appeal of the original crude form of the principle. It was too complicated, too hedged about with qualifications, too unwieldy to be a useful guide to the experimentalist. After all, the old way had worked because one always formulated a new problem by constructing a "suitable" analogy to some previously understood case, and parameters in the proper categories were "instinctively" chosen. "Perhaps," said Ehrenfest, "the old form of the principle was better just because it forced one to make these fruitful comparisons and so to train one's physical intuition." In a sense this analogy between the new situation and the old constituted the true essence of the principle. Ehrenfest concluded with the remark: "The Le Chatelier-Braun principle is surely no rule to be applied mechanically, but instead it leaves us something further to be discovered in each case: the correct sign of the inequality!"[44]

7. Practically every one of Ehrenfest's papers contains interesting ideas, but not all of his interesting ideas reached the printed page. His notebooks are full of fascinating and often tantalizing fragments. In the spring of 1904, for example, he jotted down the remark: "The Hamilton-Jacobi equation of Lagrangian mechanics corresponds to a diffractionless optics. What is the *super-Lagrangian mechanics* whose Hamilton-Jacobi equation is adequate for describing the diffracted wave?" There is nothing further on this line; it is certainly not a real foreshadowing of the work of de Broglie and Schrödinger some twenty years later, but it is striking all the same.

Another idea of Ehrenfest's, closer in spirit to a much later develop-

---

44. It is interesting to note that Lord Rayleigh wrote a brief note on the principle: "The Le Chatelier-Braun Principle," Trans. Chem. Soc. **111** (1917), p. 250. *Scientific Papers* Vol. VI (1920), p. 475. This was stimulated by Ehrenfest's work, which Rayleigh had learned to appreciate. Ehrenfest's paper also served as the starting point for later work by Planck, "Das Prinzip von Le Chatelier und Braun," Ann. Phys. **19** (1934), p. 759 and "Bemerkungen über Quantitätsparameter, Intensitätsparameter und stabiles Gleichgewicht," Physica **2** (1935), p. 1029. *Planck Papers* II, pp. 665 and 681. Planck seems to have taken Ehrenfest's concluding remark a little too seriously. (See p. 668.)

ment, appeared in one of his letters to Joffe.[45] He apparently carried it no further. One need not try to trace its relationships to information theory in order to enjoy the way Ehrenfest's mind played with the idea.

"Had a funny idea this morning as I was waking up. (You know Lichtenberg's remark 'Many things that please us very much when we think about them lying down leave us completely cold as soon as we stand up!') So you had better lie down on the couch for a while in order to enjoy the following remarks completely! My boy, the thing is this:

The entropy theorem may not be valid for the actions of organized matter. Suddenly wanted to prove this *by calculation* for an example! Thought of a Chinese craftsman making an ivory carving. How much 'improbability' does he put into his carving, and how much "improbability' does the sun put into the food that the carver consumes during the time he does the work? Had to modify the example in order to be able to calculate. Chose the following example: Give an individual 100,000 numbered cards, in a state of *complete* disorder – he is to put them in order. Here it is easy to calculate:

$$E = \frac{\text{Probability of order}}{\text{Probability of disorder}}.$$

Really seems to be a huge production of 'improbability,' and this would now have to be balanced out by the food consumed. In this phase of my reflections it didn't seem very unlikely that this man creates more improbability than he consumes. I wanted to get your help with the numerical values for the foods. Was struck by a characteristic difficulty which suddenly brought me to quite another series of questions. Have to enumerate the following remarks even though they came to me at first as a tangled skein. When you have heard them all you will easily be able to combine them again.

1. All negative-entropy is improbability but not all improbability is negative-entropy. At least not if I ask that the word negative-

45. P. Ehrenfest to A. F. Joffe. Undated. Probably c. 1909. The ideas here should be compared with some of the recent discussions of information and entropy, which are not always sufficiently critical. See also M. J. Klein, "Order, Organisation and Entropy," British Journal for the Philosophy of Science **4** (1953), p. 158.

entropy be restricted to the following use: I say that a system S in state A has a negative-entropy H_A different from that in state B, H_B, if, by taking the system from state A to state B, it is possible to put an end to an irreversible exchange of heat. If the 100,000 cards are ordered (A), they contain a definite supply of improbability compared to the completely disordered state (B), but they contain no 'atonement for thermal sins,' no 'negative-entropy,' because one really cannot pump any heat back by using them.

2. If you grant this, then you probably ought to agree to the following assertion: The difference between the logarithms of the *probabilities* for two states of a system is *greater* than or equal to the entropy difference of the two states.

3. In general then, if one wants to establish the balance between the production and consumption of improbability for any individual, one has to avoid the following mistake. It would be wrong, on the one hand, to calculate that he creates improbability by sorting the cards, but, on the other hand, to neglect the fact that in his food, and in the heating and lighting of his room, he consumes not only true negative-entropy, (which one could take into account relatively easily), but also *other* forms of improbability. There is, for example, all of the improbability that is put into his prepared meals on their way from the places where their raw materials originated to the dishes on which they are served. These other forms of improbability are (a) in no sense supplies of negative-entropy, (b) hardly susceptible of numerical evaluation, and (c) of such a nature that one cannot judge to what extent they are necessary for the work of sorting out the cards. This last is particularly sad.

4. This is the point at which I remain stuck, but I *believe* the following. It ought to be true that for any closed group of people, less improbability is produced by them, in any instant, than is consumed. Furthermore, every form of *organization*, (technical: Niagara turbines, adiabatic walls of houses, or social: address book, or intellectual: Physics Abstracts, etc.) will reduce this underproduction in ideal cases. There is an essential truth of some sort concealed here, but in the formulation I have at the moment everything is still expressed incorrectly.

5. When a single individual is considered, it would perhaps be possible to prove (?!?!!!) that here there are transient, temporary,

surpluses of improbability production over consumption, (Brownian motion).
6. It does not seem possible to make any rational connection between the orientation of human efforts towards feelings of pleasure and the characteristics of improbability.
7. The amount of improbability that is pumped into a modern steam engine does not enter the entropy balance of its functioning. The improbability of all the practice in association that has to be put into a human brain before it is capable of assorting 100,000 cards does not enter into the improbability of this activity. (The only exceptions to this are those cases in which the wear and tear plays an essential part. Example: the lubricating oil that is burned in internal combustion engines – right?!).
8. Striving towards the slowest disorganization for a given value of the improbability.

However, if, despite my request, you were not lying down, all of this was written in vain."

CHAPTER 8

The Search for a Position

Paul Ehrenfest did not have an easy time of it when he began to search for an academic position in the spring of 1911. What he hoped for was *Habilitation* at a university, that is, formal recognition and appointment as a lecturer, a *Privatdozent*. Such a position was unsalaried, but a *Privatdozent* did receive fees from his students. This was the normal first step in an academic career and, for the able, productive, and fortunate few, would eventually be followed by appointment to a professorship at the same or another university, Ehrenfest had every right to expect such an appointment as *Privatdozent* without excessive difficulty. He had taken his degree at a major university seven years earlier under an outstanding physicist, he had written some fifteen or so significant articles published in principal journals, his work was known and respected by many, and it was obvious to anyone who had ever heard him lecture that he was a born teacher. Unfortunately, however, the rational criteria which he met more than adequately were not the only ones imposed, and a difficult and discouraging time was in store for him.

The natural way to begin his job hunting would have been to write to his friends and acquaintances, inquiring about possibilities. Living in Russia for almost four years had put him a little out of touch; he did not know all his contemporaries as he would have, had he remained in Austria or Germany. Boltzmann was dead, and Felix Klein, as a mathematician, could not be of much direct help. Only two of Ehrenfest's closest friends were physicists and of these, one, Joffe, was himself in Petersburg, and the other, Walther Ritz, had died in July, 1909.

Ehrenfest did have another very close friend, the mathematician Gustav Herglotz, who had been professor of mathematics at the University of Leipzig since 1909. Ehrenfest had kept in contact with Herglotz since leaving Göttingen, and he had even visited him during the summer of 1910; they spend some time together at the Bavarian resort town of Bad Kissingen.[1] (This brief trip, which included a visit to his brothers and other friends in Vienna, was Paul Ehrenfest's only one to Western Europe since going to Russia.) Herglotz's presence in Leipzig was enough to make that university attractive to Ehrenfest. Leipzig really did not have much else to attract Ehrenfest: physics there had none of the lustre that Boltzmann had brought to Vienna, for example. Neither Theodor Des Coudres not Otto Wiener, the professors of theoretical and experimental physics, respectively, was as impressive or distinguished as some of the physicists at other universities. Nevertheless Ehrenfest tried, through Herglotz, to obtain an appointment at Leipzig. It was enough for him to know that he would have a real friend there, one with whom he could discuss his scientific problems.

Ehrenfest's inquiry brought an unexpected answer: he did not meet the qualifications for the position. The statutes of the University of Leipzig required a *Privatdozent* to have a doctorate earned at a *German* university. Foreign degrees, which would, of course, include Ehrenfest's Vienna degree, could not be accepted. (It would have been interesting to hear Boltzmann's reactions to this remarkable piece of academic chauvinism, especially since he had been professor at Leipzig himself for two years, less than a decade earlier!) The Leipzig statutes actually allowed for exceptions to this requirement on appeal, but both Leipzig professors of physics were reluctant to support such a petition for Ehrenfest.[2]

Another of Ehrenfest's early inquiries concerned a position at the University of Munich. The great attraction at Munich was Arnold

1. The "Red Book," entries for the summer of 1910.

2. Entries in Ehrenfest's diary "A", covering the period 21 June 1911 to 21 August 1911. (These dates are on the Russian calendar, which was 13 days behind the Western calendar. Russian diary entries are listed by the Russian date only, unless there is a special reason for listing them both ways. Letters from Russia have the Western date in parentheses. Letters from Western Europe have the Western date.) P. Ehrenfest to H. A. Lorentz, 6 (19) May 1912.

Sommerfeld, professor of theoretical physics there since 1906. Sommerfeld, then in his early forties, was the leading mathematical physicist of his generation, a real virtuoso in his mathematical technique, and an outstanding teacher. Under his leadership Munich had already become a major center for theoretical physics, one of the most active in Europe.

Ehrenfest had never met Sommerfeld and, perhaps for that reason, he did not write him directly. Instead he wrote to Peter Debye, a man a few years younger than himself, whom he probably considered more approachable. Debye, a native of the Netherlands, had been a student of Sommerfeld's and then a junior colleague of his at Munich, and had just been appointed assistant professor at the University of Zürich, where Ehrenfest wrote to him. Debye did inquire about an opening for Ehrenfest, and at the end of May he wrote Ehrenfest to say that he had just received Sommerfeld's answer.[3] Sommerfeld was very reluctant to take on a *Privatdozent* from outside his own school, and had in fact just refused another applicant. Precisely because he had such a thriving group of students of his own who were potential *Privatdozenten*, he did not want to block their chances by taking on outsiders. He felt that other universities–Leipzig, Göttingen and Berlin–really needed some *Privatdozenten* and ought to take on their fair share. Of course, Sommerfeld wrote, it would be very stimulating and a great pleasure to have Ehrenfest with him since Ehrenfest's papers were a sheer delight to read, but nevertheless . . . Debye tried to soften the blow a little by stressing how highly Sommerfeld thought of Ehrenfest's work and by assuring Ehrenfest that there must be other universities that would welcome him with open arms.

Ehrenfest actually considered one other possibility in the spring of 1911, a possibility very remote from these others – California. It is not clear what inspired this idea – it might even have been Boltzmann's example – but he noted it in his diary in April. During the next month or so he wrote a number of letters about a situation in California or elsewhere in America to various people, including his brother Hugo in Missouri. But nothing came of the idea.[4] By summer, only Leipzig seemed to be a live possibility, and then only under their unreasonable conditions.

---

3. P. Debye to P. Ehrenfest, 30 May 1911.
4. Entries in Ehrenfest's 1911 diary for April, May and June 1911.

The Ehrenfests proceeded to the summer house at Kanuka as usual. This summer they had the additional pleasure of having Joffe and his family at Kanuka, too. Perhaps Abram Fedorovich felt that he might not have much more time to spend with his friends before they left Russia. During the previous February, Paul Ehrenfest had noted in his diary that Joffe was now a frequent visitor, and that he was obviously sad over the fact that they would have to separate.[5] At Kanuka they had plenty of time to be together, often with other friends, and to talk endlessly about everything from photochemistry to the unrest among Russian students, from Gibbs's entropy definitions to prospects for the future.[6]

In July, Joffe and Ehrenfest took a little excursion to Finland, just across the Gulf, for five days. They crossed from Tallinn (then called Reval) in Estonia, to Lovisa in Finland on a small sailing ship carrying flour and only five passengers. The wind failed and they were becalmed for hours, spending the night on deck in the pleasant weather and watching the "splendid sunrise over the open sea". From Lovisa, where Ehrenfest telegraphed home for Galya's first birthday, they hiked part of the way to Helsinki along the Finnish coast; Ehrenfest was struck by the "Scandinavian" quality of the country, in contrast to Russia. He was proud of his new achievements as a hiker, though the party took a boat at Borgå for the rest of the trip to Helsinki – a "very, very beautiful" city. They returned to Russia by overnight steamer from Helsinki to Petersburg, and saw Kronstadt's golden dome rising from the sea, and then finally the harbor of Petersburg with the dome of St. Isaac's cathedral over the tangle of masts. Paul arrived at Kanuka that evening, full of stories of his journey to tell Tatyana as she prepared his supper.[7]

In the intervals of work, discussions, socializing, music, walking in the woods and on the beach, and playing with his little daughters, Ehrenfest pondered over the possibilities for the future. He was in constant touch with Herglotz by mail about the Leipzig situation. In August Herglotz wrote, apparently urging his friend to come to Leipzig despite the unreasonable requirements. Herglotz suggested writing directly to Sommerfeld, presumably to see whether Ehrenfest could take a doctorate at Munich quickly and then proceed to Leipzig. Herglotz's letter and suggestions put Ehrenfest out of sorts. An

5. The "Red Book", entries for February 1911.
6. Diary "A".
7. Diary "A" and a series of postcards from P. Ehrenfest to T. Ehrenfest, July 1911.

unsatisfactory conversation with Joffe about the situation did not help matters any. He accomplished nothing that day, and as a heavy rain poured down in the afternoon Paul Ehrenfest stood at the door of his summer house watching the water drip from the eaves.[8] For several days after this his diary records "No work done all day", or "Lost day". But he did write to Sommerfeld, who soon answered that it would be possible for Ehrenfest to submit a thesis for a doctorate at Munich.[9]

By mid-September the Ehrenfests had moved back to town from Kanuka. Nothing was settled; the night before leaving for Petersburg Ehrenfest had noted "How obscure the future is!!" On September 28th a letter arrived from Herglotz indicating that the Leipzig physicists would finally and definitely insist on the German doctorate.[10] Ehrenfest proceeded at once to draft a long letter to Sommerfeld in which he wrote about some possibilities for his Munich thesis topic. Ehrenfest actually described two subjects: one, in statistical mechanics, he felt he could do without difficulty. It would be a natural outgrowth of his years of work on the Encyclopedia article, but he suggested using this subject only if Sommerfeld thought it necessary. "I say 'in case of necessity' because what I would really like, if only it is possible, is to be able to work on a subject of an essentially different kind *under your guidance*."[11]

Ehrenfest must have decided that if he did have to get a second doctor's degree he might as well profit by the occasion to learn an art that he had never acquired, that of carrying out a complicated calculation; it was an art of which Sommerfeld was a real master. Ehrenfest wrote: "For two or three years now . . . I have been wishing to go to Munich so that under your direction I could learn, among many other things, *this* in particular: how one carries through a piece of work that *really* requires a display of calculation. I cannot do this; I can only accomplish those things which can be settled by intuitive, qualitative considerations, or possibly by an 'epsilontic' mathematical argument. For this reason I am condemned to do only sketchy work, and I see that I must be ruined by this if I am unable to overcome it. On the other hand, I feel that if I had proper instruction things would go somewhat better.

8. Diary "A", 11 August 1911.
9. Diary "A", 21 August 1911.
10. Diary "B", covering the period 22 August 1911 to 24 October 1911. (Russian calendar.) Entry of 15 September 1911.
11. P. Ehrenfest to A. Sommerfeld, 17 (30) September 1911.

"To be sure, Leipzig is pressing me. But I would not like to let this valuable opportunity for learning slip by unused. What I would most prefer, by far, would actually be to undertake one of the problems that you intend for your students. To be sure, I do have some questions, some of them for quite a long time now, that are perhaps quite close to that group of problems. But just because it depends upon really calculating and not just reflecting, I am completely unable to judge whether the individual questions can be mastered somehow."[11]

With his letter Ehrenfest enclosed a six page outline of one of the problems he had just been characterizing. It concerned the angular momentum of electromagnetic radiation – how it was distributed in space and how it was transferred to a body that absorbed or transmitted the radiation. The problem might require Sommerfeld's powerful mathematical technique for its solution, (Ehrenfest thought that diffraction effects might be essential), but Ehrenfest had formulated it very much in his own way: a series of sharply worded assertions that built up to an apparent paradox.

Ehrenfest expressed his hope that Sommerfeld would consider this problem both suitable and tractable, and, if not, that he would still be able to work under Sommerfeld's guidance in Munich on a problem calling for the use of the same sorts of methods. "Should this wish be fulfilled," he went on, "I would regard the pedantry of the Leipzig statutes as a very valuable stroke of good luck for me, because I am in bitter earnest about my inability to carry through a calculation."[11]

Sommerfeld was not keen on Ehrenfest's suggestion that he try to do a complicated calculation for his Munich thesis. Sommerfeld's views are evident from Ehrenfest's subsequent letter to him written a few weeks later.[12] Sommerfeld might simply have considered it unrealistic for Ehrenfest to attempt such a problem working against a deadline. He offered instead to accept an extended and developed version of the paper that Ehrenfest had just published in the *Annalen der Physik* on the essential features of the quantum theory.[13] Ehrenfest agreed to this idea with alacrity, writing that he hoped to have a "readably written" manuscript in "perhaps about two weeks", and expressing his relief at not having to make his chances depend upon a problematic calculation. He had not dropped the idea of the calculation and still looked forward

---

12. P. Ehrenfest to A. Sommerfeld, 3 (16) October 1911.
13. P. Ehrenfest, "Welche Züge der Lichtquantenhypothese spielen in der Theorie der Wärmestrahlung eine wesentliche Rolle?", Ann. Phys. **36** (1911), p. 91. *Papers*, p. 185. See Chapter 10.

to discussing it with Sommerfeld, but it could wait until he came to Munich.

Some time that autumn, Ehrenfest decided to make a grand tour, visiting a number of universities in Germany, the Austro-Hungarian Empire and Switzerland – the whole German-speaking world – to see if he could turn up a more desirable position by personal contact with his colleagues. He arranged to leave Petersburg early in January, right after a mathematical congress in which he took part. On the evening of January 6, 1912, by the Russian calendar, (it was January 19th already in Western Europe, the day after his thirty second birthday), he said goodby to Tatyana's aunt and piled into a coach with Tatyana and his daughters for the ride from their house on the Lopuchinskaya to the railroad station. It was bitter cold, as it had been for weeks – sometimes the brilliantly clear cold that Ehrenfest loved, with sun on the packed snow, sometimes the gray and gloomy days that almost invariably darkened his spirits. He paid the coach driver under the arc lights at the station. A snack at the buffet, fond farewells to wife and children, and then alone into his place on the train for the long journey to Berlin, the first stop on his two month tour.[14]

Whenever Paul Ehrenfest traveled he recorded his impressions day by day in a pocket notebook, a practice he began at least as early as 1903 on a tour of Italy.[15] The jottings he made were notes for himself: records of sights and sounds, places and people, sketches of rooms, snatches of conversation, meals that he ate and what they cost, emotional reactions to people and situations – a figurative, and sometimes literal, shorthand record of his days. These notes were the raw material for the stream of letters in which he described his experiences, often in enormous detail, for his wife and, in later years, also for his children and intimate friends. As he wrote these letters his fragmentary notes, very often meaningless by themselves, would suggest the full picture or situation to him. Ehrenfest wrote those detailed narrative letters for himself as much as for anyone else, to satisfy his need to order and work through his experiences; the more interesting or complicated he found them, the more he wrote. This trip to Western Europe gave him much to write about, as he went from city to city and

14. Diary "C", covering the period 25 October 1911 to 26 January 1912. (Russian calendar.) Entry for 6 January 1912.

15. A considerable number of these notebooks, covering most of Ehrenfest's journeys from 1903 on, form part of the Ehrenfest Collection.

met a series of physicists, many of them men whose work he had known and studied for years. All the more unfortunate then that his letters from this journey were not available; only his tantalizingly incomplete notes remain.

His visits to a dozen or so cities tended to follow a pattern. Where possible he stayed with friends, who always seemed to have room for him. Much of his time was spent in visits to the universities and to friends and colleagues. At many places he was invited to give a lecture, and there were at least three different subjects on which he spoke in the course of his trip: his recent paper on the essential features of the quantum theory,[13] his paper on the Le Chatelier principle which had been published in Russia in 1909 and had just recently appeared in German,[16] and some as yet unpublished work on the question of the dispensability or indispensability of the luminiferous aether.[16] Besides the lectures there was always talk – about his own work and the work of those he was visiting. And, of course, Ehrenfest never forgot the primary purpose of his trip – inquiring about the existence of an opening for himself. This always came up, though sometimes not until the end of an afternoon or evening spent in talking about other matters.

His first day after the train ride was a very long one – from his predawn arrival in cold, dark sleeping Berlin, through a series of visits culminating in an evening spent with Max Planck at his home. Planck was interested in hearing about the state of affairs in Russia, but he was displeased by Ehrenfest's account of the theatrical blow he had struck against the Petersburg *Magister* examinations. They talked about all the things Ehrenfest had been working on recently, many of them very close to Planck's own interests, and about Ehrenfest's current search for a position. Planck apparently did not commit himself on this subject, although Ehrenfest found him very friendly on this visit. The following day he noted that he had written a "long long letter to Tanya" about that day's experiences.[17]

At Leipzig he stayed with Herglotz and discussed the question of his *Habilitation* with Professors Des Coudres and Wiener. His impressions were strongly negative: "Leipzig obviously lost," he noted, but it is not clear whether he meant that his chances were gone or that he would not want a post there now that he had met the professors of physics.[18]

16. See Chapter 7.
17. Diary "C", 8–9 January 1912.
18. Diary "C", 9–10, 13–14 January 1912.

From Leipzig Ehrenfest proceeded to Munich for a stay of over a week. He met Sommerfeld and had plenty of opportunity for long talks with him about a great variety of problems. On the key question, Ehrenfest's coming to Munich, the outlook was a little brighter. It even seemed possible that he might receive an appointment as *Privatdozent*. The reason for the change in his prospects was that Max von Laue, who had been *Privatdozent* at Munich for several years, might be leaving; a professorship for him at another university was in the wind. If this should go through, then Ehrenfest could have Laue's Munich position, the only one that Sommerfeld was not reserving for his own students. It was all very uncertain, to be sure, but there was now some hope.[19]

Apart from the question of his future, Ehrenfest's week in Munich was a lively one. Almost everyone at Munich was interested in X-rays. Their discoverer, Wilhelm Röntgen, was director of the Institute for Experimental Physics, where a variety of problems in X-rays were under study. Röntgen himself had become rather withdrawn, and did not often take part in the frequent lively discussions of the younger physicists. Ehrenfest did have one chance to meet him and to bring him news of his former student, Joffe.[20] Sommerfeld had his own Institute for Theoretical Physics. It is perhaps characteristic of Sommerfeld that he had several experimentalists on his staff whose role was to do experiments in support of the theoretical research work. (This rather unusual arrangement was the reverse of the common situation where a theorist is attached to a laboratory in support of the experimental effort.) Sommerfeld, too, was actively working on X-rays, whose theory was far from clarified in 1911. No final evidence for the wave nature of X-rays had yet been provided, but such evidence was about to come from Munich. In just a few months after Ehrenfest's visit, Laue would propose the use of crystals as diffraction gratings for X-rays, give the theory of this crystal diffraction, and have his ideas verified experimentally by Walter Friedrich and Paul Knipping, two young men whom Ehrenfest met in Munich.[21]

A few months before Ehrenfest's visit Sommerfeld had been in Brussels for the week long meeting of the first Solvay Conference, devoted to the current problems of the quantum theory. Ehrenfest

19. Diary "C", 14–17 January 1912.
20. Diary "C", 17 January 1912.
21. See P. P. Ewald, ed., *Fifty Years of X-Ray Diffraction* (Utrecht: A. Oosthoek, 1962), pp. 15–16, 31–34, 37–56, 291–293.

had not been one of the small group of distinguished European physicists who made up the invited membership of the conference, and he was eager to learn about the discussions that had gone on at Brussels. Sommerfeld had a copy of the proceedings which Ehrenfest had a chance to look through. As he might have suspected, there was no mention of his paper on the essential features of the quantum theory,[13] a paper that had appeared just before the Solvay Conference. This was a real loss to the discussions, and a personal loss to Ehrenfest, as he was beginning to discover. For Henri Poincaré, the great French mathematician, had become interested in the quantum theory through his participation in the Solvay Conference and had independently raised and answered one of the major questions already answered in Ehrenfest's paper. Ehrenfest first became aware Poincaré's work at Leipzig when he saw a short note which the French master had published in the *Comptes Rendus*. "What will become of me?", Ehrenfest noted in his diary.[22] He wrote to Poincaré from Munich,[23] sending him a reprint of his own paper, but a week or so later, at Zürich, Ehrenfest learned that he was too late. Poincaré's full discussion of the problem had just appeared in the January issue of the *Journal de Physique*. Ehrenfest recorded his depression in his notebook.[24] Poincaré's paper played an important part in the development of the quantum theory by convincing many who had formerly been skeptical of the whole idea. That Ehrenfest received no credit for having demonstrated the same result months earlier was part of the price he paid for his isolation in Petersburg.

A week after leaving Munich, Ehrenfest was in Vienna. He had spent the intervening days in Zürich, in discussion with Debye and others that offered very little promise of a position at either the university or the Polytechnic Institute there.[25] Ehrenfest's week in Vienna involved much more than science and the university. Three of his brothers still lived there; only Hugo had emigrated to America and was practicing medicine in St. Louis. Paul spent a busy time getting to know his brothers' families, visiting with his step-mother, Tante Pepperl, and seeing other family and old friends after so many years. Not everything went smoothly: Paul quarreled bitterly with his oldest brother,

22. Diary "C", 13 January 1912. Also see Chapter 10.
23. Diary "C", 22 January 1912.
24. Diary "D", covering the period 8 February 1912 to 21 February 1912. (Western calendar.) Entry of 9 February 1912.
25. Diary "D", 8–14 February 1912.

Arthur, but they were reconciled some hours later. Paul also made use of his stay in Vienna to revisit many of the places he remembered – his childhood home on Himbergerstrasse, the Kepler school, his high school and other old familiar spots.[26]

He spent much time at the university with Philipp Frank, who was now *Privatdozent* there, and also with Boltzmann's successor, Fritz Hasenöhrl. One morning, when Ehrenfest arrived at the university, the only one present in the Boltzmann library was the assistant to Professor Franz Exner, a young man named Erwin Schrödinger who had received his degree in Vienna a year or so earlier.[27] He immediately enticed Schrödinger to go out to have some lunch with him. They talked about the opportunities for study in Vienna, and Ehrenfest stressed the need to study elsewhere, too. Then Ehrenfest dragged Schrödinger off to a coffee house where they could talk physics freely since one could scribble equations in pencil on the little white marble tops of the tables – an ideal substitute for a blackboard. He discovered that Schrödinger had not heard of the new work in the theory of magnetism done by Paul Langevin and Pierre Weiss, and proceeded to give him an impromptu lecture on the subject. Schrödinger was so swept up by Ehrenfest's enthusiasm that he went on to study the subject thoroughly, give some seminars on it, and then write a paper of his own on diamagnetism. He vividly remembered the conversation with Ehrenfest forty five years after it took place.[28]

On Friday, February 23, 1912, Paul Ehrenfest left the home of his old university friend, Heinrich Tietze, in Brünn (now Brno), where he had spent a day after leaving Vienna, and took the train for Prague.[29] One wonders how keenly he anticipated this next visit—to Albert Einstein. Ehrenfest had been reading Einstein's papers for almost a decade; he shared so many of Einstein's scientific interests that he could well appreciate the unprecedented boldness and depth of Einstein's ideas. Einstein was only a few months older than he, and had a similar Jewish background. But would there be a rapport between them? They had corresponded only very briefly, in connection with

26. Diary "D", 15–22 February 1912.
27. Diary "D", 19 February 1912.
28. Letter from E. Schrödinger to the author, 17 April 1957.
29. Diary "E", covering the period 21 February 1912 to 10 March 1912. (Western calendar.) Entry for 22-23 February 1912.

their common objection to a misinterpretation of one aspect of the theory of relativity,[30] and they had never met one another.

When Ehrenfest wrote Einstein about his intended visit, Einstein had answered at once, offering the hospitality of his home in Prague.[31] Ehrenfest had written again, probably from Zürich, expressing his growing concern over not yet having found a position. This time Einstein answered in his usual direct way: "Don't worry yourself over an *Habilitation*. Whatever I can do for you I will certainly do. What I would like best is to see you get an assistant professorship, and I believe that something like that could certainly be obtained. We will discuss all the rest when you come to see me. Let me know the day and hour of your arrival and at what station you come in, and stay at my house so that we can make good use of the time."[32] This certainly sounded promising enough—frank, friendly and informal, just as Ehrenfest would have liked.

As he left the train in Prague, there was Einstein, with a cigar in his mouth, waiting to meet him. Einstein's wife, Mileva, had come along too and accompanied them to a coffee house where they talked of Vienna, Prague and Zürich. When Mileva Einstein left to go home, their conversation turned at once to physics, as they plunged into a discussion of the ergodic hypothesis, a key question in statistical mechanics. Walking to the University's Institute of Physics through the rain a little later, the two talked constantly, arguing about everything. This discussion had to be broken off for a few hours since Einstein had a prior engagement to play string quartets. (Einstein left him with his Prague colleague, Professor Anton Lampa, an old acquaintance of Ehrenfest's, who had taught at the University of Vienna when Ehrenfest was a student.) By midnight they were arguing about physics again over a cup of tea in Einstein's house, and they continued until half past one.[33]

Two days of continual scientific dispute, (by no means one-sided, as Ehrenfest found an error in Einstein's reasoning and supplied a simple intuitive way of seeing the correct result), must have broken down any barriers there might have been between them. By Sunday they were playing Brahms violin and piano sonatas together. "Yes, we will be

30. See Chapter 7.

31. A. Einstein to P. Ehrenfest, undated postcard, January 1912. (Sent in care of Herglotz at Leipzig.)

32. A. Einstein to P. Ehrenfest, 12 February 1912.

33. Diary "E", 23 February 1912.

friends", Ehrenfest wrote in his diary, "Was awfully happy."[34] And many years later Albert Einstein, describing their first meeting, wrote: "Within a few hours we were true friends – as though our dreams and aspirations were meant for each other."[35]

Staying at Einstein's home gave Ehrenfest a chance to get to know the family, and especially Einstein's seven year old son, Hans Albert, whom Ehrenfest naturally referred to in his notes as Albert′. He hit it off quickly with little Albert, sitting next to him at meals, talking to him at length both seriously and jokingly, even taking him along to the museum one day. The boy liked to listen to his father and their lively visitor playing music together, sometimes beating time as they played. One Sunday afternoon they went out for a walk, Ehrenfest with young Albert and Einstein pushing his younger son in his baby carriage.[36]

There was time for many things during that week at Prague – time to hear about Einstein's remarkable new ideas on gravitation (the germs of the general theory of relativity), time to walk along the Moldau, time for Ehrenfest to give his memorable lecture on radiation theory – carefully preparing the blackboard in advance with the necessary equations, time to hear a seminar by Einstein on Nernst's heat theorem. And, of course, time to discuss his own future with Einstein.

Einstein himself was just beginning to achieve the recognition his genius had earned.[37] After seven years as an obscure patent examiner in the Swiss patent office at Berne, the only position he could get following a couple of years of doing scientific odd jobs, he had spent two years as assistant professor at the University of Zürich. Einstein had only been in Prague, as full professor, for about a year when Ehrenfest visited him, but he had already notified the authorities that he would be leaving at the end of that academic year to return to Zürich, this time as professor at the Federal Institute of Technology, his alma mater. The chair at Prague would then be vacant; what could be more suitable for Ehrenfest than an appointment as full professor

34. Diary "E", 25 February 1912.

35. A. Einstein, "Paul Ehrenfest in Memoriam", (1934), in his *Out of my Later Years*, (New York: Philosophical Library, 1950), p. 215.

36. Diary "E", 23–29 February 1912.

37. See P. Frank, *Einstein, His Life and Times*, trans. G. Rosen, ed. S. Kusaka (New York: Knopf, 1947), pp. 74–100. Also C. Seelig, *Albert Einstein, A Documentary Biography*, transl. M. Savill (London: Staples Press, 1956), pp. 119–130.

at Prague as Einstein's successor? This was Einstein's proposal, agreed to by his colleague Lampa.[38]

There was one difficulty: no one could be appointed to a professorship anywhere in the Austro-Hungarian Empire who had no formal religious affiliation. Einstein, who had had no religious commitments for over a decade, had conformed to the regulations and listed himself as an adherent of the "Mosaic creed", re-establishing his status as a Jew when he came to Prague. Ehrenfest felt himself to be in a different position: when he married Tatyana Alexeyevna Afanassjewa, they had officially declared themselves to be without any religion, since the law forbade the marriage of Jew and Christian. Perhaps he considered that it would be official hypocrisy to redeclare his Judaism just so as to become eligible for the Prague professorship; perhaps he felt that the state had already demanded one religious avowal, or rather disavowal, from him, and he would not give it another. In any case Paul Ehrenfest indicated at once that he was not prepared to conform to this requirement, despite assurances from Einstein that it would be strictly a matter of going through the motions.[39]

Ehrenfest was more attracted by another possibility that came up in their conversations. Einstein was going to Zürich; why shouldn't he go to Zürich too? Even a few days of Einstein's company were enough to convince Ehrenfest that any position where he could see and work with Einstein would automatically be the best one for him. There was no assurance of an opening at either the University or the Federal Institute of Technology in Zürich, but Einstein promised to look into the matter on Ehrenfest's behalf.[40]

On his last afternoon in Prague, Ehrenfest had lunch with the Einsteins, and said a fond goodby to young Albert′ when he left to go back to school. There was one more chance to play a few Brahms songs with Einstein and then the two men left the house for a leisurely trip to the railroad station. Ehrenfest had checked his heavy luggage in the morning and so they were able to stroll along the Moldau, stopping occasionally for chocolate or a cup of coffee. This time they

38. Diary "E", 23 February 1912 and several subsequent entries. Personal communication from Mrs. T. Ehrenfest–Afanassjewa. P. Ehrenfest to H. A. Lorentz, 6 (19) May 1912.

39. Personal communication from Mrs. T. Ehrenfest-Afanassjewa.

40. P. Ehrenfest to H. A. Lorentz, 6 (19) May 1912. A. Einstein to P. Ehrenfest, undated letter beginning: "Shortly after your departure, I wrote to Weiss asking him to do whatever he thought suitable for your *Habilitation* in Zürich."

talked no science at all for once, but ranged over everything from the Moscow Art Theater to the musical appreciation of the Viennese, from the Russian situation to the constancy or inconstancy of human nature. Finally it was time to go. Once again Mileva Einstein was at the station, this time to say goodby. Einstein helped him put his baggage on the train and then they said a warm farewell. "What does the future bring?!", Ehrenfest wrote in his notebook as he sat on the train taking him east again.[41]

His homeward journey took him first to Lemberg (now Lvov) in Austrian Poland, where he spent a day with Maryan von Smoluchowski in intense discussion of the problems of statistical mechanics, the subject that concerned both men most deeply.[42] From Lemberg Ehrenfest went to Czernowitz (now Chernovtsy) for a five day stay with Hans Hahn, professor of mathematics at the university there. This was another reunion with a member of the old "inseparable four".[43] Two more stops, at Kiev and Kharkov, for visits with the local physicists, and then Ehrenfest was really on his way homeward to Petersburg.[44]

He could not forego the opportunity to pay tribute to one of his heroes, and stopped off to see Tolstoy's grave at Yasnaya Polyana on his way from Kharkov to Moscow. (Tolstoy had died on November 20, 1910, only a little over a year earlier.) Ehrenfest received permission from Countess Tolstoy, the writer's widow, to visit the house as well, and he carefully noted the contents of Tolstoy's study and his bedroom. He was driven by sledge across the snow covered fields, glaringly bright in the sunshine, to a nearby village where he had to wait many hours for the next train.[45]

Ehrenfest spent a day in Moscow, arriving from Yasnaya Polyana early in the morning, and leaving in the evening on the train for Petersburg. It was a bright Sunday, with the sun accentuating the strong colors of the Kremlin, and he joined the crowds making the long climb up the tower of Ivan the Terrible for a view of the city.

41. Diary "E", 29 February 1912.
42. Diary "E", 1 March 1912.
43. Diary "E", 2–7 March 1912.
44. Diary "F", covering the period 1 March 1912 to 31 March 1912. (Russian calendar again.) Ehrenfest was at Kiev from 25 to 29 February and at Kharkov on 1–2 March 1912.
45. Diary "F", 3 March 1912.

He was at the railroad station in time to be the first one to board the Petersburg train, despite his fears of missing it and spoiling the last leg of his long journey. The following noon he was at home again, (a day less than two months since he had left), with his daughters dancing around him for joy. It was a happy reunion.[46]

Paul had a lot to talk about with his wife and friends after his trip. Future prospects continued to develop in a confusing way. Einstein wrote that he had already asked Pierre Weiss at the Zürich Polytechnic to do what he could for Ehrenfest's *Habilitation* there, and that he foresaw no serious difficulties. (With his usual enthusiasm, Ehrenfest had also asked Einstein to try to arrange a position for Joffe in Zürich, at the University. Einstein declined;he did not doubt Joffe's ability but he was not "the Grand Mogul of Zürich with all the strings in his hand.") He did recommend Ehrenfest as a possible successor to Debye at the University of Zürich, calling him "a very independent fellow who has recently written several really remarkable things." Einstein commented particularly on how outstandingly good Ehrenfest's lecture on the radiation problem had been.[47]

In the meantime there was the question of Prague. On objective grounds this was by far the most attractive possibility open to Ehrenfest. It was the only position in sight that was a full professorship, which meant that in both prestige and remuneration it far exceeded any of the other possibilities. In addition, Ehrenfest would be Einstein's successor if he went to Prague, no mean distinction in itself. The position could apparently have been his for the asking, yet he ruled it out. Ehrenfest's uncompromising opposition to declaring a religious affiliation struck his friends as unreasonable, unrealistic, and out of all proportion. In April, Einstein wrote: "I am frankly *annoyed* that you have this caprice of being without religious affiliation; give it up for your children's sake. Besides, once you are professor here you can go back to this curious whim again – and it is only necessary for a little while. Let yourself be influenced by your wife. Correction of a slip of the pen in your letter: not 'my wife and I have decided', but rather 'I have decided'."[48] Einstein was particularly annoyed that Ehrenfest persisted in refusing to consider the Prague professorship, because in the same letter he had to forward the negative reports he had received from Switzerland:

46. Diary "F", 4–5 March 1912.

47. A. Einstein to P. Ehrenfest, undated. See note 40. A. Einstein to A. Kleiner, 3 April 1912.

47. A. Einstein to P. Ehrenfest, undated. See note 40.

48. A. Einstein to P. Ehrenfest, 25 April 1912.

there was no possibility of an assistant professorship at the University of Zürich, and, furthermore, nothing could be done about Ehrenfest's *Habilitation* at the Polytechnic until Einstein himself was in Zürich.

Einstein was not alone in his efforts to change Ehrenfest's mind. Hans Hahn had written to Tatyana Ehrenfest some weeks earlier, right after Ehrenfest's visit to him in Czernowitz, to try to persuade her to convince her husband of his folly.[49] He had argued with Ehrenfest himself a number of times to no avail, and even though Ehrenfest had already warned him that his letter would have no effect, Hahn felt he had to write. For one thing he wanted to emphasize the completely formal and perfunctory nature of the requirement of religious affiliation. Neither the faculty of the University nor the Ministry of Education took any interest in it; the only reason for the requirement at all was the personal whim of the octogenarian emperor, Kaiser Franz Josef.[50] For this reason, wrote Hahn, Ehrenfest should not hesitate to go through the motions; no one would think less of him for it, no one would regard him as a hypocrite for doing so, no one but the Kaiser really cared about it at all. In Hahn's view, Ehrenfest owed it to himself, to his children, and to his scientific vocation not to turn down this first class opportunity by taking a rigid moral position over an empty formality. Neither Ehrenfest was convinced by the argument in Hahn's long and moving letter; they had ruled Prague out as a real possibility.

Paul Ehrenfest had decided to gamble on moving to Switzerland.[51] If he were very lucky there might be some sort of professorship at Basel or elsewhere; if not there was still a reasonable though not very strong chance of *Habilitation* at the Zürich Polytechnic. But even if all these chances fell through he wanted to be near Einstein. Those days in Prague had made an indelible impression.

During the month of April 1912, his future as confused and unsettled as ever, Ehrenfest received a letter from Professor H. A. Lorentz.[52] Ehrenfest had not seen Lorentz since the spring of 1903 and had written to him only once, a few years after that, for help in understanding a point in one of Lorentz's papers.[53] But he had sent Lorentz a copy of

49. H. Hahn to T. Ehrenfest, 12 March 1912.

50. Philipp Frank makes a similar remark in his *Einstein*, p. 79.

51. P. Ehrenfest to H. A. Lorentz, 6 (19) May 1912.

52. H. A. Lorentz to P. Ehrenfest, 20 April 1912. Diary "G", covering the period 1 April 1912 to 21 July 1912. (Russian calendar.) Entry of 10 (23) April 1912.

53. P. Ehrenfest to H. A. Lorentz, 2 April 1906.

the article in the *Mathematical Encyclopedia*, which had finally appeared in December 1911.

Lorentz's letter was apparently written for a rather different reason. He addressed himself to Ehrenfest as someone who could inform him about the current situation of the Moscow physicists. Their leading spirit, Peter Nikolaevich Lebedev, had died in March.[54] Lebedev had lost his professorship during the repression of the universities the year before, and had spent much of his last year abroad. He had written to Lorentz to ask for aid from the new Solvay Foundation, of which Lorentz was president, in setting up a new laboratory in Moscow. Lorentz thought this suggestion might have come from Ehrenfest, but in any case he now regretted that he had not answered Lebedev. The foundation was not yet fully established, but a letter to Lebedev might have brought him some pleasure during his last illness. Lorentz was now interested in knowing who might take Lebedev's place as the principal physicist in Moscow, and what were the prospects for physics there.

Since he was writing, Lorentz took the opportunity of thanking both Ehrenfests for the copy of their "beautiful and profound" *Encyclopedia* article, whose contents he found "highly interesting." And he went on to add a more personal note. "Our paths have not crossed again since I had the pleasure of seeing you and your friend Ritz here for a short time many years ago. I have only seen the articles written by you and your wife. But because one is interested in people, and not just in their scientific work, I should be very pleased if in your answer you were to tell me about your career and about your present situation. At that time I did not even know you were Russian."

Ehrenfest answered at once,[55] undoubtedly pleased that Lorentz remembered him. In addition to answering Lorentz's questions about Moscow he explained politely that he was an Austrian national and not a Russian, despite his present residence. He also wrote that – by no choice of his own–he did not hold any position in Russia.

Paul Ehrenfest had no way of knowing what lay behind Lorentz's letter. Lorentz had decided to retire from the professorship of theoretical physics at the University of Leyden.[56] Only fifty eight years old

---

54. See P. P. Lazarev, "My Recollections of P. N. Lebedev," Usp. Fiz. Nauk. **77** (1962), p. 571. English translation in Soviet Physics Uspekhi **5** (1963), p. 617.

55. P. Ehrenfest to H. A. Lorentz, 11 (24) April 1912.

56. See G. L. de Haas-Lorentz, "Reminiscences" in *H. A. Lorentz. Impressions of His Life and Work*, ed. G. L. de Haas-Lorentz (Amsterdam: North-Holland, 1957), pp. 97-102.

and still at the height of his powers, he was giving up the regular duties of the professorship he had held for thirty three years to take up a dual position as curator of the physics division at Teyler's *Stichting* (Institute) and also secretary of the *Hollandsche Maatschappij van Wetenschappen*, the Dutch Society for Science. He would move to nearby Haarlem, where his new position amounted to a research appointment with few formal duties. Lorentz would be free to pursue his own studies and to meet his growing responsibilities as an international leader in science, continually in demand for meetings and lectures. The one crucial task that remained to be done before he felt that he could leave Leyden was to find a successor for his professorship.

There was one obvious choice – Albert Einstein. Lorentz knew the profoundly original nature of Einstein's work, had been in correspondence with him for several years, and had recently been reminded of his unique scientific and human qualities when the two men met again at the Solvay Congress. At the time of that meeting, Einstein was considering an offer of the chair of theoretical physics at the University of Utrecht in the Netherlands.[57] He was reluctant to accept, partly because he thought that Lorentz might consider it improper for him to do so as a foreigner. Lorentz did not feel that way at all and would have been delighted to have Einstein at Utrecht, but neither man spoke freely on the subject during their conversations in Brussels. As a result Einstein declined the Utrecht offer, and then wrote to Lorentz saying that he did so "with a heavy heart, like someone who has done an injustice to his father."[58] Einstein was actually expecting an offer from the Polytechnic at Zürich, the Swiss Federal Institute of Technology, an offer that he had decided to accept when it was made.[59]

It must have been at just about this time, December 1911, that Lorentz made his decision to go to Haarlem. Only two days after he answered Einstein's letter, reassuring him that he was not at all angry, but only sorry to have lost him for the Netherlands, and wishing him joy in Zürich and success in his work,[58] Lorentz wrote again to

57. "Einstein and the University of Utrecht", Janus **47** (1958), p. 131.

58. A. Einstein to H. A. Lorentz, 23 November 1911. H. A. Lorentz to A. Einstein, 6 December 1911. These two letters are published in M. Rooseboom, "Albert Einstein und die Niederländischen Universitäten," Janus **47** (1958), p. 198.

59. A. Einstein to M. Grossmann, 18 November 1911. Published in C. Seelig, *op. cit.*, p. 132.

Einstein.[60] This time he wanted to know whether Einstein's decision to go to Zürich was really firm. Einstein was surprised by the letter, thinking that Lorentz still had the Utrecht position in mind.[61] By the time he learned that Lorentz wanted him to come to Leyden as his successor, the Zürich arrangements had been made final. Einstein wrote to an old friend that he was rather glad it had turned out that way: he could not have said no to Lorentz, but he really preferred going to Zürich. (Einstein had had no problems in declining handsome offers from Vienna and from the *Reichsanstalt* at Berlin at just about the same time.)[62]

Since Einstein could not accept a call to Leyden, Lorentz had to look for someone else. Paul Ehrenfest had just sent him the *Encyclopedia* article on the conceptual foundations of statistical mechanics, and so his name was present to Lorentz at this time. Besides, Lorentz had himself thought long and hard about the subtleties in the work of Boltzmann and Gibbs, and he appreciated the masterful way in which these subtleties were discussed and clarified by the Ehrenfests. Lorentz also read Ehrenfest's 1911 article on the essential features of the quantum theory and admired his ability to disentangle a very complex theoretical situation.[63]

At about the same time that Lorentz wrote Ehrenfest to find out about his current situation, he also wrote for first hand opinions from people who knew Ehrenfest and whose views he could trust. One of these was Arnold Sommerfeld, who had been considering Ehrenfest as a possible *Privatdozent* at Munich. Sommerfeld did not know Ehrenfest very well and said so in his letter to Lorentz.[64] But Ehrenfest had recently spent a week or so in Munich, and Sommerfeld had found him to be a sensitive and sympathetic person. He knew that Ehrenfest's friends praised him highly, and he was evidently a man of principle; Sommerfeld mentioned Ehrenfest's political and religious positions and even his vegetarianism. When it came to more professional subjects, Sommerfeld did have some strong impressions of Ehrenfest as a lecturer and as a scientist.

---

60. H. A. Lorentz to A. Einstein, 8 December 1911.
61. A. Einstein to M. Grossmann, December 1911.
62. A. Einstein to H. Zangger, February and March 1911. The letters referred to in notes 59–62 were quoted in letters to the author from Miss Helen Dukas.
63. J. D. van der Waals, Jr., "In Memoriam Prof. Dr. P. Ehrenfest," De Gids (1933) No. 4, p. 1.
64. A. Sommerfeld to H. A. Lorentz, 24 April 1912.

"He lectures *like a master.* I have hardly ever heard a man speak with such fascination and brilliance. Significant phrases, witty points and dialectic are all at his disposal in an extraordinary manner. His way of handling the blackboard is characteristic. The whole disposition of his lecture is noted down on the board for his audience in the most transparent possible way. He knows how to make the most difficult things concrete and intuitively clear. Mathematical arguments are translated by him into easily comprehensible pictures."

Sommerfeld went on to write about Ehrenfest's approach to physics. "I had the impression from personal contact with him, more than from his papers, that he cares about the *physical facts.* In his papers he is more of a logician and dialectician. Mathematics is not an end in itself for him, which is as it should be. He shows himself to be much more versatile in personal contact than he does in his publications. He follows up experimental results so long as they concern questions of principle."

Sommerfeld's opinion of Ehrenfest had obviously risen sharply since his visit to Munich. Sommerfeld wrote Lorentz that he would now be happy to have Ehrenfest as a *Privatdozent* in Munich, a fact that he had not yet communicated to Ehrenfest himself. Despite his recommendation of Ehrenfest, Sommerfeld could not resist pointing out to Lorentz that he might be able to obtain Peter Debye as his successor rather than Ehrenfest. Debye had already accepted the Utrecht position, where he had been second choice after Einstein,[57] but Sommerfeld thought there might still be time to get him for Leyden instead. Lorentz apparently ignored this suggestion, but he was happy to see that Sommerfeld's opinion of Ehrenfest supported his own views. Sommerfeld's letter also informed Lorentz that Ehrenfest could have the chair at Prague, if he were willing to compromise his (non-)religious principles, which Sommerfeld thought unlikely. As a result Lorentz decided to write to Ehrenfest again and let him know what he was thinking about.

On the day Lorentz's first letter arrived, Paul Ehrenfest was busily trying to arrange a trip he had been dreaming about for a long time – a cruise down the Volga.[65] It took another week to complete his plans, but on April 20th Paul and Tatyana left the two little girls with their great-aunt and took the night train from Petersburg to Rybinsk, where

65. Diary "G", 10 (23) April 1912.

they embarked the following day. They were gone for almost two weeks, travelling slowly down to Kazan and then back again. No holiday was ever a complete vacation for Paul Ehrenfest, and he managed to work on one or two problems during the voyage, but it was a memorable experience nevertheless.[66] This trip took him deep into central Russia, and gave him an immediate feeling for the extent and variety of the vast country in which he had been living for almost five years, and which he knew second hand from the great Russian novelists.

The Ehrenfests had left their itinerary with friends and family so that they could be reached if necessary and kept informed of any new developments in the search for a job, but the precautions proved unnecessary. There was some mail waiting for them in Petersburg, however, including a postcard from Ehrenfest's friend Paul Epstein in Munich. This brought news of Max von Laue's great discovery, not yet published, – the diffraction of X-rays by crystals. As usual, Ehrenfest noted in his diary – "depressed."[67]

The following morning a letter arrived from Munich.[68] It was the long awaited letter from Sommerfeld offering him a position as *Privatdozent*; Laue was leaving Munich to take Debye's place at Zürich, and Ehrenfest could have his place. Paul read the letter to Tanya, who was still in bed when it came. He spent the morning packing his many books in anticipation of the annual move to Kanuka and the major move to come in the fall – although its destination was still uncertain. After lunch he was lying on the sofa reading, books piled up everywhere, when the next mail came with a startling letter from Lorentz.[69]

After thanking Ehrenfest for the information he had sent, Lorentz proceeded to admit that he had had another reason for asking about Ehrenfest's personal situation. He went on to describe his own plans for retirement and to say that "the profundity, clarity and penetration" which Lorentz valued so highly in Ehrenfest's works had made him consider Ehrenfest when thinking of a possible successor. Lorentz could not be sure that the faculty at Leyden would not insist on naming a Dutch physicist to the chair, and he would not have mentioned the matter to Ehrenfest so prematurely, except that he did not want Ehrenfest to commit himself to some other university before he had

66. Diary "G", 20 April (3 May)–2 (15) May 1912.
67. Diary "G", 2 (15) May 1912.
68. Diary "G", 3 (16) May 1912.
69. H. A. Lorentz to P. Ehrenfest, 13 May 1912.

a chance to consider the Leyden possibility. He asked Ehrenfest to be kind enough to let him know if he received any other offers, and to try to defer a response until he heard from him. Lorentz hoped Ehrenfest would not consider this request to be unreasonable or indiscreet. After all, he did not know whether the Ehrenfests would even consider moving to Holland. Lorentz wrote only because he was so firmly convinced that any university would be fortunate to have Ehrenfest, even though he could not say what action the faculty at Leyden would actually take.

Ehrenfest was overwhelmed by this letter. Deeply moved, he rushed out to tell the news, first to Tanya's aunt whom he met on the way downstairs, and then to Tanya herself, who was in the garden with the children on that cool and sunny afternoon. He was almost afraid to rejoice yet.[68]

During the next few days Ehrenfest spent many hours composing a long and highly unconventional answer to Lorentz's request for more information about himself.[70] He asked Lorentz's indulgence for not putting his answer into a completely objective form, but his profound and joyful surprise at Lorentz's letter made him want to try instead, as conscientiously as he could, to give Lorentz a complete picture of his situation. After all his unsatisfactory experiences during the preceding months, something about Lorentz's letter had reached deep into Ehrenfest's soul. He simply could not answer it by keeping within the conventions and limiting his reply to an account of his academic history and prospects. Perhaps it was a fatherly tone that Ehrenfest found or read into Lorentz's letter, perhaps it was an absence of academic pomposity in Lorentz's style, perhaps it was what Einstein called "the special charm that emanated from Lorentz's personality" – whatever it was that Ehrenfest responded to, he felt that he could answer only by writing freely to Lorentz about all his hopes and his frustrations.

He explained his situation to Lorentz in detail, but without giving the names of any of the universities: the Prague professorship – ruled out by his lack of religious affiliation, the very slight hope for a professorship of some sort in Switzerland, and the newly received invitation to become *Privatdozent* in Munich. He planned to accept the Munich invitation, but would explicitly reserve the right to reject it for a professorship, provided one were to be offered. Ehrenfest wrote:

---

70. P. Ehrenfest to H. A. Lorentz, 6 (19) May 1912.

"I would *unconditionally* accept an offer from a Dutch university, except in the most unlikely event of a competing offer from a Swiss university. In the event of an offer from Switzerland the particular circumstances of a Dutch offer would have to be considered – because of the language situation." Referring to Prague, he remarked: "I want to add explicitly that I would feel *very* uncomfortable at that university, and I would unconditionally prefer a university like Leyden, despite the great difficulty of acquiring a new language."

Ehrenfest went on to describe his education and his academic experience, adding that, "Of great influence on my development were my relationships with my close friends: G. Herglotz (since Gymnasium), W. Ritz, and A. Joffe (in St. Petersburg), as well as the possibility of being able to work jointly with my wife, particularly on questions involving conceptual analysis."

Having disposed of all the official business of his letter, Ehrenfest turned to something that was much on his mind that year, when the whole pattern of his life was about to change. "My experiences in the last ten years are characterized, more than anything else, by my lack of a homeland through no will of my own. I have long been convinced that, apart from cases of exceptional talents, one can develop his powers only if he feels that the people he normally has to deal with are not foreign to him. In present day Vienna I felt and I feel more alien than practically anywhere else. I felt far more at home in the circle of my Göttingen comrades. It was the same way later on in German-speaking Switzerland. Yes, ridiculous as it may sound, even during a stay of several weeks among the fishermen of Schiermonnikoog I very quickly felt more at home, (despite my most defective knowledge of Dutch), than I could in Vienna. But Russia would undoubtedly have become my home in the most serious sense of the word, if only I could somehow have been admitted to a regular teaching position; despite my imperfect mastery of the language, I feel absolutely at home in my relations with all classes of the population (except political officials), and also, strange to say, Russians of all possible classes do not treat me like a foreigner."

Despite the satisfaction and pleasure that all their unofficial seminars and other activities had brought the Ehrenfests in Russia, he now had to face the fact that there was no academic future for him there. "Without such a *responsible* position I was afraid that I would completely stagnate; I already felt the increasing disorganization of my activities. Also, for me 'working' and communicating my thoughts verbally are inseparable. Furthermore, we had to take into consideration the fact that the income from our (inherited) capital would not be enough to bring up our two children in an unhampered way, even if we increasingly restricted our personal needs." And a little later, after describing his plan to follow Einstein to Switzerland despite the "distressing uncertainty" of the move, Ehrenfest wrote that a "final separation from Russia would be a weighty step for me and all the more so for my wife." He mentioned once more the "profound and encouraging impression" that Lorentz's letter had made, "regardless of whether or not it would lead to practical consequences", and he promised to avail himself of Lorentz's offer of advice before he made any final decision.

What did Lorentz make of this letter? Lorentz, that natural aristocrat, who always had "the remarkable gift of maintaining a certain distance between his students and himself, without appearing to do so,"[71] whose "superiority was felt by everyone,"[72] had discovered a man very different from himself. His choice of Ehrenfest despite all these evident differences, and his real understanding for the younger man's problems, give some clue as to why Lorentz meant so much as a man to so many people. Lorentz had decided that Ehrenfest was the one he wanted, the one "to dot the i's in physics," and he read his letters approvingly, quoting passages to his physicist son-in-law "with visible pleasure."[73]

Lorentz did not answer Ehrenfest's long and intimate letter in any detail. He asked only for a complete list of Ehrenfest's publications and copies wherever possible, which Ehrenfest sent off early in June from Kanuka.[74] The family had left Petersburg for the summer – for the last time – after several exhausting and depressing days spent packing all their belongings.[75] They had given up their home in Petersburg, but the future was still obscure.

Ehrenfest kept his friends informed about his situation, and he

71. G. L. de Haas-Lorentz, *op. cit.*, p. 35.
72. A. Einstein, "H. A. Lorentz, His Creative Genius and His Personality," in G. L. de Haas-Lorentz, *op. cit.*, p. 8.
73. W. J. de Haas, "Prof. dr. P. Ehrenfest†", Nieuwe Rotterdamsche Courant, 27 September 1933.
74. H. A. Lorentz to P. Ehrenfest, 30 May 1912. P. Ehrenfest to H. A. Lorentz, 22 May (4 June) 1912.
75. Diary "G", 15 (28) May–18 (31) May 1912.

soon received a letter from Einstein that must have cheered him up, at least for a while.[76] "My hearty congratulations on Lorentz's news," Einstein wrote. "Besides yourself no one would be happier than I if you were to be appointed in Holland. You are one of the few theoreticians who have not been deprived of their native intelligence by the mathematical epidemic." Furthermore, Einstein went on to acknowledge that Ehrenfest had been right in a scientific discussion they had been carrying on.

But neither Einstein's letter nor his usual active life with family and friends at the shore could ease Ehrenfest's mind that summer. "Lorentz? Sommerfeld? Prague?", he wrote in his diary soon after arriving at Kanuka.[77] He was worried because the Leyden appointment might not materialize, and in that case he would have to prepare an inaugural dissertation (*Habilitationschrift*) for Munich to qualify as *Privatdozent.* Understandably enough the work on this project was not going very well.[78] After a month at Kanuka, and no word from Leyden, Ehrenfest wrote somewhat apologetically to Lorentz to make sure that a letter had not gone astray.[79]

Lorentz answered promptly on July 7th. He must have realized how uncomfortable and uneasy Ehrenfest felt, because he wrote at some length this time.[80] No letter had gone astray, but Lorentz took the opportunity to say, confidentially, that he was really very hopeful that Ehrenfest would receive the appointment as professor at Leyden some time during the summer, an outcome that would give Lorentz great pleasure. Of course he ought to be prepared for the possibility of unexpected circumstances, but Lorentz implied this was unlikely by sending Ehrenfest details about salary, pension plans and insurance benefits. He also wrote about the other physicists at Leyden, the students, the courses, and so on. Not surprisingly Ehrenfest regained some of his confidence when this letter came. "Very happy all day", he wrote.[81]

A few weeks later, Lorentz wrote again to explain some of the delay.[82] He had hoped himself that matters would be settled by now, but the

---

76. A. Einstein to P. Ehrenfest, n.d., received 7 (20) June 1912.
77. Diary "G", 24 May (6 June) 1912.
78. Diary "G", 21 May (3 June) 1912.
79. P. Ehrenfest to H. A. Lorentz, 17 (30) June 1912.
80. H. A. Lorentz to P. Ehrenfest, 7 July 1912.
81. Diary "G", 30 June (13 July) 1912.
82. H. A. Lorentz to P. Ehrenfest, 25 July 1912.

Minister of Education was off on a month's trip, and his approval was required. The formal appointment had to be made by the government.

The summer dragged on for Ehrenfest. When he bought a new notebook early in August for the next installment of his diary, he inscribed "Leyden or Munich?" on the cover.[83] And almost daily he remarked on this subject, never far from his thoughts. "Will Holland say yes or no?" "Again no letter!!" "No letter from Lorentz!!" "Will Lorentz finally write?!"[84] Ehrenfest started to write another letter to Lorentz, but decided to give it up, unfinished.[85] "Lorentz is silent." "Another day with no news." "Lorentz's silence *entirely incomprehensible*."[86] Tatyana developed a severe cough early in September and Paul was very concerned about it; his worry over "the Lorentz affair" seemed completely abstract by comparison. But a day later he was writing, "Lorentz????"[87]

By the middle of September Ehrenfest could stand it no longer, and he sent a wire to Lorentz to ask if there had been a misunderstanding or if something had gone astray. Lorentz's return wire came within a few hours. "Unfortunately no decision yet."[88]

The summer was over, and it was time to leave Kanuka. On the afternoon of the 29th of September, in the midst of the chaos of packing up, a telegram arrived. Paul tore it open. "Ehrenfest named professor at Leyden. Sincere congratulations. Letter follows. Lorentz." Paul was in tears, hardly able to feel his full happiness yet. It was a deeply emotional moment for all four Ehrenfests. He was up late into the night writing cards and letters to all his friends to share the good news.[89]

Lorentz's letter, when it came, was full of warm good wishes for Ehrenfest's future in every respect.[90] Lorentz considered Ehrenfest's appointment to be the realization of a plan he had formed as soon as he knew that Einstein was not available, and he was personally delighted with the outcome. He and his wife offered to help in every

83. Diary "H", covering the period 22 July (4 August)–7 (20) October 1912.
84. Diary "H", 14 (27), 15 (28) August, 19 August (1 September), 23 August (5 September) 1912.
85. Diary "H", 26 August (8 September) 1912.
86. Diary "H", 28 August (10 September)–3 (16) September 1912.
87. Diary "H", 4–5 (17–18) September 1912.
88. Diary "H", 9 (22) September 1912. Telegram, H. A. Lorentz to P. Ehrenfest, 22 September 1912.
89. Diary "H", 16 (29) September 1912. Telegram, H. A. Lorentz to P. Ehrenfest, 29 September 1912.
90. H. A. Lorentz to P. Ehrenfest, 29 September 1912.

possible way, even to taking care of the Ehrenfest children while the parents looked for a house. "We would be very glad to do anything that can help to make you feel immediately at home in this country."

On October 5th Ehrenfest answered Lorentz:[91]

"I had to wait at least two or three days to answer your letter so that I could manage to express my feelings in not too tumultuous a form. Please forgive this delay and allow me to put my thanks in the form of a vow: I will really exert all my strength in order to justify, at least to some extent, the confidence of those men who found it possible to appoint me to your professorial chair. I will also strive to fashion my personal behavior in such a way that neither you nor Professor Kamerlingh Onnes nor Professor Kuenen will regret having so heartily welcomed a complete stranger."

The Ehrenfests' thanks were expressed from their "jubilant hearts."

The ten days after the arrival of Lorentz's telegram were hectic. Joffe was at the train to meet them when they returned to Petersburg. He had already heard the great news from another friend who had received Lorentz's telgram for Ehrenfest and relayed it to Kanuka. Many friends wanted to see the Ehrenfests, share their excitement and good fortune, and say farewell. Most of the preparations to leave Petersburg had already been made.

On October 8th the Ehrenfests boarded the train for the west. "Only a little while ago the appointment question was still in the dark," Ehrenfest wrote that night on the train. "Now our destiny is in Leyden!"[92]

91. P. Ehrenfest to H. A. Lorentz, 22 September (5 October) 1912.
92. Diary "H", 19–25 September (2–8 October) 1912.

CHAPTER 9

Early Years in Leyden

On the morning of October 17, 1912, Paul and Tatyana Ehrenfest got off the train in Leyden with their two young daughters, a Russian nursemaid, and Tatyana's aunt, Baba Sonya.[1] They had left Petersburg over a week earlier, and had spent the intervening week in Berlin. The time between the arrival of Lorentz's long awaited telegram and their departure from Petersburg had been so short and so hurried; a week in Berlin might help to ease the transition to the new life.

Much mail was waiting for them at the Berlin post office. Letters of congratulations on Paul's new appointment had come from many friends and from his brothers in Vienna. Otto wrote that Paul's old fellow student and almost-namesake, Ehrenhaft, "would burst" when he heard the news. Emil sent a package of gifts for the family, including a Meccano set with which Paul immediately built a windmill for little Tanya.[2] The Ehrenfests spent their week in Berlin sight-seeing (including a trip to the zoo), shopping, and visiting their many acquaintances, especially in scientific circles. One evening Max Planck came to call on them for half an hour. After his visit, Ehrenfest reflected that Planck had been appointed as Kirchhoff's successor in Berlin only after Boltzmann had refused the position, in much the same way that he himself was succeeding Lorentz at Leyden because the obvious first choice, Einstein, was not available. It left him feeling a little empty.

1. Diary "H", 17 October 1912.
2. Diary "H", 10 October 1912.

With younger Berlin colleagues, like Gustav Hertz and James Franck, there were many discussions about the current state of physics and their own research work.

In Berlin Paul and Tatyana made the acquaintance of another young scientific couple. They were Lorentz's daughter and her husband, and it was Lorentz who had the suggested the visit to Ehrenfest.[3] W. J. de Haas, Lorentz's son-in-law, was an experimental physicist who would eventually become Ehrenfest's colleague in Leyden, and his wife, G. L. de Haas-Lorentz, had just written her dissertation on the theory of the Brownian motion. The two couples found much to discuss, as Lorentz had expected they might. Ehrenfest was particularly struck by Mrs. de Haas's parting words: "Tell my father that I envy you because you will be able to be with him."[4]

Paul had already spent many hours of a long and largely sleepless night on the train west from Petersburg wondering about the future. The many questions in his mind went down in his notebook as usual, perhaps to help keep them from looming too large. What would his new colleagues in Leyden be like, and how would he get on with the experimentalists? Where would they live? How would Tanya take to the new life, and would they all find Leyden a healthful place to live? What about the language? How would he get through his inaugural lecture? Would he see much of his old friend, Herglotz, and his new friend, Einstein? But above all, how would things go with Lorentz?[5]

His meeting with Lorentz's daughter and son-in-law must have been reassuring on this last point. And at about this time Ehrenfest also received a warm congratulatory letter from Smoluchowski in Lemberg.[6] Smoluchowski wrote, "I really envy you one aspect of this whole move, and that is the frequent contact you will have with H. A. Lorentz. I have just recently gotten to known him personally, and I must confess that I am completely enchanted by him. He is a wonderful person and it is a pleasure to be able to talk about scientific matters with him. I also found Kamerlingh Onnes thoroughly pleasant and interesting in many respects, but his is another kind of talent. In any case you will be living in a thoroughly stimulating and congenial atmosphere there."

3. H. A. Lorentz to P. Ehrenfest, 10 October 1912. See also G. L. de Haas-Lorentz, *op. cit.*, p. 103.

4. Diary "H", 16 October 1912.

5. Diary "H", 8 October 1912.

6. M. v. Smoluchowski to P. Ehrenfest, 9 October 1912.

On their first day in Leyden, after putting up at a hotel near the station in thoroughly unsatisfactory quarters and wandering around town a bit, Paul could wait no longer. He decided to go to Haarlem with Tatyana that very day to visit Lorentz. From Leyden to Haarlem is only a short run by train, and they were soon standing in front of the house at 76 Zijlweg, noting the "Lorentz" inscribed over the door. The drawing room was upstairs, and there was Lorentz, "enormously calm – noble", greeting them with "sincere warmth." For several hours the Ehrenfests sat talking quietly with Lorentz and his wife about some of the many questions they had in their new situation. Lorentz showed them his study and its books. It was a room that would come to mean much to Ehrenfest. When it was time to leave, Lorentz went to the station with them, and on the way Ehrenfest spoke of his hopes and plans for close scientific contact between them. Lorentz accompanied them to the platform, saw to it that they were on the correct train, and waited to wave goodby as the train pulled out. Despite the "impossible" accommodations at the hotel, Ehrenfest could sleep well that night.[7]

The early weeks in Leyden were busy ones. Finding a place to live was the first and most urgent matter. The family moved to another hotel the next day, and Tanya set out to look for a house to rent. With the help of Mrs. Nieuwenhuis, the wife of one of the Leyden professors, she located one immediately, on Groenhovenstraat.[8] It was small, Ehrenfest noted, but in a very pretty location near the Witte Singel, the gracefully curving canal that goes by the University Observatory. Although the house was located quickly, it took almost a month before it could be made ready for them to move in. The house had been without electricity, and Ehrenfest worked with the electrician in planning the disposition of the wiring to be installed. New plumbing, more furniture, and a telephone were also required, as well as a vast amount of scrubbing, before the Ehrenfests considered it really habitable.

Lorentz and his wife sent flowers to the Ehrenfests to welcome them to the new house. In his note to Lorentz acknowledging the gift, Ehrenfest described their situation.[9] "Our house is still in a semi-

7. Diary "H", 17 October 1912.

8. Diary "H", 18 October 1912. Also Diary "I", covering the period 21 October 1912 to 30 October 1913. Entries of 21 October, 5–13 November 1912.

9. P. Ehrenfest to H. A. Lorentz, undated. (November 1912).

chaotic state. Only the bare necessities are present, and not always even those! Your rosy flowers naturally have a particularly cheering effect against this dreary background!" Ehrenfest was inviting Lorentz to spend the next Monday afternoon and early evening with them, since Lorentz had to be in Leyden in the morning for his regular Monday lecture and then in the evening for a meeting. He described to Lorentz what he could expect to find. "My workroom is in order, to a first approximation. Permit me to enumerate its virtues: it has five windows but it can be kept quite warm, nevertheless, by means of a gas stove. There are blotting paper, writing paper, pen and ink; also two tables, two chairs, a couch and an electric table lamp – even a blackboard. If you are willing to put up with this inventory, and with the rather camp-like meals (in an absolutely bare room, decorated only by your flowers!), well, if you are willing to put up with this, we should be very very pleased to have you. My room, that is really very tolerable, but I am afraid that the meals will unfortunately be *very* bad. But perhaps in spite of this you would rather stay with us than travel back and forth to Haarlem again."

Ehrenfest naturally met a great many people during those first weeks. He liked Heike Kamerlingh Onnes, the professor of experimental physics and director of the low temperature laboratory, "a friendly gentleman", but thought he might be a little like Chwolson of Petersburg, which was not a recommendation. He also liked J. P. Kuenen, "a nice fellow". Kuenen was the third professor of physics at Leyden, an experimentalist who was responsible for much of the elementary physics teaching at the university. Ehrenfest visited Amsterdam one day and called on J. D. van der Waals, whose theory of the equation of state of gases had been famous since 1873. Van der Waals had just turned seventy five and was retired, but it was only two years since he had received the Nobel Prize for his work. Ehrenfest also called on J. D. van der Waals, Junior, who now occupied his father's chair in theoretical physics at Amsterdam, but was not impressed by his conversation. Pieter Zeeman was much more to his liking.[10]

Another visit that had to be paid was to the Ministry of Education at the Hague, since Ehrenfest's appointment as professor at a national university came officially from the government of the Netherlands. Ehrenfest, on Lorentz's advice, had purchased a top hat and a coat suitable to the occasion! The visit went off smoothly enough. Ehrenfest

10. Diary "I", 31 October 1912.

spent the rest of the afternoon admiring the Holbeins, Vermeers and Rembrandts in the Mauritshuis Museum and wandering through bookstores. Back at the hotel in Leyden two year old Galya was allowed to try on his top hat. It was a "very, very lovely day."[11]

On another day Ehrenfest went to Utrecht to call on Peter Debye, who had recently become professor of theoretical physics there. Debye and Ehrenfest talked about their plans for courses and colloquia. Because of the closeness of the Dutch cities it was, and is, quite possible for people at one university to attend a colloquium or lecture held at another. Debye went on to tell his visitor about his recent work on the theory of specific heats of solids and on several other problems. Ehrenfest was much impressed by Debye. He "felt his infinite superiority in ability and drive". And, although Ehrenfest liked Debye and liked the others he met that day in Utrecht, especially the astrophysicist W. R. Julius, he was depressed by the feeling that he was no match for Debye as a physicist.[12] The next day Lorentz was in Leyden to give his weekly lecture. Ehrenfest told Lorentz about his visit to Debye and his reactions afterwards. Lorentz was quick to reassure Ehrenfest that there was no reason for him to feel that way.[13] But no objective arguments could ever give Paul Ehrenfest the self-assurance he needed to resist the depressing effects of the comparisons he forced on himself.

During his first week in Leyden, Ehrenfest had the opportunity to attend the inaugural lecture of one of his new colleagues, W. J. M. van Eysinga, professor of international law. He was of course, very interested in what such an occasion was like, since his own ordeal was to come in only six weeks. Ehrenfest carefully observed the formal entrance of the faculty and the curators of the university, noting that all the while "poor Eysinga was standing, standing, standing". The talk itself seemed painfully slow to Ehrenfest and for the most part rather boring. He may well have resolved not to follow this example in his own forthcoming lecture. Afterwards there was the typical reception at a nearby hotel – many people in a large room, cigars, drinks, congratulations and conversation; not something that Ehrenfest would look forward to.[14]

His own day came soon enough. The inaugural lecture had been on Ehrenfest's mind almost since he had known of his appointment. In

11. Diary "I", 30 October, 2 November 1912.
12. Diary "I", 3 November 1912.
13. Diary "I", 4 November 1912.
14. Diary "I", 23 October 1912.

November he mentioned in a letter to Lorentz that he was "working uninterruptedly on the horrible inaugural lecture". Ehrenfest added, "I haven't yet reached the state where I see myself giving the inaugural lecture every night in my dreams, but that will come!"[15] He had been doing the actual writing for several weeks, but suddenly on the eve of the great day he was plunged into despair over his manuscript. He cut away much of his text and changed what remained. Tanya was with him, trying to set his mind at rest. She finally went to bed at half past twelve, but Paul went on writing until two o'clock. It was long before he could fall asleep at all, and by five thirty he was wide awake again. He worked on his manuscript until ten, when it was finally ready. Then Tanya accompanied him as he went out for a last minute visit to the barber, and a look at the auditorium where he was to speak, to check that the lectern was where he wanted it. Now it was time to get into his formal clothing, which he did very slowly as if he were preparing to march against an enemy battery. Some final revisions of his text with Tanya's help, a bite of lunch, his top hat in place, and he was ready to be picked up by his colleagues.

The first part of Ehrenfest's installation as professor took place at the main University building, the Academy, on the Rapenburg. He passed through the little "Sweating-out room", where generations of students have waited for their oral exams and then waited again to hear the results. The walls are covered with graffiti that commemorate these occasions. One can imagine Ehrenfest's thoughts as he went by. His scientific colleagues were already waiting for him in one of the faculty rooms. They proceeded to the Faculty Senate Chamber, leaving him briefly to stare out the window into the courtyard. Finally he was called in. The President of the Board of Curators administered the formal oath of office, and then spoke briefly. The whole company then departed, to walk the few short blocks to the auditorium for Ehrenfest's lecture.

Lorentz, with his usual tact and human insight, made sure he walked with Ehrenfest, engaging him in conversation about French physicists as they went by the Pieterskerk, distracting him from his anticipations. Tanya was already at the hall, waiting with her aunt, who had come despite a toothache, along with Mrs. Lorentz and other faculty wives. Ehrenfest stopped in an anteroom to put on his academic robe, as did all his colleagues. He was ushered into the auditorium by the university

---

15. P. Ehrenfest to H. A. Lorentz, undated. (November 1912).

proctors, and discovered that the audience was rather small. This cheered him up, and helped dispel his anxiety. While waiting on the podium for the curators to be ushered in, Ehrenfest carefully arranged his glass of water, lumps of sugar, and his manuscript just so, to the sympathetic amusement of Lorentz and Tanya. At last all the curators were seated, the proctors were gone. He was alone with his audience, and suddenly quite sure of himself.

The speech went well, as we have already seen in Chapter 1. Lorentz in particular was very pleased. As soon as it was all over he took Ehrenfest aside and "spoke very, very fondly" to him. At the reception Ehrenfest received congratulations from everyone, including the delegates of the student Corps, which he disliked. In responding to congratulations from faculty wives he made sure to mention that Tanya was his scientific collaborator. Eventually the reception ended. Paul and Tanya were in high spirits, dancing across the little bridge over the Witte Singel on their way home. There, still in his top hat and tails, Paul recounted the whole afternoon's proceedings to little Tanya and her nursemaid, while Galya listened, too young to know quite what had happened.[16]

Paul Ehrenfest missed some things in Leyden. One of the questions he had asked himself on that long train ride from Petersburg was "Will I find a comrade like Joffe in Leyden?"[17] It did not take very long before he decided, regretfully, that the answer was "No."[18] There was no one in Leyden with whom he could share all his ideas, not only on physics but on all possible subjects, none who responded to life in a way that matched his own. No one, in short, about whom he felt as he did about Joffe, "It would be a pity if we had never known each other!!"[19]

He did what he could to form relationships with the younger scientists in Holland. Within a week or so of Ehrenfest's arrival he had organized a physics discussion club, much like the one he had started in Petersburg some years earlier. It met regularly at the homes of the members. During this earliest period in Leyden, Ehrenfest thought he might have found a "Joffe-substitute" in the person of J. A. Vollgraff.[20] Vollgraff, a few years older than Ehrenfest, had taken his doctorate at

16. Diary "I", 3 and 4 December 1912.
17. Diary "H", 8 October 1912.
18. Diary "H", 19 October 1912.
19. P. Ehrenfest to A. F. Joffe, 20 February 1913.
20. Ehrenfest had met Vollgraff when he was at Leyden in the spring of 1903.

Leyden in 1903 with Lorentz. He had been teaching the history of physics and mathematics at the university since a year before Ehrenfest's arrival. The two men hit it off well at first, so well that Ehrenfest included this remark addressed to Vollgraff at the end of his inaugural lecture: "Both of us are already convinced that our common interest in the conceptual foundations of physics, which we approach with a similar attitude, brings us together out of an inner necessity. I attach great importance to the exchange of ideas with you, Herr Vollgraff, and with the other scientific collaborators at this university and at your institute."[21] For one reason or another this attraction based on common scientific interests seems never to have developed into a close friendship. Vollgraff remained in Leyden until 1916, when he left to accept a professorship of mathematics at Ghent. He later directed the publication of Huygens' *Works*, but he did not play any further part in Ehrenfest's life.

Among the other members of Ehrenfest's circle of young Leyden scientists were H. J. Backer, the chemist, who later became professor at Groningen, and two of Lorentz's students, J. Droste and A. D. Fokker. Droste, whose interests were divided between physics and mathematics, helped Ehrenfest with his Dutch. The two met regularly for a while, and Ehrenfest would practice by discussing some current work in physics in his new language.[22] It did not take very long before he could speak Dutch freely, but his Dutch was never grammatically correct, or accentless. German words or expressions, or even sentences, often appeared in his Dutch lectures. His students would sometimes joke about Ehrenfest's mixed language or his pronunciation,[23] but they never had any doubt about his meaning, no matter what language he spoke. Ehrenfest never felt at home writing anything but German. A number of his letters to Dutch friends start in Dutch, but they often turn into German very quickly as he warms to his writing.

Fokker, who later succeeded Lorentz in Haarlem and also became an adjunct professor at Leyden, was already working on his dissertation problem. This was a study of the equilibrium between free electrons and electromagnetic radiation, a singularly perplexing problem which was very close to Ehrenfest's own interests. A tall and impressive young man, Fokker had been an active member of Leyden's Student Corps. Ehrenfest did not hesitate to express his hearty dis-

---

21. See *Papers*, p. 326.
22. Diary "I", 25 and 29 October 1912.
23. Personal communication from Professors D. J. Struik and A. J. Rutgers.

like for this elite group and what it stood for, and Fokker later gave Ehrenfest credit for freeing him from this narrow tradition. Ehrenfest's arrival in Leyden was "like a bolt of lightning" to Fokker.[24]

The new Dutch acquaintances were no substitute for old friends, and the Ehrenfests were delighted to have Gustav Herglotz come for a ten day visit at Christmas that first year in Leyden. There was plenty of time to catch up with what had been going on in their separate lives during the past few years. Ehrenfest proudly showed Herglotz the sights of his new country, and even more proudly brought him to visit Lorentz in Haarlem. By the time Herglotz left he had met practically all of Ehrenfest's new circle, been to Amsterdam, the Hague, Delft, and the seaside towns, Katwijk and Noordwijk, and heard about Ehrenfest's ideas on everything from the new reading room for students to the current status of a variety of research problems.[25]

In the spring of 1913 the Academy of Sciences at Göttingen organized a week-long series of lectures and discussions on "The Kinetic Theory of Matter and Electricity".[26] This was to be a meeting on a rather larger scale than the first Solvay Conference held in 1911, which had dealt with closely related subject matter. The Solvay Conference had involved only a small, especially invited group of outstanding European physicists. The Göttingen meeting, held under the auspices of the Wolfskehl Foundation, consisted of invited lectures by Max Planck and Walther Nernst of Berlin, Arnold Sommerfeld of Munich, Marian von Smoluchowski of Lemberg, and two representatives from the Netherlands – H. A. Lorentz and Peter Debye. In addition to the lecturers there were some invited listeners, presumably those who were particularly wanted for the discussions. The meetings were, however, in contrast to the Solvay Conference, to be open to all who were interested. The subjects to be discussed included the current status of statistical mechanics and various aspects of the quantum theory, on both of which Paul Ehrenfest had recently published important work. He might well have expected an invitation, if not as a lecturer, then at least as an "official listener", but none came. He commented on this situation in a letter, written in February 1913, to

24. Personal communication from Professor A. D. Fokker.
25. Diary "I", 25 December 1912–3 January 1913.
26. *Vorträge über die kinetische Theorie der Materie und der Elektrizität*, M. Planck, P. Debye, W. Nernst, M. v. Smoluchowski, A. Sommerfeld, and H. A. Lorentz, (Leipzig: B. G. Teubner, 1914).

Joffe, who was coming to Göttingen from Petersburg for the meetings.[27]

"Without you I would not have traveled to Göttingen. You can easily imagine how endlessly I am going to bewail my fate to you during those days in Göttingen. But it really vexes me very much that I wasn't even invited as a listener. That I wasn't invited as a speaker, I consider fair, (if also bitter to have to admit it to myself), but they really could have invited me as a listener. Well – you will be there – you'll help me through it. Naturally, I have to go in order to learn, or else I will fall infinitely far behind."

Ehrenfest arrived at Göttingen on Saturday, April 19th, and was up until 2 the next morning talking physics with some of the others who had come early. Before Joffe's arrival the next day Ehrenfest had a chance to call on Felix Klein, to whom he owed so much, and to walk by the house in which he and Tanya had lived some seven years earlier. On Monday morning the lectures began. The crowd around the entrance to the lecture hall reminded Ehrenfest of a swarm of bees, although it could not have been terribly large. Ehrenfest took part in some of the formal discussions after the lectures and was in almost constant informal discussion with other physicists at meals, in coffee houses, and walking on the streets of Göttingen. It was a very warm April week, and he could not help looking out the window of the lecture hall during one unexciting lecture, noticing the fresh green of the trees and listening to the birdsong.[28]

In the course of the week Ehrenfest met a number of physicists he had not known before, and improved his acquaintance with others. Among the near contemporaries he met were Max Born and Theodor von Karman who were then working together on an elaborate and rigorous theory of the specific heats of crystals. There were also Richard Courant and Hermann Weyl, the mathematicians, Ludwig Hopf, who had recently collaborated with Einstein, and many others. One of the older physicists Ehrenfest came to know better was Max Planck, with whom he talked often, and also, on several occasions, played Brahms.

Ehrenfest spent one evening walking with Frederick Lindemann, and talking with him at length about his future development as a physicist. They made a strange pair. Lindemann[29] was a young English physicist

27. P. Ehrenfest to A. F. Joffe, 20 February 1913.

28. Diary "I", 18–26 April 1913.

29. F. W. F. Smith, Earl of Birkenhead, *The Professor and the Prime Minister. The Official Life of Professor F. A. Lindemann, Viscount Cherwell.* (Boston: Houghton Mifflin Company, 1962).

who had been working with Nernst in Berlin for several years, and was about to leave for Chicago to give a course of lectures at the University. He was the son of a very wealthy Alsatian engineer and amateur astronomer who had settled in England as a young man. Lindemann was accustomed to wealth and luxury: he lived at a fashionable Berlin hotel while he was a graduate student, and played championship tennis at the best clubs. But he also did outstandingly good physics, and several of his papers on low temperature physics were much discussed at that time. Lindemann and Ehrenfest were in sharp contrast in their backgrounds, their appearances, (Lindemann was tall and thin), and in their whole life styles. Lindemann would spend the years of the First World War working at the Royal Aircraft Establishment at Farnborough, confirming his theory of how the deadly spin of an airplane could be avoided by becoming a test pilot himself, and go on from that to an Oxford professorship. Later on, during the Second World War, he was Winston Churchill's trusted adviser and became the redoubtable Lord Cherwell. But in April 1913 he was still a somewhat diffident young man, happy to hear what Paul Ehrenfest had to say about his future as a physicist.[30]

Ehrenfest returned to Leyden with Joffe, who stayed on for a week and was shown much the same round of sights and visits as Herglotz, a few months earlier. Joffe had to return to Petersburg to finish his second, Russian, doctorate without which his prospects for an academic appointment were slight.[31]

The Ehrenfests had not been in their rented house for more than a month before Tanya began drawing up sketches for a new home of their own.[32] They had a small capital sum, whose income had supported them during the Russian years. Some of it came from Tanya; the remainder was Paul's inheritance from the old grocery business in the Favoriten district. They had decided to put it into a house that would meet their needs and suit their tastes. Tanya took on the task of preparing the plans, and the house was basically her design, although a professional architect eventually worked with her. Paul and Tanya were quite satisfied with the neighborhood they were living in, and located a suitable building site just one street away from their present

30. Diary "I", 22 April 1913.
31. P. Ehrenfest to A. F. Joffe, 22 May 1913.
32. Diary "I", 16 December 1913.

location on Groenhovenstraat. The new home would be built at the end of Witterozenstraat and its garden would overlook a small canal.

Construction started at the end of the summer of 1913. (Tanya's mother was visiting in Leyden that summer. Her aunt had returned to Russia in the spring.)[33] By fall Ehrenfest noted that the house was "growing like a mushroom", but it was not until July 1914 that the Ehrenfests moved into their own home.[34]

The house that Tanya had designed was a most unusual addition to the urban landscape of Leyden. It differed from the rest of Witterozenstraat as much as the Ehrenfests and their ways differed from their Dutch neighbors. Large and spacious, set apart in its garden, decorated in a beige-yellow stucco, the house stood out on a street consisting largely of typical small, red brick, Dutch row houses. Furthermore, it did not present its front to the street. The principal room on the main floor was not a living room in any ordinary sense, but it did suit the living habits of the Ehrenfests. It was a long room, running across the width of the house, fitted out as a study, library, and work-room. Perhaps it is best to describe it in the words of Jan Burgers, one of Ehrenfest's first students.[35]

"How much was I impressed by his study, a large room with three windows in one of the walls, looking out on a part of the garden, and a large couch at the other wall, where I have been sitting so very often. On the shorter wall there was a large blackboard, a strange form of decoration even to me; (it was in this room that also the colloquium met in those years.) In the bookcases I saw many books, amongst them some Russian books, and I remember Hilton's 'Mathematical Crystallography'. There were pictures between the windows at the wall: Boltzmann, who had been Ehrenfest's teacher in Vienna, Maxwell, Walther Ritz with whom Ehrenfest had had a close friendship, but who had died young; Tolstoy, Dostoyevsky (of whom I had never heard). And one or two views from the surroundings of St. Petersburg, from which Ehrenfest had pleasant recollections."

Ehrenfest's "study" was not the work-room of the lonely scholar. He worked best in collaboration, at the blackboard, arguing and discussing, and his study was the site of much of this work. "To me," Burgers goes on, "this room has been more impressive than the study

33. Diary "I", 24 October 1913.
34. Diary for 1914, entry of 26 July 1914.
35. J. M Burgers, *Autobiographical Notes* (Unpublished), pp. 27–28. (American Institute of Physics Collection, New York City).

rooms of any other scientist by whom I have been received. It had a simplicity and austerity, and for me it has a grandeur, since it speaks of the many chapters of modern physics which have been discussed in it, with Ehrenfest's colleagues [including his wife!] as well as with the many, many visitors who have come to see him. The room has spoken to me the stronger since I was received there as a close friend, almost as a close relative, while also my best friends were at home there."

The open quality of Ehrenfest's home struck all his students. Naturally, Tanya played an important part in this. "Soon I was also received in the dining room, which so often had full sunshine coming from the garden. Mrs. Ehrenfest had a very pleasant way of treating us, and of talking; she was almost as inquisitive as Ehrenfest was himself, and she often engaged us for work in the garden of which she was very fond." And Burgers concluded, "All this deepened the impression I had from Ehrenfest's study room; I have found myself almost at home there as it had been in my father's house."[36]

Life at the Ehrenfests' had some peculiar features. We have already seen Ehrenfest apologizing to Lorentz in advance for the food he might be served. This culinary situation could not all be blamed on the fact that they had just moved in. Meals were never elaborate, often irregular, and furthermore, the Ehrenfests were vegetarians. Alcoholic drinks were never served. A Dutch colleague who wanted a glass of *jenever* (gin) or beer would have to get it elsewhere. The Ehrenfests did not smoke and did not permit smoking in their home, although guests who stayed with them were allowed to smoke upstairs in the privacy of their own rooms.

And there were often guests! The upstairs guest room saw a remarkable succession of visitors over the years. Edmond Bauer, a friend of the Ehrenfests since 1905, described their novel way of keeping a "guest register".[37] "In Leyden, on a white wall behind a hanging, in a house belonging to two physicists, one can read the signatures, inscribed in pencil, of all the scientists who stayed there during a period of twenty years, who discussed and dug into all the problems of modern physics there with the owner of the house and his students. Englishmen, Scandinavians, Frenchmen, Russians, Germans, Americans,

36. The Burgers home in Arnhem was itself a curious museum, as described by Burgers, *op. cit.*

37. E. Bauer, *L'Électromagnétisme hier et aujourd'hui* (Paris: Albin Michel, 1949), p. 324.

Italians,–this assembly of names, the most illustrious of the time, are a moving symbol of that great scientific family which, never dying, always re-forms after each war." Bauer did not exaggerate: fifteen Nobel Laureates were among those who signed the wall in the Ehrenfests' guest room over the years.

Paul Ehrenfest himself had mixed feelings about the house, as he did about so many things. He was uncertain about his future in Leyden, still feeling that he must prove to himself that he could do what he considered justice to Lorentz's professorship. As he wrote to Joffe, "My freedom is seriously endangered. If the time should come when I ought to give up my position to someone else, then this house will tie me down more than is good for me. I now think about it this way: in any case I will only find the moral strength to give up my professorship here, if I can find *some* other appointment (e.g. perhaps still in Russia). But in that case our means would certainly suffice, even with the possible large losses we would suffer on selling the house."[38] Ehrenfest never lost the feeling that someone worthier than he ought to occupy the chair at Leyden.

Jan Burgers, whose impressions of the Ehrenfest home have already been quoted, came to Leyden to study physics on October 1, 1914.[39] He was the first to receive a doctorate under Ehrenfest's sponsorship, and therefore also the first to experience the intellectual and emotional ties that bound Ehrenfest and his students. Burgers was only nineteen when he came to Leyden from his home at Arnhem. Until that time his father had been the great influence in Burgers' life. Burgers senior was a clerk at the railway post office in Arnhem, a self-educated man with an absorbing interest in science of all kinds. He gave lectures on scientific subjects to family, friends and other interested people in the area, and his interest and openness of spirit were communicated to his sons, Jan and Willy (younger by two and a half years). Both sons studied at the H.B.S. (*Hogere Burger School*), the secondary school that emphasized science, mathematics, and modern languages, but did not, at that time, prepare its students for the university. After graduation, Burgers studied the necessary Latin and Greek at home for two years with the help of a classics teacher from the Gymnasium, and continued his science and mathematics on his own during this time

38. P. Ehrenfest to A. F. Joffe, undated.
39. J. M. Burgers, *op. cit.*, and personal communication.

from books and university lecture notes made available to him by his physics and mathematics teachers at the H.B.S.

When he came to Leyden, Burgers already knew the names of Lorentz and Kamerlingh Onnes, and he began to attend Lorentz's Monday morning lectures, which dealt with physical optics that year. On the first Monday, October 5th, Burgers arrived at the lecture room early, not by the appropriate side entrance, which he did not yet know, but by way of the Low Temperature Laboratory with its corridors full of pumping machinery and elaborate pipe lines, to await the lecturer. He noticed that a short dark man accompanied Lorentz to the lecture. This was his first glimpse of Ehrenfest. Burgers, in turn, was noticed by Ehrenfest. The audience at Lorentz's lectures was small enough so that new faces were quite evident, and it was rather unusual to find a first year student in the group. Later on, this same new student was also often to be found studying in "Bosscha", the reading room for students of the physical sciences set up by Ehrenfest. (It was actually located in the same room that was used for lectures.)

Ehrenfest made it his business to get to know young Burgers, invited him to a special lecture he was giving, and then asked him to join the regular Wednesday evening colloquium at his own home. (This was a special distinction for a brand-new student. But once one was invited to join the colloquium, Ehrenfest expected regular attendance. He kept a record of who came, a practice he had started with his informal colloquia in Petersburg, as we have seen.[40]) Ehrenfest also saw to it that Burgers met the other students with similar interests – Dirk Struik, Dirk Coster, Hans Kramers, and the van Leeuwen sisters. Burgers became a regular member of the Christiaan Huygens society for scientific students, the society which owed its revival to Ehrenfest.

Since Burgers was obviously both gifted and well prepared in physics and mathematics, Ehrenfest urged him not to drag out his studies, as so many Dutch students did.[41] On Ehrenfest's prompting, Burgers successfully took his *Candidaatsexamen* (roughly equivalent to an American B. S. degree) in May, 1915, less than a year after his arrival at the university. This was made possible by the background in physics and mathematics which Burgers had acquired during the two years he studied Latin and Greek at home.

---

40. See Chapter 5.

41. It was common for students to take the *Candidaatsexamen* after four years at the university and the *Doctoraalexamen* after three more years, and only then to begin a dissertation. H. A. Lorentz to P. Ehrenfest, 7 July 1912.

It goes without saying that Burgers took Ehrenfest's courses. Ehrenfest gave a regular sequence consisting of electromagnetic theory and statistical mechanics (including what was known about the quantum theory), in alternate years. There were also special courses on various chapters of classical and modern physics. In 1914, for example, Ehrenfest announced a course on theoretical mechanics, which included a discussion of the problem of small oscillations and normal coordinates.[42] Ehrenfest took this as point of departure for an introduction to the theory of integral equations, stressing the analogy between the integral equation and a finite system of linear algebraic equations. Ehrenfest had learned this method from Hilbert, but it was far from common knowledge at that time; "Hilbert space" had not yet become fashionable.

Ehrenfest always tried to get his students actively involved in physics as soon as possible. During the Christmas holiday of Burgers' first year, Ehrenfest wrote him a note with some questions about crystallography, a subject in which Burgers had already developed an interest and done some reading.[43] It was a typical informal Ehrenfest note, ending with a greeting to Jan's parents, whom he had not yet met but had probably heard about. It must have been quite an occasion for a first year student to get such a note, addressing him straightforwardly as a collaborator in scientific inquiry. At other times Ehrenfest would ask Burgers or one of the other students to give a lecture or two in one of his courses, carefully explaining what he wanted covered and how. Thus, in January 1916 he wrote to ask Burgers to talk about elliptic coordinates in three dimensions, "since you have surely thought these things over carefully already." He asked Burgers to treat all the degenerate cases carefully, to stick to a standard notation, and to put "brutally schematic sketches on the blackboard!!"[44]

In 1916 and 1917 Burgers was extending Ehrenfest's earlier work on the adiabatic principle to include the new developments in the quantum theory made by Sommerfeld and his disciples. (This work will be discussed later on.) During the summer of 1917, when Burgers went home to Arnhem for a vacation, Ehrenfest wrote a note to his parents, urging them to be sure "that Jan does absolutely no calculating or

---

42. J. M. Burgers, *op. cit.*, p. 29. Ehrenfest's lecture notes are to be found in the Ehrenfest Collection, along with some notes taken by students. Notes on this course by D. J. Struik are available on microfilm in the archive, Sources for History of Quantum Physics, at the American Philosophical Society, Philadelphia.

43. P. Ehrenfest to J. M. Burgers, 17 December 1914.

44. P. Ehrenfest to J. M. Burgers, 11 January 1916.

theorizing in this whole time," although he could do anything else as much as he liked, except "not too much reading."[45] At the same time Ehrenfest wrote to Burgers himself, criticizing .... his handwriting. It had been getting more legible, (no doubt after Ehrenfest's complaints), but it was now "threatening to pass through a maximum of clarity" and decline, because of some new twists given to the letters. "Excuse me for this meddling, but look: I hope to read a lot more of your handwriting in my lifetime and so I am strongly interested!!"

This half-joking concern with Burgers' penmanship does indicate how Ehrenfest involved himself in his student's lives. He offered much and demanded much. A student had to meet him on his own ground, at least to the extent of feeling fully "possessed" by his work. Sometimes Ehrenfest felt this to be a serious limitation in his ability as a research director. He once wrote to Burgers, by then a professor himself, about another student who was giving him a great deal of concern.[46] "You know from your own experience that I can only help someone by drawing his attention to problems and then providing criticism. The other person must show initiative, and in a sensible direction." He had suggested a problem to this student, but "he absolutely doesn't bite into it. Since I don't know how to help him, his presence here is a martyrdom to me." But a student like Burgers saw Ehrenfest's role differently. "Ehrenfest taught us how to read scientific papers, to look for the assumptions made by the authors, and to hunt them out when they were not given explicitly. His powerful analytical mind opened our eyes to many subtleties in physical theory. He always strove to find interpretations of new thoughts, and had striking ways for the illustration of their peculiarities .... He encompassed and taught theoretical physics as a whole ...."[47]

Ehrenfest's students were brought into direct contact with many things besides physics. He did not compartmentalize his life. Burgers remembered Ehrenfest playing his favorite Bach fugues and Beethoven bagatelles, as well as he remembered his ideas on the quantum theory. One May morning Ehrenfest took his students after class to the garden of the Leyden Ethnographic Museum where there was "a set of five wonderful statues of the Buddha, standing under a magnolia tree in full flower, with bright sunshine over everything."[48]

Many of his students could echo Burgers' words: "No longer was it

45. P. Ehrenfest to J. M. Burgers, 15 August 1917.
46. P. Ehrenfest to J. M. Burgers, 26 December 1919.
47. J. M. Burgers, *op. cit.*, p. 29.
48. *Ibid.*, p. 30.

my father's maxims which took the first place in my thinking; Ehrenfest's influence became the stronger one."

When Jan Burgers' younger brother, Willy, came to Leyden to study chemistry, Ehrenfest saw to it that he, too, took an active part in the scientific life as soon as possible. On one occasion Ehrenfest suggested rooms for the Burgers brothers, also offering them a study room in his own home. Burgers met his future wife at the meetings of the Huygens society, and of course Ehrenfest knew her too.

There came a time when Burgers had to fight to free himself of Ehrenfest's influence, which had become oppressive to him. Some of Ehrenfest's other students would also feel this way – (one of them, Dirk Coster, once felt him to a "vampire")[49] – but others never felt the need to break free in the same way. Ehrenfest knew that he put too much of himself into his students' lives, though he could not control this tendency. He wrote about it to Burgers, just after Burgers had taken his degree and gone to Delft to become professor of hydrodynamics at the Institute of Technology there.[50]

"Listen, Jan – you mustn't let yourself be hurt by quarreling with me. I feel that I am now *rapidly* losing all contact with the younger people and am growing old – which is cause and which is effect I do not know. Old, anxious, tired – and, more than anything else, because I am always being reproached for having hurt younger people by my uninvited interference. (The evening before last Coster made this clear to me once more – note: not spitefully, but rather with a great deal of warmth.) Now I should be quite willing to stay quietly in my corner, 'ready for particulars', if only the older students would see to it that everything remained lively. But everything is out of joint. How about the mathematics colloquium and the physics colloquium? And will the Huygens club be ruined again? (Self criticism: mothers-in-law, grown old, always believe that their children and grandchildren would absolutely go to pieces if they didn't put their noses into everything. Since mothers-in-law who have grown old have to think this way and can't do anything different, I believe it is far better for them to be dead than to grow still older). To freeze, dry up, grow stupid and end in petty hatred."

Paul Ehrenfest did not stay in such bitter moods for long. A few weeks later he was writing to Burgers in a very different vein, about

---

49. Personal communication from Professor D. J. Struik.
50. P. Ehrenfest to J. M. Burgers, 12 December 1918.

some photographic plates located at the Teyler Foundation in Haarlem.[51] "And which of us is to get the plates from Haarlem? It would be as noble as helium of *you* to do it, especially since I am almost twice as old as you, and furthermore you know the catacombs of the Teyler Foundation better than I do – and, in general! N.B. It would be interesting if an electron occasionally fell into the atomic nucleus 1. Energy. 2. Displacement of the atom to the 'left' in the periodic table. Now imagine some time what an incredibly noble gas would be formed this way from hydrogen. And *that's* how noble it would be if you would get the plates."

When Paul Ehrenfest wrote to Lorentz, early in October 1912,[52] to accept the Leyden professorship, his "heart full of rejoicing", he mentioned his fears of making "very bad blunders" at first, due to his "unfortunately all too lively temperament". He asked for Lorentz's guidance and forebearance. No doubt Ehrenfest's ways of going about things caused a good many raised eyebrows in Leyden. His conversation, which would range quickly over a wide variety of subjects, treating none of them in a conventional way, left his colleagues "tired". He was "different," and he would remain so no matter how long he lived in Holland. But the difference in style and temperament between Ehrenfest and his Dutch colleagues by no means precluded mutual respect and admiration.

Not long after Ehrenfest's arrival Lorentz remarked about him to his Amsterdam colleague, the younger van der Waals:[53] "He has already managed, in this short time, to accomplish what I tried in vain to do during all the years of my professorship. He has gotten the students talking." Lorentz appreciated Ehrenfest's efforts to set his students on their own feet, to convince them that there was still physics to be done and that they were the men to do it.

Lorentz himself was so utterly different from Ehrenfest. He had lived happily in Arnhem and then in provincial Leyden, completely without personal contact with foreign scientists, never even attending the scientific meetings of his neighboring countries until he was forty years old. His daughter recalled some occasion in the early 90's when a visitor arrived in Leyden, one described as "looking like a foreign

51. P. Ehrenfest to J. M. Burgers, 2 January 1919.
52. P. Ehrenfest to H. A. Lorentz, 5 October 1912.
53. J. D. van der Waals, Jr., "In Memoriam Prof. Dr. P. Ehrenfest", De Gids, (1933), No. 4, p. 1.

professor" – (!!) – and her father's only reaction was a heartfelt "I hope he will not turn out to be a physicist."[54] (The young Lorentz who never even tried to meet his contemporaries – Maxwell, Hertz, Fitzgerald, Helmholtz – had later come out of his isolation and very rapidly assumed an uncontested position as "the leading spirit among theoretical physicists of all nations", to quote Einstein.[55]) Yet Lorentz could see and sympathize with the needs of Paul Ehrenfest, for whom doing physics meant lively discussion rather than Lorentz's own quiet contemplation.

Ehrenfest constantly turned to Lorentz for support and advice in his efforts to get physics moving in Leyden. One can easily imagine Lorentz being a little amused and also moved by the whole-hearted and vigorous way in which Ehrenfest advocated the cause of his Dutch students. In June 1914, for example, Lorentz received these heated words in a letter from Ehrenfest.[56]

> "I hear (accidentally!!) that not only will we be unable to get hold of a stipend for Struik, but we will also lose the stipend necessary for Buning. Professor Lorentz! That is really *completely* impossible!! Buning absolutely must receive a stipend—he is a *particularly* clear and energetic worker, with clearly expressed initiative, and besides that an uncommonly fine person. I am sure that if you were to see him you too would have to value him highly. And I am also considering the following point in this connection: next year there will also be a physicist (Buning is a chemist) who will be finished with his *Candidaats* examination, whom Kuenen and I think highly of: Coster. He is a former teacher who is living on his savings which, so far as I know, will just last him through the year. At my instigation he has already passed three of his preliminary examinations, and done them very well, and this during his *first* year as a student; he will do the *Candidaats* exam *very* soon. But then he has to be able to work in peace; (he is *very* industrious and works with a warm interest in physics). And Struik also seems to deserve our attention (I don't know him personally.). So the faculty has to arouse itself and be more forceful. You understand, Professor Lorentz, that I express myself about these things more frankly only to you and De Sitter

54. G. L. de Haas-Lorentz, *op. cit.*, p. 89.
55. *Ibid.*, p. 5.
56. P. Ehrenfest to H. A. Lorentz, 17 June 1914.

because I am afraid of the others, but I must really say that this business is really being settled in an unusually bureaucratic way, *that is, if I see it all correctly*.

Naturally people can say to me: don't constantly put your nose into everything possible. But, really and truly, I don't do it presumptuously. So often, though, when I look into why it is that one or another *evidently* talented student doesn't accomplish as much as one would obviously expect from him, I immediately run into the same few typical causes: civil service examinations, student commuting, no communication with fellow students in the same field, (the latter is now rapidly improving and bearing good fruit among the youngest students – the Huygens Club is now a very lively society for the mathematics, physics, chemistry and astronomy students), the examination system, and economic pressure (whose very worst consequence is student commuting).

Maybe I am completely wrong, maybe everything is really splendid, or if not the causes lie somewhere else than where I seem to see them. (Because I naturally form my opinions of everything only by making analogies with Vienna, Petersburg, Göttingen.) And I don't forget for an instant that one can say to me: Just you do your own work well and deeply and you will see that, without any organizational preparations, the young people all around you also will be working.

Well, maybe I am all wrong, but I ask you to believe me that I only feel compelled to poke my nose into something when I *believe* I clearly see that the thing concerned is closely connected with the development of the students close to me.

And the question of stipends, at all events, belongs to that category; and if nothing more is done than what I am able to see, I really believe that one has to set about things much more energetically. I mean not so abstractly, bureaucratically, but rather concretely: the people who have the decisions in their hands must be made to see the situations and the people involved so clearly, that they can form their opinions on the realities and not purely formally. I hope this demand is not too completely anarchistic. (What I would really like to say is this: so clearly that it begins to trouble them with even one ten thousandth of the intensity with which their 'own' affairs trouble them. But I wouldn't venture to say something so barbaric and uncivilized.)

Please excuse me for continually creating a disturbance."

Ehrenfest was often apologizing to Lorentz, especially in his early years at Leyden – for the "very bad and nonuniform" way that he gave the oral exams in his courses, or for his behavior and tone at one occasion or another.[57] But the most frequent and most deeply felt apologies were for what Ehrenfest considered the inadequacy of his research – "that since last summer I haven't had or followed up a single idea."[58]

At the same time, however, Ehrenfest took great pleasure in sharing his ideas and feelings with Lorentz. He wanted Lorentz to see his new house, while the weather was still good and the house was "still surrounded with flowers and greenery, and flooded by sunlight!!!!!"[59] He described to Lorentz his excitement at seeing one of Kamerlingh Onnes' experiments on superconductivity in which a current, once established, continued to flow on its own for over 24 hours.[60] "It is uncanny to see the influence of these 'permanent' currents on a magnetic needle. You can feel almost tangibly how the ring of electrons in the wire turns around, around, around – slowly, and almost without friction." And, of course, there is a steady stream of reports to Lorentz on whatever Ehrenfest found especially interesting in his own work on theoretical physics, and that of his students and friends.

Ehrenfest knew very well that his own restless striving contrasted sharply with Lorentz's serenity. And for all his lack of self assurance and his tangled feelings about being Lorentz's successor, he was confident of having Lorentz's understanding, sympathy and support. He could write to him, at the end of a long letter in 1919:[61] "Dear, dear Professor Lorentz, we younger men really don't have an easy time of it. We *cannot* simply perform our current duties, uniformly, calmly, patiently. To accomplish great and beautiful things – how rarely that succeeds, and in any case, it is reserved for very few. So what remains for us is only restless, self-tormenting searching about, and somewhere in a corner of our soul the hope of eventually being able to redeem ourselves by self-sacrifice."

In one of the descriptive sketches of Lorentz that Ehrenfest wrote, he discussed Lorentz's way of providing counsel in quite personal

57. P. Ehrenfest to H. A. Lorentz, 10 August 1913.
58. P. Ehrenfest to H. A. Lorentz, 14 March 1913.
59. P. Ehrenfest to H. A. Lorentz, 4 September 1914.
60. P. Ehrenfest to H. A. Lorentz, 11 April 1914.
61. P. Ehrenfest to H. A. Lorentz, 26 February 1919.

matters. Ehrenfest was evidently describing a situation he knew well.[62]

"Lorentz really had to do an extraordinary number of things of many sorts. The only way he could keep on top of them all was by careful apportionment of his time. If you asked to have a detailed conversation with him, whether it was about scientific or about personal matters, he inquired first of all whether there was any hurry about the matter, in order to be able to determine a suitable time. As soon as he perceived that it would really be better if the advice you needed were given soon, he somehow knew how to arrange for a conversation within the next few days or possibly even on the same day, despite the fact that his time was already overburdened. And just especially when a difficult personal matter was involved, you were particularly happy to accept the invitation to come to his home in Haarlem for the conversation. You came then at the appointed time, having already thought over as well as possible precisely what you wanted to ask Lorentz, so as at least not to take up any more of Lorentz's time than was absolutely necessary. Lorentz came downstairs to meet you, brought you in to see his wife, invited you to have a cup of tea, and, in nice contrast to the whole rush to save time with which you had come to the appointment, you now had to sit quietly over another cup of tea, and you had to tell how your wife and children were and talk about all sorts of other things. Not until Lorentz had the impression that you had relaxed from all your haste and tension would he suggest that you go upstairs to his study with him.

The study was not very large, but very comfortable and cheerful. Book shelves covered all the walls. Also a number of separate cases completely full of manuscripts and reprints. A large completely plain table in the middle of the room, almost completely covered, but not untidy. Over the mantel a large photographic portrait of Einstein.

Now Lorentz, with smiling formality, pushes the large comfortable armchair nearer to the table for his guest, sees to it in all other ways that you feel as comfortable as possible, and, while he seats himself in his desk chair, suppresses any too business-like or too tense beginning of the conversation by some joking remark or other, and really achieves what he wants: that you not

62. P. Ehrenfest to A. D. Fokker, 2 February 1929.

only feel comfortable sitting in a soft armchair, but you also feel surrounded by a cheerful atmosphere, without rush, and full of seriousness and at the same time radiantly warm peace.

You begin to talk about your problem, and about the difficulties confronting you. By the remarkably sensitive questions he asks as you talk, Lorentz gradually begins to disentangle, to deepen, and to illuminate the picture. Any over-dramatizing remarks evaporate before his smile. You feel that you are expressing yourself as you have only rarely before in your whole life. Finally you have said everything. And so now at last should come the moment when you ask Lorentz for advice, that advice for which you came to him in the first place. But to your surprise you notice you no longer need to 'ask for advice': by means of his questions this wonderfully wise man has helped us clarify our motives and weigh their importance, and we feel ourselves *assured in our decision.*"

Fig. 7 Two views of the Ehrenfest home on Witterozenstraat, Leyden.

Fig. 8 Invitation to the 25th annual meeting of the Christiaan Huygens Society. Ehrenfest appears as a guide.

Fig. 9 Paul Ehrenfest.

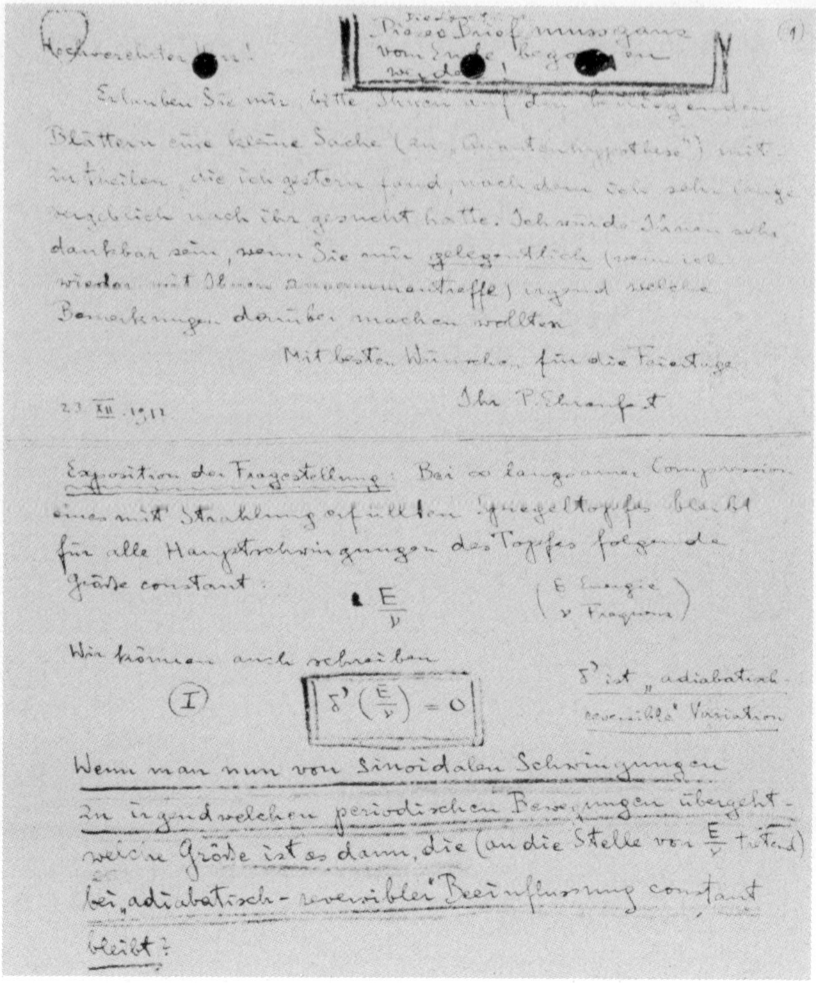

Dieser Brief muss ganz vom Ende begonnen werden!

(1)

Hochverehrter Herr!

Erlauben Sie mir, bitte Ihnen auf den beiliegenden Blättern eine kleine Sache (zu „Quantenhypothese") mitzuteilen, die ich gestern fand, nachdem ich sehr lange vergeblich nach ihr gesucht hatte. Ich würde Ihnen sehr dankbar sein, wenn Sie mir gelegentlich (wenn ich wieder mit Ihnen zusammentreffe) irgend welche Bemerkungen darüber machen wollten

Mit besten Wünschen für die Feiertage

Ihr P. Ehrenfest

23. XII. 1912

Exposition der Fragestellung: Bei so langsamer Compression eines mit Strahlung erfüllten Spiegeltopfes bleibt für alle Hauptschwingungen des Topfes folgende Grösse constant:

$$\frac{E}{\nu} \qquad \begin{pmatrix} E \text{ Energie} \\ \nu \text{ Frequenz} \end{pmatrix}$$

Wir können auch schreiben

(I) $$\boxed{\delta'\left(\frac{E}{\nu}\right) = 0}$$ δ' ist „adiabatisch-reversible" Variation

Wenn man nun von sinoidalen Schwingungen zu irgendwelchen periodischen Bewegungen übergeht – welche Grösse ist es dann, die (an die Stelle von $\frac{E}{\nu}$ tritt und) bei „adiabatisch-reversibler" Beeinflussung constant bleibt?

Fig. 10 Two pages from a letter by Paul Ehrenfest to H. A. Lorentz, 23 December 1912 (p.t.o.)

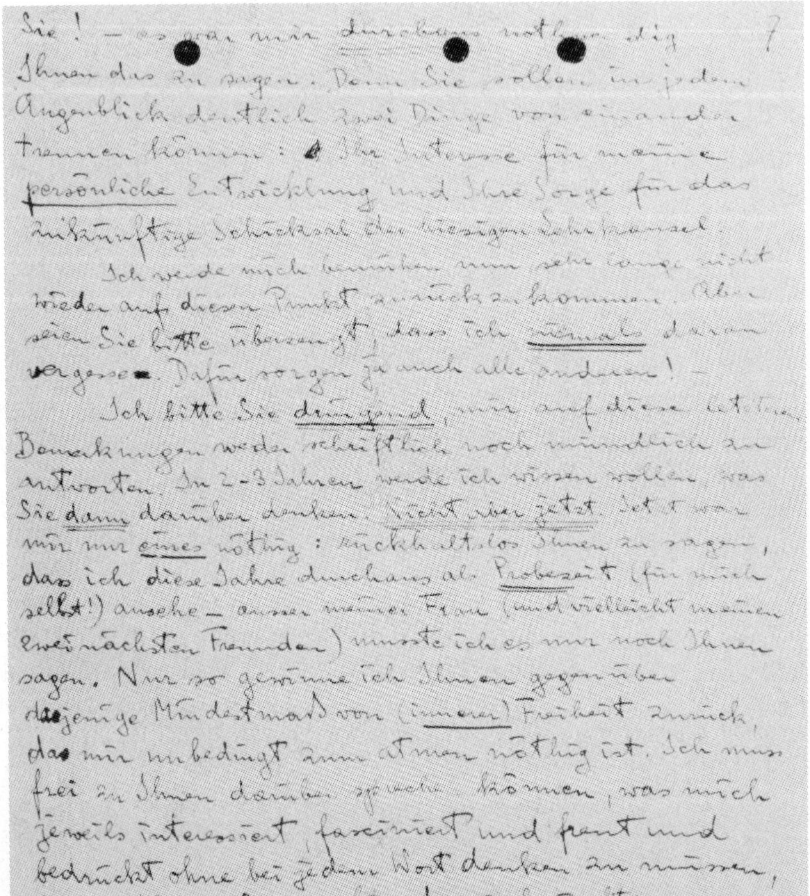

Sie! – es war mir durchaus nothwendig 7
Ihnen das zu sagen. Denn Sie sollen in jedem
Augenblick deutlich zwei Dinge von einander
trennen können: Ihr Interesse für meine
persönliche Entwicklung und Ihre Sorge für das
zukünftige Schicksal der hiesigen Lehrkanzel.
Ich werde mich bemühen um sehr lange nicht
wieder auf diesen Punkt zurückzukommen. Aber
seien Sie bitte überzeugt, dass ich niemals daran
vergessen. Dafür sorgen ja auch alle anderen! –
Ich bitte Sie dringend, mir auf diese letzten
Bemerkungen weder schriftlich noch mündlich zu
antworten. In 2-3 Jahren werde ich wissen wollen was
Sie dann darüber denken. Nicht aber jetzt. Jetzt war
mir nur eines nöthig: rückhaltslos Ihnen zu sagen,
dass ich diese Jahre durchaus als Probezeit (für mich
selbst!) ansehe – ausser meiner Frau (und vielleicht meinen
zwei nächsten Freunden) musste ich es nur noch Ihnen
sagen. Nur so gewinne ich Ihnen gegenüber
dasjenige Mindestmaass von (innerer) Freiheit zurück,
das mir unbedingt zum atmen nöthig ist. Ich muss
frei zu Ihnen darüber sprechen können, was mich
jeweils interessiert, fasciniert und freut und
bedrückt ohne bei jedem Wort denken zu müssen,
dass es Ihnen Sorge für das zukünftige
Schicksal der hiesigen Lehrkanzel bereiten
könnte. Jetzt nachdem ich es Ihnen ein für allemal

CHAPTER 10

The Essential Nature of the Quantum Hypothesis

1. Paul Ehrenfest became a professor at Leyden at just the time that the questions raised by the concept of energy quanta began to be the central concern of theoretical physicists. Over a decade had gone by since Max Planck had introduced this concept, but physicists had been slow in grasping the very serious implications of his work. When Ehrenfest made a passing reference, in his Leyden inaugural lecture, to the "tangled skein of problems" associated with the quantum hypothesis, he knew just how complex these problems were. Few people had done so much to try to clarify the issues, and fewer had had so much success.

Ehrenfest's work on the quantum theory was, like almost all his work, primarily in the critical mode. It was the most productive criticism he ever undertook. After almost ten years of probing into the meaning of Planck's strange new idea of quantization, and the related but even stranger ideas of Albert Einstein, Ehrenfest proposed a generalization of these ideas that proved to have major significance. This generalization, the adiabatic hypothesis, was one of the few principles that could serve as reliable guides to progress in the years before the new quantum mechanics was developed. It was certainly Ehrenfest's most significant positive contribution to physics, but he won his way to it by practicing his characteristic kind of penetrating criticism. Even though these early critical papers had little influence on his contemporaries, the entire development aptly illustrates a remark

Wolfgang Pauli made about Ehrenfest's scientific work as a whole: "It appears to us as a living testimony to the lasting truth: objective criticism in science, no matter how sharp, is always stimulating and fruitful, if only it is thought through consistently and completely."[1]

The material to which Ehrenfest applied his critical methods was the writings of Planck and Einstein. In contrast to the circumstances of his work in statistical mechanics, Ehrenfest's concern with quanta dates almost from the inception of this new idea. He was among the first to undertake the serious study of this work and, if we are to appreciate what Ehrenfest did, we have to go back to the beginning and follow the development of Planck's ideas.

2. Max Planck formed himself as a physicist by his self-study of the writings of Rudolf Clausius, and it was Clausius' thermodynamics that set the pattern of Planck's scientific career.[2] He devoted the first fifteen years or so of that career to clarifying, expounding and applying the second law of thermodynamics. It was the concept of irreversibility and its relationship to the monotonic increase of entropy that especially fascinated Planck. As a disciple of Clausius, Planck looked upon the second law of thermodynamics as having absolute validity; processes in which the total entropy decreased simply did not occur in nature. And so, despite their common concern with entropy and irreversibility, Planck most certainly did not subscribe to Ludwig Boltzmann's reformulation of the second law as a statistical law. He could not accept the view that the increase of entropy is a highly probable rather than an absolutely certain feature of natural processes.

When Planck wrote the Preface to his *Treatise on Thermodynamics*, in April 1897, he characterized his own method of handling the subject as one that "does not advance the mechanical theory of heat, but, keeping aloof from definite assumptions as to its nature, starts direct from a few very general empirical facts, mainly the two fundamental principles of thermodynamics". He admitted that a molecular theory "penetrates deepest into the nature of the processes considered", but thought that "obstacles, at present insurmountable, seem to stand in the way of its further progress". These obstacles were "due not only to the highly complicated mathematical treatment, but principally to essential difficulties .... in the mechanical interpretation of the funda-

1. W. Pauli, "Paul Ehrenfest†", Naturwissenschaften **21** (1933), p. 843.
2. M. Planck, *Scientific Autobiography and Other Papers*, transl. F. Gaynor (New York: Philosophical Library, 1949).

mental principles of thermodynamics." Despite the existence of these obstacles, Planck thought that his own "more inductive treatment" would not be the last word, but might well "have in time to yield to a mechanical, or perhaps an electromagnetic theory", one that would fulfill "our aspiration after a uniform theory of nature".[3]

Planck's reference to a possible electromagnetic foundation for the second law of thermodynamics was quite deliberate. He was really alluding to research of his own which was already in progress. In March 1895 he had presented a paper to the Academy of Sciences at Berlin on the resonant scattering of plane electromagnetic waves by an oscillating dipole of dimensions small compared to the wave length.[4] This study in Hertzian electrodynamics only seemed to represent a shift away from Planck's normal thermodynamic pursuits, because at the end of his paper Planck explained that his real interest was in the problem of black-body radiation. This had been only a preliminary consideration of the scattering process as a mechanism for maintaining the equilibrium state of the radiation in an enclosure. The following February, Planck had reported on an analysis of the radiation damping of his charged oscillators. He was much impressed by the difference between radiation damping and damping by means of the ordinary resistance of the oscillator. Radiation damping was a completely conservative mechanism involving no transformation of energy into heat and involving no new characteristic constant for the oscillator. Planck saw this as a possible way of explaining irreversibility by a conservative mechanism. "The study of conservative damping seems to me to be of great importance, since it opens up the prospect of a possible general explanation of irreversible processes by means of conservative forces – a problem that confronts research in theoretical physics more urgently every day."[5]

A year later, in February 1897, Planck communicated the first of what would become a long series of papers on irreversible phenomena

---

3. M. Planck, *Treatise on Thermodynamics*, transl. A. Ogg. Reprinted (New York: Dover Publications, n. d.) Preface dated April, 1897, p. viii.
4. M. Planck, "Absorption und Emission elektrischer Wellen durch Resonanz", Wied. Ann. **57** (1896), p. 1. Planck's scientific papers have been reprinted in three volumes: M. Planck, *Physikalische Abhandlungen und Vorträge* (Braunschweig: Friedr. Vieweg & Sohn, 1958). They will be cited here as, e.g. *Planck Papers* I, p. 445.
5. M. Planck, "Über elektrische Schwingungen, welche durch Resonanz erregt und durch Strahlung gedämpft werden", Wied. Ann. **60** (1897), p. 577. *Planck Papers* I, p. 466.

in radiation.[6] The extended introduction itself indicated that Planck was planning a major work. He began by asserting that no one had yet explained successfully how a system governed by conservative interactions could proceed irreversibly to a final state of thermodynamic equilibrium. Planck explicitly rejected Boltzmann's *H*-theorem as an unsuccessful attempt in this direction, and cited the criticisms recently raised by his own student, E. Zermelo, against Boltzmann's analysis.[7] He then announced his own program for deriving the second law of thermodynamics for a system consisting of radiation and charged oscillators in an enclosure with reflecting walls. He would introduce no damping other than radiation damping, but would take the basic mechanism for irreversibility to be the alteration in the form of an electromagnetic wave by the scattering process – its apparently irreversible conversion from incident plane wave to outgoing spherical wave. The ultimate goal of this program was the explanation of irreversibility for strictly conservative systems, with no appeal to statistical considerations, and a valuable by-product would be the determination of the spectral distribution of black-body radiation.

Planck had to reconsider his high hopes when Boltzmann[8] promptly pointed out a basic flaw in his proposal: the equations of electrodynamics are invariant under time reversal. The scattering process, which was to have served as the mechanism for irreversibility, does have an inverse process which is an equally legitimate solution of the electrodynamic equations. Electrodynamics alone cannot be the basis for a proof that there is an irreversible approach to equilibrium in a system composed of oscillators and radiation, any more than mechanics alone can be the basis for a proof that a gas approaches equilibrium irreversibly. In both cases the underlying theory (mechanics or electrodynamics) must be supplemented by an appropriate statistical assumption.

Planck finally granted the need for a statistical assumption if he were to carry out the rest of his program, and introduced what he

6. M. Planck, "Über irreversible Strahlungsvorgänge. Erste Mitteilung", Berliner Berichte (1897), p. 57. *Planck Papers* I, p. 493.

7. See above, Chapter 6.

8. L. Boltzmann, "Über irreversible Strahlungsvorgänge", Berliner Berichte (1897), pp. 660, 1016. *Wiss. Abh.* III, pp. 615, 618.
L. Boltzmann, "Über vermeintlich irreversible Strahlungsvorgänge", Berliner Berichte (1898), p. 182. *Wiss. Abh.* III, p. 622.

called the hypothesis of "natural radiation".[9] This was an assumption of randomness, of the absence of correlations between the various Fourier components of the radiation. It was the analogue of Boltzmann's assumption of "molecular chaos" in the theory of gases. With its help Planck was able to reach his goal, although not exactly in the way he had hoped.

Let us look more closely at what Planck was doing.[10] The basic quantity that describes black-body radiation in thermodynamic equilibrium within an enclosure whose walls are kept at temperature T, is the spectral distribution function, $\rho(\nu, T)$. Here $\rho(\nu, T)\mathrm{d}\nu$ is the energy of the radiation, per unit volume, which lies in the frequency interval from ν to $\nu+\mathrm{d}\nu$. It had been shown by Gustav Kirchhoff that the character of the radiation, as expressed by the function $\rho(\nu, T)$, depends only on the temperature and not at all on the nature or properties of those material bodies which are in equilibrium with the radiation. In 1894 Wilhelm Wien[11] had combined thermodynamic methods with the electromagnetic theory of light to show that $\rho(\nu, T)$ must necessarily have the form

$$\rho(\nu, T) = \nu^3 f(\nu/T), \tag{10.1}$$

where $f(\nu/T)$ is a function of the single variable (ν/T). This result, Wien's displacement law, shows that the spectral distribution law is determined for all temperatures once it is known at a single temperature. The displacement law also implies that the total energy per unit volume, the integral of $\rho(\nu, T)$ over all frequencies, must be proportional to T^4, a result already found experimentally by Josef Stefan and then deduced theoretically by Boltzmann.

---

9. M. Planck, "Über irreversible Strahlungsvorgänge. Vierte Mitteilung", Berliner Berichte (1898), p. 449. *Planck Papers* I, p. 532.

10. For a detailed study of Planck's work see the following:

(a) M. J. Klein, "Max Planck and the Beginnings of the Quantum Theory", Archive for History of Exact Sciences **1** (1962), p. 459.

(b) M. J. Klein, "Planck, Entropy, and Quanta, 1901–1906", The Natural Philosopher (New York: Blaisdell) **1** (1963), p. 81.

(c) M. J. Klein, "Thermodynamics and Quanta in Planck's Work", Physics Today, **19**, No. 11 (Nov. 1966), p. 23.

(d) L. Rosenfeld, "La première phase de l'évolution de la Théorie des Quanta", Osiris **2** (1936), p. 149.

(e) M. Jammer, *The Conceptual Development of Quantum Mechanics* (New York: McGraw-Hill, 1966), p. 10.

11. W. Wien, "Temperatur und Entropie der Strahlung", Wied. Ann. **52** (1894), p. 132.

In addition to these theorems about black-body radiation Planck had one other result to guide him, but it was a result of a rather different kind. Wien[12] had proposed an explicit form for the distribution law in 1896, and while Wien's derivation left something to be desired, his distribution did account very well for the available experimental results. Wien's distribution had the form

$$\overset{\text{(Wien)}}{\rho(\nu, T)} = \alpha\nu^3 \exp(-\beta\nu/T), \tag{10.2}$$

where α and β are constants. Planck had every reason to believe that the fundamental theory of the thermodynamics of radiation, which he had set out to create, should culminate in a sound derivation of the Wien distribution, (10.2), as the distribution law for radiation in thermodynamic equilibrium.

Planck's aim was to understand irreversibility in radiation processes. He made use of the latitude granted by Kirchhoff's theorem, and considered the radiation as interacting only with oscillating dipoles. These oscillators were not intended to represent real material systems, but were just the simplest systems one could treat in looking for a result one knew to be independent of the nature of these systems anyway. One fundamental result that Planck obtained[13] was the proportionality of $\rho(\nu, T)$ and the average energy, $E_\nu(T)$, of an oscillator of frequency ν at temperature T. This theorem, derived by equating the rates at which energy is emitted and absorbed by an oscillator at equilibrium, has the form

$$\rho(\nu, T) = (8\pi\nu^2/c^3)\, E_\nu(T) \tag{10.3}$$

where c is the velocity of light.

The two crucial points that remained were first, the proof that radiation processes took place irreversibly, and second, the determination of $E_\nu(T)$ and therefore of the radiation distribution law. Planck solved these together by a method he later called "thermodynamic", although that term hardly suggests what he actually did. In order to demonstrate irreversibility he had to study the behavior of the entropy of his system of oscillators and radiation. The entropy, however, was not

12. W. Wien, "Über die Energievertheilung im Emissionsspectrum eines schwarzen Körpers", Wied. Ann. **58** (1896), p. 662.

13. M. Planck, "Über irreversible Strahlungsvorgänge. Fünfte Mitteilung", Berliner Berichte (1899), p. 440. *Planck Papers* I, p. 560.

given, and Planck first had to determine it. He did this by simply defining it; the entropy S of an oscillator of frequency ν and energy $E_\nu(T)$, (to be written as E for brevity), was defined by the equation

$$S = -(E/\beta\nu)[\ln(E/\alpha'\nu) - 1]. \qquad (10.4)$$

Planck did not arrive at this definition arbitrarily. This expression for S is exactly what one obtains from the Wien distribution, (10.2), if one uses (10.3) and the second law of thermodynamics in the form, $T^{-1} = (\partial S/\partial E)$. The constant α' of (10.4) is then equal to $\alpha c^3/8\pi$, α and β being the constants in (10.2). It seems very likely that Planck's definition of the entropy of an oscillator was obtained by arguing backwards from Wien's law in this way. Using (10.4) and an analogous expression for the entropy of the radiation itself, Planck calculated the time rate of change of the total entropy of the system and showed that it could never be negative, thereby demonstrating irreversibility. He also derived the equilibrium distribution by requiring that the total entropy be stationary at equilibrium with respect to a virtual exchange of energy between oscillators of different frequencies. The result was the Wien distribution, as required by the assumed form for the entropy.

Naturally, Planck realized that his results depended on the particular form he had assigned to the entropy by his definition, (10.4). He argued that this definition was the only one compatible with the second law of thermodynamics: if one defined the entropy in any other way, one could not prove that entropy increases monotonically in time. This seemed to put the Wien distribution on very solid ground. "I believe," Planck wrote, "that it must therefore be concluded that the definition given for the entropy of radiation, and also the Wien distribution law for the energy, which goes with it, is a necessary consequence of applying the principle of entropy increase to the electromagnetic theory of radiation, and that the limits of validity of this law, should there be any, therefore coincide with those of the second law of thermodynamics. Further experimental test of this law naturally acquires all the greater fundamental interest for this reason."[14]

These remarks appeared in the fifth and last paper of the series presented to the Prussian Academy, and they were repeated verbatim in an article Planck submitted to the *Annalen der Physik* on November

14. *Ibid. Planck Papers* I, p. 597.

7, 1899.[15] This article, presenting the results of several years of work, must have seemed like the final stage of Planck's program. He had had to make some important changes along the way, such as admitting that a statistical assumption was essential and invoking the second law as part of his argument, but he had apparently constructed the theory he wanted.

3. By the time Planck corrected the proofs of his *Annalen* paper he had seen the first results of some of the "further experimental tests" of the Wien distribution which he had called for. He felt obliged to add a note to the proofs remarking that deviations from Wien's law were observed. Since Planck had asserted that the Wien distribution and the second law of thermodynamics had the same domain of validity, these deviations had to be taken very seriously indeed. Even though the experimental situation was not yet completely clear, Planck undertook a careful review of his arguments and sent off another paper to the *Annalen* in March 1900.[16] He decided that his earlier claim had been too strong: it was not really necessary to have the Wien distribution in order to satisfy the second law. The quantity $(\partial^2 S/\partial E^2)$, which is simply proportional to the negative reciprocal of E if the Wien distribution holds, need not have this special form. It has only to be negative to insure that the total entropy of the system increases monotonically with time. Nevertheless, Planck was still sure that the Wien distribution was correct, though his new arguments rested only on a plausible assumption on the functional dependence of the entropy change. By October, however, new experimental results at long wavelengths left no room for doubt – the radiation distribution was definitely in disagreement with Wien's distribution law.[17]

Planck realized[18] that the assumption which had seemed so plausible

15. M. Planck, "Über irreversible Strahlungsvorgänge", Ann. Phys. **1** (1900) p. 69. *Planck Papers* I, p. 614.

16. M. Planck, "Entropie und Temperatur strahlender Wärme", Ann. Phys. **1** (1900), p. 719. *Planck Papers* I, p. 668.

17. (a) O. Lummer and E. Pringsheim, "Über die Strahlung des schwarzen Körpers für lange Wellen", Verh. d. Deutsch. Phys. Ges. **2** (1900), p. 163.
(b) H. Rubens and F. Kurlbaum, "Über die Emission langer Wellen durch den schwarzen Körper", Verh. d. Deutsch. Phys. Ges **2** (1900), p. 181. Only the title of Kurlbaum's talk is given, but there is a remark that Planck's work (note 18) was presented during the "ensuing lively discussion".

18. M. Planck, "Über eine Verbesserung der Wienschen Spektralgleichung", Verh. d. Deutsch. Phys. Ges. **2** (1900), p. 202. *Planck Papers* I, p. 687.

had to be given up. He had no theoretical guideline to follow, but the new experimental data indicated that $\rho(\nu, T)$ was proportional to T at low frequencies. Planck found that if he made the quantity $-(\partial^2 S/\partial E^2)^{-1}$ proportional to the sum of a linear and a quadratic term in E, the next simplest form for this expression after Wien's law, he could match this result. His new expression for $\rho(\nu, T)$ had the form

$$\overset{\text{(Planck)}}{\rho(\nu, T)} = \frac{\alpha\nu^3}{\exp(\beta\nu/T) - 1}, \tag{10.5}$$

reducing to the Wien distribution for high frequencies and becoming proportional to T for low frequencies. Planck constructed this expression within a few days after hearing of the experimental results obtained by his Berlin colleagues, H. Rubens and F. Kurlbaum. It successfully accounted for all available data – then, and ever since. But the question raised at once for Planck was what did it mean? As he wrote years later: "But even if the absolutely precise validity of the radiation formula is taken for granted, so long as it had merely the standing of a law disclosed by a lucky intuition, it could not be expected to possess more than a formal significance. For this reason, on the very day when I formulated this law, I began to devote myself to the task of investing it with a true physical meaning."[19]

On December 14, 1900 Planck reported the results of his efforts to the German Physical Society.[20] Only eight weeks had gone by since he had first proposed the new distribution law at a meeting of the same Society, but they were "weeks of the most strenuous work of my life".[19] He had succeeded in finding a meaning for his new law and constructing a derivation of it, but neither he nor anyone else could have foreseen the consequences of his work.

The crucial step needed to complete Planck's earlier work was the determination of the entropy of an oscillator as a function of its energy. Twice already Planck had been misled into thinking he had sound reasons for accepting (10.4) as the form of this relationship. This time he was sure he knew what the entropy-energy relationship had to be.

---

19. M. Planck, Nobel Prize Lecture 1920, "The Origin and Development of the Quantum Theory", in his *A Survey of Physical Theory*, transl. R. Jones and D. H. Williams. Reprinted (New York: Dover Publications, 1960), p. 102. *Planck Papers* III, p. 121. (I have revised the translation).

20. M. Planck, "Zur Theorie des Gesetzes der Energieverteilung im Normalspektrum", Verh. d. Deutsch. Phys. Ges. **2** (1900), p. 237. *Planck Papers* I, p. 698.

If the distribution law had the form (10.5), then from (10.3) and the second law he could conclude that the replacement for (10.4) would have to be the equation

$$S=\frac{\alpha'}{\beta}\left[\left(1+\frac{E}{\alpha'\nu}\right)\ln\left(1+\frac{E}{\alpha'\nu}\right)-\frac{E}{\alpha'\nu}\ln\left(\frac{E}{\alpha'\nu}\right)\right], \tag{10.6}$$

where $\alpha' = \alpha c^3/8\pi$. Since, however, Planck's attempts to derive an entropy-energy equation by electrodynamic and "thermodynamic" arguments had failed to produce satisfactory results, he had to turn to the one available method he had not yet tried – Boltzmann's statistical mechanics.

Planck had gradually been forced to give up some of his opposition to Boltzmann's ideas. As a result of Boltzmann's criticisms he had had to introduce the statistical hypothesis of "natural radiation" into his own work. Now he would have to go much further and make use of Boltzmann's connection between entropy and probability. Planck actually had begun to consider this in the first paper[18] introducing his new distribution, when he referred to the logarithmic dependence of S on E implied by the distribution, saying that such a logarithmic dependence was suggested by the theory of probability. The basic relationship Planck referred to was the equation he proceeded to write in the form

$$S_N = k \ln W \tag{10.7}$$

which relates the entropy S_N of a set of N oscillators of the same frequency whose total energy is E_N, to the quantity W, which is proportional to the probability that the set of oscillators have this value for their total energy. The quantity k is a universal constant known now (but not then) as Boltzmann's constant.

The problem was now to determine W. For this purpose Planck had to do two things. First, Planck assumed that the quantity W could be set equal to the number of ways in which the energy E_N could be shared among the N oscillators. Using Boltzmann's terminology Planck called this the number of complexions, and, by taking this quantity as proportional to the probability, he was assuming that each of these complexions was equally probable. Second, in order to calculate the number of complexions, Planck had to "treat E_N not as a continuous, infinitely divisible quantity but rather as composed of an integral number of equal finite parts". He called these "parts" elements of energy, ϵ, and wrote the equation

$$E_N = P\epsilon \tag{10.8}$$

where P is an integer, a large integer.

The number of complexions W was now fixed as the number of distinct ways the P energy units could be distributed among the N oscillators, and Planck took this number from combinatorial analysis,

$$W = \frac{(P+N-1)!}{P!\,(N-1)!} \cong \frac{(P+N)^{P+N}}{P^P N^N}. \tag{10.9}$$

The second, approximate, form comes from dropping the 1 compared to the large numbers N and P and using Stirling's approximation for the factorials.

From (10.7) one now obtains the entropy of the set of N oscillators by taking logarithms,

$$S_N = k\,[(N+P)\ln(N+P) - P\ln P - N\ln N]. \tag{10.10}$$

If one now makes use of the fact that S_N is just N times the entropy of one oscillator, and that P can be written as NE/ϵ from (10.8), (where E is again the average energy of one oscillator, E_N/N), the final result has the form

$$S = k\left[\left(1+\frac{E}{\epsilon}\right)\ln\left(1+\frac{E}{\epsilon}\right) - \frac{E}{\epsilon}\ln\left(\frac{E}{\epsilon}\right)\right]. \tag{10.11}$$

Up to this point the elements of energy ϵ were completely arbitrary, an artifice for making it possible to calculate W. On the other hand, the frequency of the oscillators had not appeared in the calculation. Planck now invoked the Wien displacement law in a new and particularly simple form – the requirement that S must be a function of (E/ν) only.[21] This meant that ϵ must depend on the frequency of the oscillators in a linear way,

$$\epsilon = h\nu, \tag{10.12}$$

21. It follows from (10.1) and (10.3) that $E = \nu f(\nu/T)$, absorbing the constants. If we invert this equation we obtain $T = \nu F(E/\nu)$. Now, using the second law in the form $T^{-1} = (\partial S/\partial E)$, this can be integrated to obtain $S = g(E/\nu)$.

where h is the second universal constant in the theory; (the first was k). When this last expression for ϵ, the energy unit, was substituted into (10.11), the resulting expression for the entropy had just the form Planck was seeking, that of (10.6). Identifying the constants, he obtained the Planck distribution law in its now standard form

$$\overset{\text{(Planck)}}{\rho}(\nu, T) = \left(\frac{8\pi\nu^2}{c^3}\right)\frac{h\nu}{\exp(h\nu/kT) - 1}. \qquad (10.13)$$

This argument is essentially the one Planck gave in the paper he wrote for the *Annalen* early in January 1901.[22] (He had given only a sketch of his reasoning when he spoke to the German Physical Society a few weeks earlier.) By his references to Boltzmann's 1877 memoir[23] on the relationship between probability and the second law of thermodynamics, as well as by his actual calculations, Planck indicated clearly that he considered his work to be an adaptation of Boltzmann's methods to the radiation problem.

4. Planck had achieved his immediate goal – he had derived a radiation distribution law. But in order to do so he had taken a fateful step by requiring the energy of one of his oscillators always to be a discrete multiple of an energy unit. Reading his papers today, we are struck by how little emphasis he gave to this radical idea. It is true that in his first announcement of the new derivation[20] he referred to his use of the discrete energy units as "the most essential point of the whole calculation". The context of this phrase suggests, however, that he may only have meant that it was essential for calculating a *finite* number of ways of dividing up the energy among the N oscillators. Nowhere else in these papers of 1900 and 1901 did he emphasize, or even discuss, his shockingly revolutionary way of treating energy.

Such a normally cautious and thorough physicist as Planck must have had a reason for taking this work as seriously as he obviously did. It was not enough that it gave him the correct form of the radiation spectrum; there must have been some additional reason for placing his confidence in this new idea of discrete energy units. Planck found this necessary corroborative support in the universal nature of the two constants, k and h, which appeared in his equations.

In 1899, when he thought that his thermodynamic-electrodynamic

22. M. Planck, "Über das Gesetz der Energieverteilung im Normalspektrum", Ann. Phys. **4** (1901), p. 553. *Planck Papers* I, p. 717.
23. See notes 16 and 17 of Chapter 6.

theory gave a rigorous derivation of the Wien distribution, Planck had attached great significance to the two constants α' and β which appear in (10.2) and (10.4). Although he could not give them specific physical interpretations at that time, Planck evaluated their numerical values from the available data on black-body radiation, and emphasized their universal quality. He proposed that these two constants, $\alpha' = 6.885 \times 10^{-27}$ erg sec and $\beta = 4.818 \times 10^{-11}$ sec °K, be made the basis of a new system of physical units, along with c, the velocity of light in vacuum, and the gravitational constant, G. The units of length, time, mass and temperature so constructed would be truly "independent of particular bodies or substances", would "necessarily retain their significance for all times and for all cultures, including extraterrestrial and non-human ones". They would deserve the name of "natural units", and would "retain their natural significance as long as the laws of gravitation and the propagation of light in vacuum, and the two laws of thermodynamics retain their validity".[24]

In his new theory the two constants retained this absolute significance by which Planck set such great store, and one of them, at least, acquired an immediate and impressive physical significance. This was the constant k, the ratio of α' and β, which, as Planck pointed out, also had to be equal to the gas constant R, divided by Avogadro's number N_0, the number of atoms in a gram atomic weight. Planck's determination of k from radiation data gave him an accurate value of Avogadro's number and with it the mass of the individual atom. This was a major achievement. Planck's value of Avogadro's number was far more accurate than any of the then existing indirect estimates based on the kinetic theory of gases. He used it not only to get the mass of the hydrogen atom but also, in combination with the Faraday electrochemical constant, to determine the charge on the recently discovered electron, the natural unit of electric charge. At this stage, however, Planck had no interpretation to offer for his second constant, h.

Planck referred to these very important by-products of his theory in his first report[20], alluding to "other relationships" deduced from the theory "which seem to me to be of considerable importance for other fields of physics and also of chemistry". And he concluded that paper, and also the separate one in the *Annalen*[25] which he devoted exclusively

24. M. Planck, *op. cit.* note 13, *Planck Papers* I, p. 599. The same quotations appear in M. Planck, *op. cit.* note 15, *Planck Papers* I, p. 666.

25. M. Planck, "Über die Elementarquanta der Materie und der Elektrizität", Ann. Phys. **4** (1901), p. 564. *Planck Papers* I, p. 728.

to his determination of the fundamental physical constants, by writing: "All these relationships can lay claim to absolute, not approximate validity, so long as the theory is really correct .... Their test by more direct methods will be a problem for further research, a problem as important as it is necessary."

There was still one more step to be taken before Planck could consider that the radiation problem was even tentatively settled. He had to show that his new entropy function, as given by (10.11), behaved in the proper fashion in irreversible processes. Planck did this in a supplement to his papers on irreversible processes in radiation, which he sent off to the *Annalen* in October 1901.[26] He proved that if the entropy of his oscillators were given by (10.11), and if the entropy of the radiation were defined analogously, then the total entropy of the system could only increase as a result of the scattering processes that take place. This theorem corresponded exactly to what he had proved earlier for the entropy associated with the Wien form of the distribution law. It was also a particular case of his assertion that all one really needed to carry out the proof was an oscillator entropy function for which $(\partial^2 S/\partial E^2) < 0$. Planck accounted for his inability to fix a unique form for the entropy by irreversibility arguments by referring to the fact that he had not treated the most general processes. He thought that a completely general treatment would fix the entropy uniquely, but also that "in the present state of our knowledge this way of determining the expression for the entropy is not practicable yet."

5. Planck's fellow theorists did not rush to explore the significance or the consequences of his work on black-body radiation; they did not even attack it. Those experimentalists who had been making measurements on radiation confirmed the correctness of his distribution law[27], but they showed no interest in the theory underlying it nor, for that matter, did they follow up his new method of determining the fundamental physical constants.

The first theoretical physicist who referred to Planck's work in print was, apparently, H. A. Lorentz, who was also thinking about the problems of radiation at the turn of the century. In a paper on "The

26. M. Planck, "Über irreversible Strahlungsvorgänge. Nachtrag." Ann. Phys. **6** (1901), p. 818. *Planck Papers* I, p. 744.

27. H. Rubens and F. Kurlbaum, "Über die Emission langwelliger Wärmestrahlen durch den schwarzen Körper bei verschiedenen Temperaturen", Berliner Berichte (1900), p. 929.

Theory of Radiation and the Second Law of Thermodynamics"[28], given to the Academy of Sciences at Amsterdam shortly after Planck's theory first appeared, Lorentz mentioned that "Wien and Planck have endeavored to discover the form" of the radiation distribution law, and made reference to Planck's papers at the German Physical Society. Lorentz did not discuss these attempts at all, although he did mention Planck's proposal of a new system of "natural units" in connection with his attempts to demonstrate Wien's distribution law.

A few years later, in 1903, Lorentz had more to say about Planck's work.[29] This time he was writing on the emission and absorption of long wave length radiation by metals, calculating the ratio of emission to absorption coefficients by an argument based on his theory of electrons. This ratio is very simply related to the energy density in black-body radiation corresponding to the same frequency interval, so that Lorentz was, in effect, independently calculating the long wave length (or low frequency) limit of the radiation distribution. What he found was a result of the form,

$$\overset{\text{(Low frequency)}}{\rho(\nu, T)} = \gamma\nu^2 T, \tag{10.14}$$

where γ is a constant. As Lorentz pointed out, this is just the form assumed by the Planck distribution, (10.13), when $h\nu$ is small compared to kT. Lorentz showed that his value for γ, determined from the electron theory, agreed precisely with the coefficient $(8\pi k/c^3)$ obtained from the Planck distribution. He also confirmed that the k obtained from black-body radiation data agreed with the cruder estimates of k from the kinetic theory. Lorentz commented: "There appears therefore to be a full agreement between the two theories [his own and Planck's] in the case of long waves, certainly a remarkable conclusion, as the fundamental assumptions are widely different."

Earlier in his paper Lorentz had had something to say about the assumptions underlying Planck's theory. He referred to Planck's way of introducing "the notion of probability" into his theory, remarking that it was "not the only one that may be chosen". Lorentz emphasized Planck's assumption that his oscillators could gain or lose energy

28. H. A. Lorentz, "The Theory of Radiation and the Second Law of Thermodynamics", Proc. Acad. Amsterdam **3** (1901), p. 436. *Collected Papers*, Vol. VI (The Hague: Martinus Nijhoff, 1938), p. 265.

29. H. A. Lorentz, "On the Emission and Absorption by Metals of Rays of Heat of Great Wave-Lengths", Proc. Acad. Amsterdam **5** (1903), p. 666. *Collected Papers*, Vol. III (The Hague: Martinus Nijhoff, 1936), p. 155.

only in discrete units, these "units of energy" being proportional to the frequency of the oscillator in question. And Lorentz added, "It appears... that the hypothesis regarding the finite 'units of energy', which has led to the introduction of the constant *h*, is an essential part of the theory." Even though Planck had not investigated "the mechanism by which the heat of a body produces electromagnetic vibrations in the aether," Lorentz found the results "most remarkable".

Lorentz lectured on the theory of radiation in the spring of 1903, and one of the students attending those lectures was Paul Ehrenfest, visiting Leyden for a time before returning to Vienna, as we have already seen. Ehrenfest read extensively on the radiation problem while he was in Leyden, studying Planck's papers as well as Lorentz's own works, but he did not really work on this subject until after he had received his degree. On November 9, 1905 Boltzmann presented to the Academy at Vienna Ehrenfest's first contribution to the theory of radiation[30], a study of the physical assumptions underlying Planck's theory of irreversible processes in radiation. In this paper Ehrenfest used a method developed by Lorentz, but the questions he raised and and the point of view he adopted showed him as Boltzmann's disciple.

Ehrenfest saw Planck's long series of papers on irreversible processes in radiation as an attempt to create a parallel to Boltzmann's H-theorem. (This was certainly correct and, although Ehrenfest did not intend it that way, also ironic, in view of Planck's earlier attitude towards Boltzmann's work.) But Ehrenfest pointed out one fundamental difference between the two theories. Boltzmann's H-theorem really had two aspects to it: the function $(-H)$ must increase with time, becoming stationary for the equilibrium or Maxwell-Boltzmann distribution; but also, if the time derivative of H is zero, then one can conclude that the velocity distribution has the Maxwell-Boltzmann form. The function $(-H)$ could really be interpreted as the entropy of the gas in the sense that it could only increase or remain constant, *and*, if it remained constant, the velocity distribution was uniquely fixed as the Maxwell-Boltzmann distribution.

Planck's radiation theory did not have this dual aspect. In the course of his work Planck had used two different functions to represent the entropy of his system, as we have seen. For both of these he was able to prove that the function could never decrease with time, but in

30. P. Ehrenfest, "Über die physikalischen Voraussetzungen der Planck'schen Theorie der irreversiblen Strahlungsvorgänge", Wiener Berichte, **114** (1905), p. 1301. *Papers*, p. 88.

neither case did the stationarity in time of the supposed entropy function allow Planck to determine the spectral distribution function. Ehrenfest then showed that this weakness was inherent in Planck's model of charged oscillators plus radiation field. All of Planck's fundamental equations, Maxwell's equations for the radiation and the oscillators' equations of motion, were linear in the respective amplitudes. Using an argument based on similarity considerations adapted from one of Lorentz's papers, Ehrenfest showed that this linearity implied that the requirement of stationarity could not lead to a unique spectral distribution without contradicting the Wien displacement law. Infinitely many stationary distributions were possible.

Ehrenfest went on to analyze the additional assumptions Planck must have made in order to obtain an answer for the spectral distribution of the radiation, since he had in fact obtained such an answer – twice. In his earlier work Planck had required that the function introduced as entropy by definition be stationary, not only for real processes occurring in time, but also for all virtual processes. (He had actually considered a virtual exchange of energy between oscillators of different frequencies.) Ehrenfest pointed out that this was a new and independent condition that Planck had imposed on his function, and it was this extra condition that had enabled Planck to complete his argument (incorrectly, as it had turned out). Ehrenfest made it clear that, if one knew in some independent way that the function in question was really the entropy, then the requirement of stationarity with respect to virtual processes would not be an independent assumption but would have to hold automatically.

Up to this point in his discussion Ehrenfest did not have to consider Planck's new approach of 1900 involving the "elements of energy". The role of this paper in the development of Planck's work on the radiation problem did not seem obvious. It amounted to an argument for introducing a new form for the entropy function or, equivalently, for the distribution law, and Ehrenfest had, therefore, been able to set it aside for the purposes of his earlier analysis. He now turned to it explicitly and pointed out how it fit into the picture. It, too, could be viewed by way of an analogy to Boltzmann's work.

"Namely, after it had been shown by means of the H-theorem that the quantity $(-H)$ behaves like the entropy, and that, in every case where the entropy of a hot body is defined thermodynamically at all, its functional form coincides with this entropy, Boltzmann then succeeded in proving that the value of H in the stationary final state is also distinguished by a maximal property of a combinatorial kind." Ehrenfest was referring to Boltzmann's memoir of 1877, the very memoir to which Planck had turned in the fall of 1900, when he decided he had to resort to Boltzmann's methods. And, as Ehrenfest wrote, "This article [Planck's derivation of the distribution law using energy elements] is devoted to an attempt to carry over this combinatorial method of deriving the entropy to the thermodynamics of radiation." In other words, Planck had turned to Boltzmann's statistical or combinatorial method when he was unable to find the entropy by using Boltzmann's kinetic method, the method of the H-theorem.

There were serious difficulties associated with this step. For, as Ehrenfest emphasized, this maximal property of the entropy "stands and falls along with a completely definite hypothesis concerning the domain of states which are treated combinatorially as having equal weight". The choice of those states to be assigned equal *a priori* probabilities is crucial to any combinatorial argument. Planck had made such a choice, and it differed from Boltzmann's. And Planck had made the additional assumption "that the radiant energies of various colors are composed of smallest particles of energy of magnitude $E_\nu = (\nu) \times (6.55 \times 10^{-27}$ erg sec), where ν is the frequency of the corresponding color". Ehrenfest was misinterpreting Planck here, since Planck's discreteness assumption concerned the energy of his oscillators and not that of the free radiation at all. The misinterpretation had no immediate consequences, however, since Ehrenfest said nothing more of substance about Planck's energy elements, although he announced his intention of returning to this subject. He did add one sentence which says a great deal about the reception of Planck's ideas. Referring to the quantum hypothesis Ehrenfest wrote, "This hypothesis, which is obviously meant to be taken only formally in its present form, requires a further reduction. So far as I can see, it has no analogy in Boltzmann's theory."

6. Paul Ehrenfest was not the only one trying to puzzle out the relationship between Planck's work and Boltzmann's in 1905. Lord Rayleigh wrote a Letter to the Editor of *Nature* in May of that year on "The Dynamical Theory of Gases and of Radiation.[31] Rayleigh had discussed the problem of the black-body radiation spectrum earlier,

31. Lord Rayleigh, "The Dynamical Theory of Gases and of Radiation", Nature **72** (1905), p. 54.

in a short article published in June 1900[32], a few months before Planck proposed his new distribution law. In two pages he had shown that if the equipartition theorem of statistical mechanics could be applied to "the modes of aetherial vibration", then the radiation distribution law would have to have a form very different from Wien's distribution function. Rayleigh was well aware of the "difficulties which attend the Boltzmann-Maxwell doctrine of the partition of energy". He recognized that "the doctrine fails in general", but thought nevertheless that "it seems possible that it may apply to the graver modes". Rayleigh did not need Planck's oscillators for his argument. He simply pointed out that the number of electromagnetic modes of vibration in an enclosure, having frequencies in the interval between ν and $\nu + d\nu$, must be proportional to $\nu^2\, d\nu$. According to the equipartition theorem the average energy of every one of these modes, regardless of its frequency, would be proportional to T at thermal equilibrium, with a universal proportionality constant. The distribution function $\rho(\nu, T)$ would, therefore, have to be proportional to $\nu^2 T$.

In his 1900 note, Rayleigh did not trouble to point out explicitly something that must have been obvious to him: a distribution of the form $\nu^2 T$ could not possibly apply to all frequencies since it would mean an infinite energy density, due to the infinite number of high frequency modes. Rayleigh only referred to "the graver modes", as quoted above and then blunted the point of his own argument by writing: "If we introduce the exponential factor, the complete expression will be $\rho(\nu, T) \propto \nu^2 T \exp(-\beta\nu/T)$", and suggested that this form be compared with experimental results. This last expression for $\rho(\nu, T)$ was evidently intended only as a guess as to how the rigorous but unacceptable consequence of applying the equipartition theorem might be modified into a convergent distribution law, but it was the only aspect of Rayleigh's paper that was noticed. Rubens and Kurlbaum[27] compared their data with this law containing "the exponential factor", and found it inadequate. They never commented on the fundamental importance of the fact that their results did have the form $\nu^2 T$ at low frequencies, which was all that Rayleigh really expected.[33] Planck, who must also have been aware of Rayleigh's result, at least through Rubens

32. Lord Rayleigh, "Remarks upon the Law of Complete Radiation", Phil. Mag. **49** (1900), p. 539.

33. See the footnote Rayleigh added to his paper in 1902 when it was reprinted: Lord Rayleigh, *Scientific Papers* Vol. IV (Cambridge: Cambridge University Press, 1903), p. 483.

and Kurlbaum, never mentioned it (or the equipartition theorem) in his writings before 1906.

When Rayleigh returned to the radiation problem in his letter of May 1905, he repeated his earlier arguments in considerably more detail. This time he began by pointing out the infinite result that follows from applying the equipartition theorem to all radiation modes. He also calculated the explicit form of the distribution law including constants. Students of theoretical physics can take heart from Rayleigh's calculation: the author of *The Theory of Sound* made an error of a factor of 8 in calculating the density of normal modes. The error was immediately called to his attention by James Jeans. Rayleigh's corrected result had the form

$$\overset{\text{(Rayleigh-Jeans)}}{\rho(\nu, T)} = (8\pi k/c^3)\, \nu^2 T. \tag{10.15}$$

This is exactly the result found by Lorentz in 1903 by a completely different method, but there is no reason to believe that Lorentz had seen Rayleigh's 1900 article, or that Rayleigh had read Lorentz's work. Rayleigh also noticed that Avogadro's number could be found from the low frequency radiation data, where Planck's law reduced to his own.

"A critical comparison of the two processes would be of interest," Rayleigh went on, "but not having succeeded in following Planck's reasoning I am unable to undertake it." Since his own equation (10.15) followed so directly from "the Boltzmann-Maxwell doctrine" Rayleigh found it hard "to understand how another process, also based on Boltzmann's ideas, can lead to a different result".

Rayleigh's request for clarification[34] was answered, at least in part, by James Jeans.[35] Jeans was not at all sympathetic to Planck's ideas, and was doing his utmost to account for the black-body spectrum by assuming that the classical laws held, but that the radiation was not in thermodynamic equilibrium. He did, however, point out clearly two essential ways in which Planck's reasoning departed from the Boltzmann pattern. Jeans attacked Planck's use of W as a "probability", since no population was given from which probabilities could be calculated, and since one could not introduce such a population with an

34. Lord Rayleigh, "The Constant of Radiation as Calculated from Molecular Data", Nature **72** (1905), p. 243.
35. J. Jeans, "A Comparison between Two Theories of Radiation", Nature **72** (1905), p. 293.

a priori weighting consistent with Planck's arguments. Jeans also claimed that Planck had no right to keep his energy elements ϵ finite, and that he was obliged to consider the limit $\epsilon = 0$. If Planck had done this, Jeans showed, he would have obtained the same result as Rayleigh, (10.15). While it was true that Wien's displacement law required ϵ to be of the form $h\nu$, it did not fix the value of h, "whereas statistical mechanics gives us the further information that the true value of h is $h = 0$". Jeans thought that since Planck had also used "the methods of statistical mechanics and of the theorem of equipartition of energy", he had no right to refrain from setting h equal to zero. Actually, the equipartition theorem had never even entered Planck's considerations.

7. Planck's book. *Lectures on the Theory of Heat Radiation*[36], appeared late in the spring of 1906. It did not advance the theory of radiation directly, but his new ordering and systematic exposition of the material helped to clarify just what he had done five years earlier. The structure of Planck's argument in his book brought out something that had been less than obvious in the original papers: the new radiation law did not depend on Planck's incomplete analogue of the H-theorem. As Ehrenfest had already emphasized, the problems of treating irreversible processes in radiation did not have to be settled before discussing the combinatorial derivation of the radiation law. Planck explicitly referred to Ehrenfest's paper in the final section of his book, as he recognized that the interaction of radiation with his oscillators could not produce a unique spectral distribution, since such an interaction did not affect the frequency of the radiation.

In his *Lectures* Planck discussed the low frequency limit of his distribution law (something he had not done before), and showed that it agreed precisely with the result derived in very different ways by Lorentz and by Rayleigh and Jeans. Since Planck was very sure of the correctness of his own law, he found no difficulty in asserting that the equipartition theorem simply could not apply to the high frequency oscillations. He had never developed a constraining respect for the central importance of this theorem, and did not hesitate in limiting its domain of applicability to situations where kT was large compared to $h\nu$.

Planck also offered a new way of looking at his fundamental assumption of discreteness. He looked at the phase plane in which the state

---

36. M. Planck, *Vorlesungen über die Theorie der Wärmestrahlung* (Leipzig: J. A. Barth, 1906).

of an oscillator of frequency ν is given by a point whose abcissa and ordinate are, respectively, the displacement x and its canonically conjugate momentum, p. The energy, E, of the oscillator can be written in the form

$$E = \tfrac{1}{2}Kx^2 + \frac{2\pi^2\nu^2}{K}p^2, \tag{10.16}$$

where K is a constant characteristic of the oscillator. The curve representing the locus of phase points with energy E is, therefore, an ellipse whose semi-axes are $(2E/K)^{1/2}$ and $(EK/2\pi^2\nu^2)^{1/2}$. The area enclosed by this ellipse is just (E/ν). The elementary cells, or the regions of equal probability corresponding to energy intervals ΔE, are elliptical rings of area $(\Delta E/\nu)$. In Planck's theory these cells must have areas equal to his new constant h, which then fixes the area of a phase cell for all oscillators, regardless of their frequencies. Only if h could be considered infinitely small would the theory reduce to the classical case. Planck's constant, h, has the dimensions of energy multiplied by time, or displacement multiplied by momentum, the dimensions of the quantity known as action in mechanics, and in his *Lectures* he proposed calling it the element or unit of action.

Very soon after Planck's book appeared, Ehrenfest sent off another paper on the theory of radiation[37]. This was in the form of a commentary on some general aspects of the theory in the light of Planck's *Lectures*. Ehrenfest began by emphasizing once again that Planck's model of linear oscillators scattering the radiation in an enclosure contained no mechanism for producing the state of true thermodynamic equilibrium. The repeated scatterings would make the radiation more disordered, but would not make it "blacker"; i.e. the scattering process could not affect the spectral distribution of the radiant energy. This time Ehrenfest described an analogous situation in a gas to bring out his point more sharply. Suppose the molecules of a gas were transparent to one another, so that they did not really collide with each other, but these molecules could make elastic collisions with a collection of scatterers fixed at random positions on the walls of the container. Then, in the course of time, the spatial distribution of the molecules would tend to become uniform, and the molecular velocities would acquire a random distribution of directions. The distribution of molecu-

37. P. Ehrenfest, "Zur Planckschen Strahlungstheorie", Phys. Z. 7 (1906), p. 528. *Papers*, p. 120.

lar speeds, however, would retain its initial shape, since the molecules keep their original energies, and there would be no tendency to approach the Maxwell distribution. The entropy would increase to a maximum value, but this maximum would not be the absolute maximum for the given energy, attained only in the Maxwell distribution. The case of radiation was exactly parallel, with the true black-body distribution, the Planck distribution, playing the part of the Maxwell distribution.

Ehrenfest then turned to what he called Planck's second theory of radiation, the "complexion theory", but he did not follow Planck's treatment. Instead Ehrenfest fixed his attention directly on the electromagnetic proper vibrations of the enclosure, just as Rayleigh and Jeans had done. These modes of vibration could be treated as independent in the same sense that the molecules of a gas are independent. Let $N(\nu)\,\mathrm{d}\nu$ be the number of modes having frequencies in the interval ν to $\nu+\mathrm{d}\nu$. Then Ehrenfest introduced the fundamental distribution function $F(\nu,\xi,\eta)$ defined so that $N(\nu)\,\mathrm{d}\nu F(\nu,\xi,\eta)\,\mathrm{d}\xi\,\mathrm{d}\eta$ is the number of modes (in the given frequency interval) whose generalized coordinates lie in the range ξ to $\xi+\mathrm{d}\xi$, and whose canonical momenta lie in the range η to $\eta+\mathrm{d}\eta$. The condition for equilibrium is that the entropy S of the radiation be a maximum, where S is defined by the equation

$$S=-k\int_0^\infty N(\nu)\,\mathrm{d}\nu\int_{-\infty}^{\infty}\int_{-\infty}^{\infty} F(\nu,\xi,\eta)\,\mathrm{d}\xi\,\mathrm{d}\eta. \qquad (10.17)$$

The maximization must be done subject to two conditions: the first is simply a normalization condition on F, and the second is the requirement that the total energy of the radiation is fixed. Such a calculation leads to the Boltzmann distribution, equipartition, and the Lorentz-Rayleigh-Jeans result, which is unacceptable. And so Ehrenfest raised the question: "By what method did Planck's 'complexion theory' avoid this difficulty?"

In principle, one could always adjoin other constraints to try to achieve any desired distribution. But, as Ehrenfest remarked, such a calculation using an additional variational condition could only deserve to be called a theoretical derivation if one could give a physical basis for the extra condition. He had not succeeded in doing this. Planck himself had not put the calculation in this form, of course, but he had imposed an extra condition by requiring that the energy of his oscillators be quantized.

Ehrenfest's description of the quantization condition is very curious. It looks as though he had forgotten that he was using the Rayleigh-

Jeans approach, where the electromagnetic vibrations of the enclosure are considered, rather than Planck's approach, using the auxiliary oscillators. In any event, Ehrenfest described the auxiliary condition "chosen by Planck" as the requirement that "the proper vibrations of frequency ν should only be able to possess energy in those amounts ϵ_ν which are integral multiples of the energy unit $\epsilon_\nu^0 = h\nu$." He also transferred Planck's phase plane discussion from the material oscillator to the electromagnetic vibration, saying that the phase point for a proper vibration of frequency ν could not lie anywhere in the phase plane, but only on one of the discrete series of ellipses, of the type of (10.16), corresponding to energy $mh\nu$ ($m = 0, 1, 2, ..$). This transference was not a slip of the pen. I think it was merely a reflection of the attitude Ehrenfest had already expressed in his earlier paper, the view that the hypothesis of energy quanta was "obviously meant to be taken only formally".[38] If one were dealing with only a formal device, then it was perfectly legitimate to use that device in such a way that the structure of the argument was clearest. For Ehrenfest that meant using it within the Rayleigh-Jeans approach, where the normal methods of statistical mechanics could be employed, rather than within Planck's peculiar complexion counting procedure. Here one could see, at least formally though not yet physically, how Planck's theory had made "the infinitely many ultraviolet overtones of the enclosure innocuous, so that they did not suck away all of the energy as they did in the Rayleigh-Jeans calculation". It was simply because Planck's "energy atoms" increase without any limit as ν increases. Rather than all of the energy going to the very high frequency modes, practically none of it does, since no finite amount of energy can make up even one of the "energy atoms" needed to excite such high frequencies.

8. Planck's radiation theory produced a variety of responses, as already described, but it had its most remarkable effect on Albert Einstein. Einstein referred to Planck's work for the first time in 1904, in the course of the first of his profound studies[39] of the limits to the validity of thermodynamics imposed by the molecular structure of matter. He

38. A rather different view of this paper was expressed by G. Wentzel in his article, "Quantum Theory of Fields (until 1947)" in the volume: *Theoretical Physics in the Twentieth Century. A Memorial Volume to Wolfgang Pauli.*, ed. M. Fierz and V. F. Weisskopf (New York: Interscience, 1960), p. 48.

39. A. Einstein, "Zur allgemeinen molekularen Theorie der Wärme", Ann. Phys. **14** (1904), p. 354.

had chosen the black-body radiation for a test of the validity of his energy fluctuation equation, thereby obtaining a new insight into the significance of the constant k as a measure of the "thermal stability" of a system. By the spring of 1905 Einstein had been familiar with Planck's work for over a year. The paper[40] he wrote in March reported the results of the reflections that had been stimulated by Planck's puzzling theory.[41]

Einstein had apparently been struck more forcibly by what Planck had not done in his theory of black-body radiation than by what he had done. He noticed that Planck's relationship between the spectral distribution $\rho(\nu, T)$ and the average energy of an oscillator led at once to the result

$$\rho(\nu, T) = (8\pi\nu^2/c^3)(kT), \tag{10.15}$$

since the equipartition theorem required that the average energy of a linear oscillator be kT. Einstein also saw at once that the divergence implied by this result had unfathomable consequences, reaching to the very foundations of physics. It must be stressed that Einstein was evidently not familiar with Rayleigh's paper of 1900 or with Lorentz's of 1903. He had arrived at the same distribution law by a completely independent argument, and he was the first to point explicitly to its catastrophic implications.

Einstein saw the origin of the trouble in the "profound formal distinction between the theoretical ideas which physicists have formed concerning gases and other ponderable bodies, and the Maxwell theory of electromagnetic processes in so-called empty space". The existence of this distinction between discrete particles and continuous field marked a fundamental inhomogeneity in the foundations of physics. The unreconciled tension between Maxwellian and Newtonian ideas was symbolized by the failure of a result like (10.15), where the first factor came from electromagnetic theory and the second from statistical mechanics. Although Einstein could not offer a solution to

---

40. A. Einstein, "Über einen die Erzeugung und Verwandlung des Lichtes betreffenden heuristischen Gesichtspunkt", Ann. Phys. **17** (1905), p. 132. Engl. transl. by A. B. Arons and M. B. Peppard, Am. J. Phys. **33** (1965), p. 367.

41. (a) M. J. Klein, "Einstein's First Paper on Quanta", The Natural Philosopher (New York: Blaisdell) **2** (1963), p. 57.

(b) M. J. Klein, "Thermodynamics in Einstein's Thought", Science **157** (1967), p. 509.

(c) L. Rosenfeld, *op. cit.* note 10 (d).

(d) M. Jammer, *op. cit.* note 10 (e), p. 28.

the problem of unifying the foundations of physics, he could and did propose a dependable way of learning something about the nature of radiation: the use of the second law of thermodynamics and its statistical interpretation.

Considering only the high frequency form of the radiation spectrum, the well-established Wien distribution, (10.2), Einstein showed purely thermodynamically that the entropy of black-body radiation in a given frequency interval depends upon the volume of the enclosure in exactly the same way that the entropy of an ideal gas depends upon its volume. By interpreting the entropy statistically Einstein recognized that it was simply the independence of the motions of the gas molecules that produced this particular, logarithmic, form for its entropy. His next step was to make the logical leap and conclude that the radiation entropy has this same form because it, too, consists, in some sense, of independent particles of energy. By identifying the appropriate coefficient in the radiation entropy with the number of these particles, Einstein determined the energy of such a particle as $(R/N_0)(\beta\nu)$, where β is the coefficient in the exponential of Wien's law. This is just $h\nu$, although Einstein did not use the constants h and k to which Planck gave such weight. Einstein concluded, therefore, that: "Monochromatic radiation of low density (within the region of validity of Wien's distribution law), behaves with respect to the thermal phenomena as if it were composed of independent energy quanta of magnitude $(R/N_0)(\beta\nu)$".

This conclusion led Einstein to suggest a new "heuristic viewpoint" from which to consider processes involving the emission and absorption of radiation, the hypothesis of light quanta. He proposed that the electromagnetic wave theory of light might not be the last word on the subject and that one should explore the idea that light behaves like a collection of independent, localized particles of energy – the light quanta. He showed how simply this completely heretical suggestion could be used to understand several phenomena which were utterly unintelligible from the point of view of the wave theory, phenomena like the lowering of the frequency in photoluminescence and the existence of a lower limit to the frequency of light that can produce photoionization. He also showed how the light quantum hypothesis immediately accounted for the strange result that the energy of electrons produced in the photoelectric effect is independent of the intensity of the incident light. Einstein predicted the precise form of the frequency dependence of the photoelectrons' maximum energy, a prediction that Millikan verified in detail a decade later.

In 1906, in another paper[42] on the radiation problem, Einstein discussed the relationship between his work and Planck's. He had just realized that Planck's theory was not in contradiction to his own, although he had thought there was such a contradiction a year earlier. He pointed out that Planck's theory implicitly presupposed the hypothesis of light quanta, and that as a result Planck's relationship between the spectral distribution function and the average energy of an oscillator had to be considered as a separate assumption. Planck's derivation was based on the classical electromagnetic theory of the oscillator, a theory inconsistent with quantized energy for the oscillator.

A year later Einstein carried his analysis a good deal further[43], pointing out that if one took Planck's theory seriously – "if Planck's theory strikes to the heart of the matter" – then one must make a fundamental change in the foundations of statistical mechanics. The oscillators involved in exchanging energy with radiation have a narrower range of possibilities open to them than bodies with which we have direct experience, i.e. their energies can take on only the values $nh\nu$ ($n = 0, 1, 2, \ldots$). But if these oscillators are so restricted, what about all other oscillators that appear in molecular theories? Einstein had no doubt that they too must obey the quantum restrictions, and that there must be phenomena which contradict classical statistical mechanics, phenomena whose explanation demands quantization[44]. He found such a phenomenon in the departure of the specific heats of the light elements, in solid form, from the rule of Dulong and Petit.

Einstein showed that these departures from the Dulong-Petit rule could be looked upon as failures of the equipartition theorem. Using the simplest possible model of a solid, treating it as a collection of N_0 independent atomic oscillators, each capable of independent vibrations in the three coordinate directions, all vibrations being of the same frequency ν, Einstein could write down the energy of a mole of the solid in the form

$$U = 3N_0 \frac{h\nu}{\exp(h\nu/kT) - 1}. \tag{10.18}$$

42. A. Einstein, "Zur Theorie der Lichterzeugung und Lichtabsorption", Ann. Phys. **20** (1906), p. 199.

43. A. Einstein, "Die Plancksche Theorie der Strahlung und die Theorie der spezifischen Wärme", Ann. Phys. **22** (1907), p. 180.

44. For further discussion and references see M. J. Klein, "Einstein, Specific Heats, and the Early Quantum Theory", Science **148** (1965), p. 173.

The second factor is just the average energy of an oscillator of frequency ν, whose possible energies are restricted to the integral multiples of $h\nu$, (See (10.13) above.) The specific heat is then the temperature derivative of U. Einstein used this result to explain the known departures from the Dulong-Petit rule and the previously puzzling failure of electrons to contribute to the specific heat. He also determined the characteristic vibration frequency ν, in many cases, as the frequency at which optical absorption occurred, thereby giving a completely new connection between optical absorption and specific heat. And Einstein predicted a new general property of matter: the specific heats of all crystalline solids must vanish as the absolute zero of temperature is approached. All these results followed directly from his assumption that Planck's quantization of the energy of an oscillator was not just a formal device introduced in order to be able to derive the radiation formula, but that instead it represented a new limitation on possible motions at the molecular level.

Einstein's predictions were confirmed. Walther Nernst's extended series of specific heat measurements at low temperatures, planned in 1906 to test his own new heat theorem, showed a universal approach of the specific heats to zero at low temperatures. Realizing that this had been predicted by Einstein, Nernst turned to a study of the quantum theory and became an enthusiastic advocate of it[44].

Still another unexpected relationship among the properties of matter came out of the quantum theory when Einstein showed[45] early in 1911 that the compressibility of a crystal could be related to its optical absorption frequency, and, therefore, also to its specific heat. All these tangible results, capable of being tested experimentally and passing these tests whenever they were made, began to convince the community of theoretical physicists that something about the idea of quantization would have to be taken seriously.

Einstein himself had had no doubts on this point for a long time and devoted extraordinary efforts to further probings of the significance of Planck's radiation law. In 1909 he described two independent arguments[46], both based on statistical fluctuations but in very different

45. A. Einstein, "Eine Beziehung zwischen dem elastischen Verhalten und der spezifischen Wärme bei festen Körpern mit einatomigem Molekül", Ann. Phys. **34** (1911), p. 170.

46. (a) A. Einstein, "Zum gegenwärtigen Stand des Strahlungsproblems", Phys. Z. **10** (1909), p. 185.

(b) A. Einstein, "Über die Entwicklung unserer Anschauungen über das Wesen und die Konstitution der Strahlung", Phys. Z. **10** (1909), p. 817.

ways, from which he concluded that the structure of radiation had a new and unprecedented character. His earlier analysis of the Wien distribution law had led him to propose a picture of radiation as composed of particles of energy, but these new studies of the full Planck distribution convinced him that both particle and wave properties were characteristic of radiation[47]. This duality of wave and particle aspects could not be avoided, and he announced his opinion that "the next phase of the development of theoretical physics will bring us a theory of light that can be interpreted as a kind of fusion of wave and and emission [corpuscular] theories". These conclusions, which Einstein considered to be unquestionable consequences of the statistical form of the second law applied to the experimentally confirmed Planck distribution, did not convince his colleagues. Few, if any, of them were ready yet for the "profound change in our views of the nature and constitution of light" that Einstein saw as "indispensable".

9. Paul Ehrenfest did not express himself directly on this question when he returned to the problem of quanta in 1911[48], but he was obviously aware that the status of the quantum hypothesis had changed sharply since he had last written on it, five years earlier. It had been applied to "a rapidly expanding circle of questions, some of which were related only very loosely to the problem of thermal radiation". The fate of the quantum hypothesis would probably be decided by experimental tests in these new areas of application. Since those tests had not yet been made, Ehrenfest thought it would be valuable to report on some work he had done which showed that at least some of the features of the quantum hypothesis could be viewed as established once and for all by the properties of black-body radiation. He proposed, in other words, to answer the question which he used as the title of his paper: "Which features of the hypothesis of light quanta play an essential role in the theory of thermal radiation?"

Planck had introduced his energy quanta in order to derive the black-body radiation spectrum. But the fact that a result can be derived from a particular assumption does not imply that it cannot be derived with-

47. For further discussion see M. J. Klein, "Einstein and the Wave-Particle Duality", The Natural Philosopher (New York: Blaisdell) **3** (1964), p. 1.

48. P. Ehrenfest, "Welche Züge der Lichtquantenhypothese spielen in der Theorie der Wärmestrahlung eine wesentliche Rolle?", Ann. Phys. **36** (1911), p. 91. *Papers*, p. 185. This paper is discussed in L. Rosenfeld, *op. cit.* note 10 (d), and in M. J. Klein, "Ehrenfest's Contributions to the Development of Quantum Statistics I", Proc. Acad. Amsterdam **B62** (1959), p. 41.

out that assumption: sufficient conditions for a proof are not always necessary conditions. Not only that, Planck's derivation was such a confusing mixture of classical electrodynamics, (used to relate the spectrum to the energy of an oscillator and to derive the displacement law), with his own idiosyncratic version of statistical mechanics as well as the new quantum hypothesis, that it was extremely hard to tell how these disparate pieces really fit together. Ehrenfest undertook to sort it all out and arrange the essential elements into a logical structure.

As he had in his earlier papers, Ehrenfest used Rayleigh's method of working directly with the normal modes of the enclosure, rather than with Planck's oscillators. The first task was to set down the language within which the problems could be formulated, the language of statistical mechanics adapted to the system of radiation modes. In a given enclosure of volume V there would be a definite number of normal modes having frequencies in the interval between ν and $\nu + \mathrm{d}\nu$. This number, $N(\nu)\,\mathrm{d}\nu$, had already been determined by Rayleigh and Jeans to be

$$N(\nu)\,\mathrm{d}\nu = (8\pi V/c^3)\nu^2\,\mathrm{d}\nu. \tag{10.19}$$

One had to imagine the amount of energy $\mathscr{E}$ distributed over all possible electromagnetic modes within the enclosure. This distribution of energy would be characterized by giving the number of modes of frequency ν to $\nu+\mathrm{d}\nu$ having energy between E and $E+\mathrm{d}E$, $n(\nu, E)\,\mathrm{d}\nu\,\mathrm{d}E$. (The phases with which the various modes oscillate are considered to be random.) The state of thermodynamic equilibrium would then be identified with the most probable state in Boltzmann's sense, i.e. that distribution of energy, that function $n(\nu, E)$, which could be achieved in the largest possible number of ways.

At this point Ehrenfest had to generalize Boltzmann's procedure, but in a manner different from the one he had conjectured in 1906. He had thought then that what was needed to rationalize Planck's method was another constraint on the possible distribution function, besides the evident ones of fixed number of modes and fixed total energy. He now realized that what was needed could be expressed better as a generalization of the "*a priori* probability" or weight function assigned to the basic phase space. Boltzmann had always assigned equal weights to equal volumes in phase space, and while this was the simplest assignment it was not the only possible one. Ehrenfest assigned the weight $\gamma(\nu, E)\,\mathrm{d}E$ to the energy interval from E to $E + \mathrm{d}E$ for a proper vibration of frequency ν.

With this weight function $\gamma(\nu, E)$ the probability of a distribution in which, of the $N(\nu)\, d\nu$ modes, n_1 of them have energy between E_1 and $E_1 + dE_1$, n_2 have energy between E_2 and $E_2 + dE_2$, etc. would be proportional to the expression

$$[\gamma(\nu, E_1)\, dE_1]^{n_1} [\gamma(\nu, E_2)\, dE_2]^{n_2} \ldots \frac{[N(\nu)\, d\nu]!}{n_1!\, n_2! \ldots}. \qquad (10.20)$$

From this point the procedure was a straightforward application of Boltzmann's methods. Taking the logarithm, using Stirling's formula, and summing over all frequencies, the expression to be maximized had the form

$$\begin{aligned} \ln W = \text{const.} + \int_0^\infty d\nu \int_0^\infty dE\, n(\nu, E)\, [\ln \gamma(\nu, E) - \ln n(\nu, E) + 1] \qquad (10.21) \\ + \int_0^\infty d\nu\, N(\nu)\, [\ln N(\nu) - 1]. \end{aligned}$$

The two constraints on $n(\nu, E)$ were the fixed number of modes in the interval $d\nu$ and the fixed total energy, and allowing for these it could be shown in the usual way that the most probable distribution $n(\nu, E)$ was given by the equation

$$n(\nu, E) = N(\nu) \frac{\gamma(\nu, E) \exp(-\mu E)}{\int_0^\infty dE\, \gamma(\nu, E) \exp(-\mu E)}, \qquad (10.22)$$

the Boltzmann distribution modified by the weighting function $\gamma(\nu, E)$. The quantity μ was the Lagrange multiplier for the energy constraint, to be determined, in principle, by the use of this constraint.

The weight function thus determined the character of the equilibrium radiation distribution, and Ehrenfest's strategy was to analyze the restrictions placed on the weight function by the experimentally established properties of the black-body radiation. The most probable distribution had to correspond to black-body radiation; what did that imply about $\gamma(\nu, E)$? Some of the most interesting aspects of black-body radiation showed up when the volume of the enclosure was varied adiabatically and reversibly. In such a process, for example, the entropy of the radiation did not change, whether or not it was in the equilibrium distribution to begin with[49]. Ehrenfest found that a study of

49. See Planck, *Vorlesungen*, Chapter 2.

such reversible adiabatic processes could solve one very important problem in a thoroughly satisfactory way.

Wien's displacement law was surely one of the essential properties of the equilibrium distribution – it had, for example, played an essential part in Planck's work – but it was very hard to see why. For the derivation of this law, as given in Planck's book and elsewhere, made use of the Doppler effect on the frequency of a ray reflected from a moving mirror, and other concepts that seemed to have very little to do with black-body radiation. Lorentz[50] had given another derivation in 1901 which worked directly with the electromagnetic field; this was less artificial looking, but still too complicated to bring out the crux of the displacement law to Ehrenfest's satisfaction. He wanted a derivation that focussed attention directly on the electromagnetic normal modes of the enclosure, in terms of which the distribution law must ultimately be expressed.

Ehrenfest found the tool he was looking for in a theorem that he distilled out of a paper by Lord Rayleigh[51]. This theorem, suggested by Rayleigh's arguments on radiation pressure as a concept applicable to all vibrating systems, was a statement about the behavior of the energy of a normal mode under a sufficiently slow variation of the dimensions of the enclosure: if the reflecting walls of the enclosure are made to approach each other infinitely slowly, the energies of all proper vibrations change in exactly the same proportion, and are directly proportional to their frequencies. In other words the ratio E_ν/ν is invariant in such a process, during which ν varies inversely as the dimension of the enclosure. The motion of the walls must be slow enough for the normal mode to retain its identity, so that many oscillations occur during any appreciable displacement of the walls.

By combining this theorem with the condition that $\ln W$, representing the entropy, must be unchanged in such an infinitely slow compression of the enclosure, Ehrenfest proved that the weight function $\gamma(\nu, E)$ had to have the form

$$\gamma(\nu, E) = G(E/\nu). \tag{10.23}$$

The weight function could depend only on the ratio (E/ν), (apart from a possible factor which is a function of frequency only and could not affect $n(\nu, E)$, according to (10.22)). But this result led directly to the

50. H. A. Lorentz, "Boltzmann's and Wien's Laws of Radiation", Proc. Acad. Amsterdam **3** (1901), p. 607. *Collected Papers* Vol. VI, p. 280.

51. Lord Rayleigh, "On the Pressure of Vibrations", Phil. Mag. **3** (1902), p. 338.

displacement law. For, the spectral distribution function $\rho(\nu, T)$ is just the integral over energy of $En(\nu, E)$, except for a factor V^{-1} to allow for the fact that $\rho(\nu, T)$ is energy per unit volume per unit frequency interval. The appropriate equation has the form

$$\rho(\nu, T) = \frac{1}{V}\left(\frac{8\pi V}{c^3}\right)\nu^2 \frac{\int_0^\infty \mathrm{d}E\, EG(E/\nu)\exp(-\mu E)}{\int_0^\infty \mathrm{d}E\, G(E/\nu)\exp(-\mu E)}. \qquad (10.24)$$

A change of variables to $x = E/\nu$ shows that $\rho(\nu, T)$ is proportional to ν^3 times a function of $\mu\nu$. But, since μ must be proportional to T^{-1} from the way it entered the whole argument, this means that (10.24) is equivalent to the displacement law. The displacement law is the expression of the second law of thermodynamics for a system of oscillators having the invariant E_ν/ν, and therefore a weight function $\gamma(\nu, E)$ of the form $G(E/\nu)$.

With this point established Ehrenfest went on to the additional restrictions on the weight function imposed by the more specific properties of the black-body radiation. He enumerated these properties in terms of the function $f(\beta\nu/T)$ in the displacement law; $\rho(\nu, T) = \alpha\nu^3 f(\beta\nu/T)$, where the β is introduced to make the argument of f dimensionless.

(a) Since the Rayleigh-Jeans law holds for long wave lengths, there is the "red-condition" which requires $f(\beta\nu/T)$ to behave like $(\beta\nu/T)^{-1}$ for small values of its argument. Thus, one must have

$$\lim_{\sigma\to 0}\{\sigma f(\sigma)\} = 1. \qquad (10.25)$$

(b) Since the total energy must be finite, $\rho(\nu, T)$ must be integrable, and therefore the function $f(\beta\nu/T)$ must fall off faster than the inverse fourth power of its argument as $\beta\nu/T$ gets large. This is the "violet-condition",

$$\lim_{\sigma\to\infty}\{\sigma^4 f(\sigma)\} = 0. \qquad (10.26)$$

(Ehrenfest put a sub-heading over the paragraph devoted to this condition, in which he introduced a phrase that has become a cliché among physicists, although like most clichés it is almost never attributed to its source. He spoke of "Avoidance of the Rayleigh-Jeans Catastrophe in the Ultraviolet". Only very few among the thousands of physicists who have since used the terms "Rayleigh-Jeans catastrophe" or "ultraviolet catastrophe" have realized that they were quoting a typical Ehrenfest phrase, expressing the drama inherent in the theoretical problem.)

(c) One could actually make a stronger statement than (10.26) about the asymptotic behavior of $f(\beta\nu/T)$ for large values of its argument, if one were willing to make more explicit extrapolations of the experimental results. Ehrenfest investigated what he called the "strengthened violet-condition", (requiring that $\sigma^n f(\sigma)$ go to zero as σ becomes infinite for all positive n), and the "Wien-Planck violet-condition," (requiring that $f(\sigma)$ exp $(L\sigma)$ approach a finite, non-zero constant, as σ becomes infinite, for some finite, non-zero value of L).

For this part of his study Ehrenfest had to generalize the idea of the weight function $G(E/\nu)$ to include the possibility of a point spectrum, i.e. the possibility that certain values of E/ν might have a statistical weight equivalent to that normally assigned to an interval of E/ν values. Thus an integral like $\int_0^\infty dq\, G(q) \exp(-\sigma q)$, $(q = E/\nu, \sigma = \beta\nu/T)$, would either have to be viewed as a Stieltjes integral or else have a sum over the discretely weighted points q_n added on to it, a sum of the form $\sum_n G_n \exp(-\sigma q_n)$.

Ehrenfest derived a series of results about this generalized weight function from the asymptotic properties required of the spectral distribution. The most important of these results was the theorem that the point q $(= E/\nu) = 0$ must have a special weight assigned to it if the total radiant energy is to remain finite. More precisely, not only must $q = 0$ have a special weight, but $G(q)$ must vanish at least as fast as q^2 near the origin to satisfy the "violet-condition" and faster than any power q^n to satisfy the "strengthened violet-condition". And if the distribution is to satisfy the exponential Wien-Planck condition at high energies, then there must be an interval starting at the origin and extending to $q = L$ which has zero weight, the point $q = L$ again being assigned a special weight. (The "red-condition" requires only that the weight function not fall off too rapidly with increasing values of E/ν).

These results, rigorous theorems about the way statistical weights must be assigned in describing the electromagnetic normal modes, (or the Planck oscillators), were consequences of only the asymptotic behavior of the spectral distribution. If the distribution is to avoid the ultraviolet catastrophe then each mode of frequency ν cannot be set into vibration unless it is supplied with a finite amount of energy, proportional to its frequency. It was possible to go a step further. If one assumed that the Planck distribution holds exactly for the entire frequency domain, then the corresponding weight function could be specified more completely as the solution of an integral equation. In this case the weight function $G(q)$ was purely discrete, having equal non-zero values only for the points $q=0, 1, 2 \ldots$. This theorem, stated without proof by Ehrenfest, who remarked only that he had gotten it from "a friendly quarter" (possibly Gustav Herglotz, his mathematical counselor, or one of his mathematician friends in Petersburg), is actually the statement that Planck's quantization condition is necessary as well as sufficient for the Planck radiation distribution. Ehrenfest did not emphasize this result at all, mentioning it only in passing. He chose to stress the much more general consequences of the weaker, asymptotic properties of the spectral distribution.

10. Ehrenfest's paper was completed in July 1911 and it appeared in October. On the 21st of that month Einstein commented on it briefly in a letter to a friend[52], but the paper did not attract much notice. At the end of October the first Solvay Congress took place in Brussels.[53] Some twenty of Europe's most eminent physicists came together for a week of intense discussion of the problems of radiation and the quantum theory. It is clear from the printed proceedings of the meeting, which contain the "official" discussion remarks as well as the prepared reports on a variety of related subjects, that Ehrenfest's paper did not play any part in the meeting, despite the fact that he had already clarified several of the issues debated in Brussels. Ehrenfest himself was not present; he was still only an unemployed physicist in Petersburg at that time. Even though Einstein, at least, had read Ehrenfest's paper, it seems unlikely that he mentioned it during the meeting.

The Solvay Congress had a great impact on many of the participants. Planck had been hesitant about holding such a meeting, when Walther Nernst first suggested the idea to him in 1910. He felt that there would be no point to it until theorists all felt that the current state of their subject had been made "*intolerable*" by his own quanta, and were

52. A Einstein to M. Besso, 21 October 1911.

53. The proceedings were published as *La théorie du rayonnement et les quanta*, ed. P. Langevin and M. de Broglie (Paris: Gauthier-Villars, 1912). A description of the Congress can be found in M. de Broglie, *Les premiers Congrès de Physique Solvay* (Paris: Albin Michel, 1951), and in M. J. Klein, *op. cit.* note 44.

impelled to do something about it. When the Congress was over, if not before, that situation did prevail. Most, if not quite all, of the participants left Brussels certain of two things – Planck's discrete units of energy had entered physics to stay, and they could not be accommodated within the existing structure of the science.

The greatest impact of the Solvay discussions was felt by Henri Poincaré, the distinguished French mathematician, mathematical physicist and philosopher of science[54]. Poincaré came to Brussels ignorant of the quantum theory, but took a vigorous part in the discussions, raising a number of penetrating questions about the new ideas he was hearing. Poincaré's freshly awakened interest in quanta did not end with the Solvay Conference. In December he announced some new results of his own to the Academy of Sciences in Paris[55], and by January a long memoir appeared in the *Journal de Physique*[56]. Poincaré had posed and answered some of the same questions that Ehrenfest had treated the preceding year.

Poincaré proved that Planck's oscillators must be restricted to discrete energy levels $n\epsilon$ if Planck's distribution law holds for the radiation. Quantization of the energy is a necessary as well as a sufficient condition for Planck's law. He also proved, as Ehrenfest had, that there must be a singular weight attached to the zero of energy in order to avoid the classical problem of infinite energy. That qualitative aspect of discontinuity is independent of the particular nature of the radiation distribution law.

Poincaré's arguments were quite different from Ehrenfest's. His physical model was the exchange of energy by collision among Planck's oscillators, rather than the electromagnetic proper vibrations of an enclosure. One fundamental aspect of Ehrenfest's investigation was completely missing from Poincaré's work. Ehrenfest had shown that the reason why energy quanta were proportional to frequency was nothing less than the requirements of the second law in its statistical form. Poincaré's discussion, which omitted any analysis of the effect of an adiabatic process on the system, could only point to the demands of the displacement law without probing any further. He recognized this as an incompleteness, writing that "it is not well understood by what mysterious, pre-established harmony these resonators have been

54. See especially R. McCormmach, "Henri Poincaré and the Quantum Theory", Isis **58** (1967), p. 37.
55. H. Poincaré, "Sur la théorie des quanta". Comptes Rendus **153** (1911), p. 1103.
56. H. Poincaré, "Sur la théorie des quanta", Journal de Physique **2** (1912), p. 5.

endowed with the only mechanical property which could be suitable."

Poincaré had obviously not seen Ehrenfest's paper. I have already described how Ehrenfest ran across Poincaré's note in the *Comptes Rendus* and then found his full length paper, in the course of his trip through Western Europe in search of a job. Ehrenfest sent Poincaré a copy of his own work which Poincaré acknowledged at once,[57] saying that he was happy to see that Ehrenfest had arrived at results in agreement with his own by a different method. He expected to find Ehrenfest's work useful in preparing his next article. Poincaré did not comment on the additional results Ehrenfest had obtained, nor did he register any awareness of Ehrenfest's clear priority of several months. His death a few months later cut off any possibility that he would ever give Ehrenfest public credit for his closely related work.

As it turned out, Poincaré's paper deeply influenced the attitudes of his contemporaries. Whereas Ehrenfest could be ignored and Einstein not accepted. Poincaré's authority could hardly be questioned. His arguments were accepted as *the* proof that discontinuity in the energy was absolutely necessary for the existence of a finite energy in the black-body radiation. Even James Jeans, a holdout against the quantum for a decade and still an unbeliever at the Solvay Congress, yielded to Poincaré's mathematics. In his *Report on Radiation and the Quantum Theory*[58], Jeans made it clear that he had been convinced and that it was Poincaré's work that had convinced him. Jeans's *Report* was highly influential, since it served for some years as the principal document in English on its subject, and Poincaré's theorem had a crucial position in Jeans's arguments. Ehrenfest's work received no mention at all.

11. At the end of his 1911 paper Ehrenfest tried to use the results of his analysis in a study of the status of Einstein's hypothesis of light quanta. The relationship between Planck's quantized oscillators and Einstein's proposal that radiation consists, at least in some sense, of independent particles of energy, was still very obscure[59]. Ehrenfest summarized the hypothesis of light quanta, Einstein's hypothesis, in

57. H. Poincaré to P. Ehrenfest, 30 January 1912.

58. J. H. Jeans, *Report on Radiation and the Quantum Theory* (London: The Electrician, 1914).

59. This may be an appropriate place to say something about the Latin word "quantum" in physics. As a German word it had no special or novel connotation. It was in fairly common scientific use to mean quantity, as, for example, in the title of Boltzmann's paper in 1883. "Über das Arbeitsquantum, welches bei chemischen Verbindung gewonnen

three statements: (a) an oscillator of frequency ν can exhibit only the energies $0, h\nu, 2h\nu, \ldots$; (b) these energy values come about by putting together a number of elementary, mutually independent units of energy, $h\nu$; and finally, (c) these light quanta not only function as "atoms" in all emission and absorption phenomena, but they also have their separate existence in empty space.

Ehrenfest pointed out that his general theorems provided a basis for (a), a complete basis if one accepted the Planck distribution law as exact for all frequencies. There was no real distinction between Planck and Einstein on this statement. The distinction seemed to lie in statement (c), affirmed by Einstein but denied by Planck. Planck himself had stressed this distinction just the year before in his first public statement on the quantum theory since his book was published[60].

werden kann" – "On the Quantity of Work Which Can be Obtained by Chemical Combination".

When Planck first treated energy as discrete in his papers of 1900 and 1901, he referred to the quantities ϵ, or $h\nu$, as "Energieelemente", "elements of energy". In these same papers, however, he did use the word "Quantum" in referring to the charge on the electron as "das Elementarquantum der Elektricität" – "the elementary quantity of electricity". When he issued the discussion of the natural constants as a separate paper in 1901, Planck entitled it "Über die Elementarquanta der Materie und der Elektricität", still using "Quanta" to refer to the fundamental units of mass and electricity – the mass of the hydrogen atom, m_H, and the electronic charge, e. The first time Planck referred to "Quantum" in the new sense was in his *Lectures* (*op. cit.* note 36, p. 154), where he described h itself as "elementares Wirkungsquantum", the elementary quantity (or unit) of action. It was h itself as a basic natural unit parallel to m_H and e that Planck called a "Quantum".

Einstein also used "Quantum" in the common sense, referring in his 1905 paper (*op. cit.* note 40) to Planck's determination of the ' Elementarquanta", and meaning the determination of m_H and e. But even in this first paper Einstein also referred to "lokalisierten Energiequanten" and to "Lichtquanten", meaning his new light corpuscles, offered as a heuristic suggestion about the structure of radiation.

The word "Quantum" seems to have been taken over from the German into both French and English, probably after the Solvay Congress. Poincaré used it in his papers and Jeans in his *Report*, (although in 1905 Jeans had referred to Planck's ϵ as a "sort of indivisible atom of energy"). It is interesting that J. J. Thomson, who independently proposed another sort of discrete structure for light, described it as "The Unit Theory of Light", and "unit" would be an appropriate translation of "Quantum" in this context. (See R. McCormmach, "J. J. Thomson and the Structure of Light", British Journal for the History of Science **3** (1967), p. 362, especially p. 383.) Erwin Schrödinger often referred to energy quanta as "portions of energy". (See, for example, his essay "Are There Quantum Jumps", British Journal for the Philosophy of Science **3** (1952), pp. 109, 233.) But no translation of "Quantum" has ever replaced the word itself in English.

60. M. Planck, "Zur Theorie der Wärmestrahlung", Ann. Phys. **31** (1910), p. 758. *Planck Papers* II, p. 237.

Ehrenfest took pains to emphasize that this question of the existence of free quanta in space was not the first point of difference between Planck and Einstein. He argued, and he was the first to do so, that statement (b) was also affirmed by Einstein but denied by Planck. Despite the appearances of Planck's combinatorial arguments (sketched in section 3 of this chapter), Planck had not assumed that the energy he distributed over his N oscillators was composed of P independent particles of energy ϵ. If one did assume the energy to consist of such independent particles, one could derive "an infinite set of other radiation formulas", but not Planck's.

The details of this argument were not given in Ehrenfest's paper. But in one of his research notebooks there is an entry dated 21 March 1911, and labeled, "On Planck's Entropy", in which Ehrenfest says that if one distributes P balls in N urns there are N^P different distributions. He then notes that Planck asserts there are $(N+P-1)!/(N-1)!(P-1)!$. (Ehrenfest's version of the formula has a small but interesting error in it.) "That must be wrong," he concluded.

Ehrenfest returned to this point several years later in a short but lively paper[61] devoted to Planck's combinatorial formula for the number of ways P energy units ϵ can be distributed over N oscillators. Planck had simply quoted the formula in his papers and in his book, referring to it as a standard result of combinatorial theory, "the number of combinations with repetitions of N elements of class P". Ehrenfest was not satisfied with the derivation given in algebra texts, since the proof was by mathematical induction and it gave no insight into the peculiar structure of the result with its $(N-1)!$ and $(P+N-1)!$ factors. Ehrenfest wanted to know where these factors really came from, and he finally found out.

In October 1914 Ehrenfest sent a note[62] to Lorentz along with the manuscript of a short paper written in collaboration with Kamerlingh Onnes. He asked Lorentz to have a glance at the manuscript, adding "I have no doubt that you will have to laugh at it as you would at a joke. We have had a lot of fun out of it, as has everyone else I've shown it to." The joke was Ehrenfest's unforgettable proof of Planck's

61. P. Ehrenfest and H. Kamerlingh Onnes, "Simplified Deduction of the Formula from the Theory of Combinations Which Planck Uses as the Basis of his Radiation Theory", Proc. Acad. Amsterdam **17** (1914), p. 870. *Papers*, p. 353. This paper, along with many of Ehrenfest's other papers of this period, was published in a number of journals. See the Table of Contents of Ehrenfest's *Papers* for details on the other publications.

62. P. Ehrenfest to H. A. Lorentz, 24 October 1914.

combinatorial formula, now given everywhere, though never attributed to its originator.

The total energy of the N oscillators is $P\epsilon$. In how many ways can this energy be distributed? A distribution is specified by giving the "energy-grade", as Ehrenfest called it, of each oscillator. Such a distribution can be expressed symbolically. If, for example, $N = 4$ and $P = 7$, one of the possible distributions is that in which the first oscillator has energy 4ϵ, the second 2ϵ, the third no energy and the fourth ϵ. The symbolic form of this distribution is shown below:

$$\| \epsilon \; \epsilon \; \epsilon \; \epsilon \mid \epsilon \; \epsilon \mid \mid \epsilon \|$$

The end symbols $\|$ are the "fixed boundaries", the symbols ϵ represent the "energy-grades", and the dividers $|$ separate the energy symbols for the different oscillators, numbered from left to right. There will be P symbols ϵ, and there will also be $(N-1)$ divider symbols $|$, when there are N oscillators. Considering the $(P+N-1)$ quantities ϵ and $|$, there are $(P+N-1)!$ ways in which they can be ordered between the fixed end symbols. But any of the $P!$ permutations of the ϵ's among themselves, and any of the $(N-1)!$ permutations of the dividers among themselves does not change the symbol for the distribution. The number of distinct distributions is, therefore, given by Planck's expression, $(P+N-1)!/P!(N-1)!$.

This witty proof did more than make the $(N-1)!$ in Planck's formula intuitively clear. It also confirmed Ehrenfest in his old opinion that the elements of energy in Planck's calculation were only a formal device. The permutations of the ϵ symbols had no more physical significance than the permutations of the divider symbols. In an appendix to the paper Ehrenfest argued that, if one really wanted to take seriously the idea that quanta were independent particles of energy, then one would have to say something different about the counting of their distributions. One would have to say, as Einstein had in 1905, that the number of ways one can distribute P particles over N cells is N^P, since each particle has N independent choices, as it were. Some of the distributions counted as distinct in this procedure are identified as one and the same in the Planck counting. This is so whenever two distributions differ only in the *particular* quanta assigned to the different cells, the *number* of quanta in each cell being the same.

Since the number of distributions as counted by Planck differed from the number obtained for quanta treated as independent particles, the entropy change in any appropriately specified process would also have to be different. In particular, as Ehrenfest remarked again, Einstein's formula, suitable for independent particles, was equivalent to the Wien distribution law and not the Planck distribution law for radiation[63].

Ehrenfest had seized on a fundamental aspect of Planck's quanta, although it would not be possible to incorporate this aspect into a suitable theoretical structure for over a decade. If one is to describe radiation by a particle representation, then the "particles" one uses must have properties fundamentally different from those of any particles previously used in physical theory. The fact that Planck's distribution law is the correct one for such particles implies that they are not independent in any normal sense of the word, but must show a kind of correlation which is utterly incomprehensible in classical physics. And, although it was far from obvious how this was related to the non-classical correlations, these particles must also have lost that most basic of properties, their individuality, and become as indistinguishable as the ϵ symbols of the calculation.

This blurring of the concept, particle, when applied to light quanta was already implicit in the results for the energy and momentum fluctuations in black-body radiation which Einstein had obtained in 1909[46]. Starting with the Planck distribution, Einstein had shown that the fluctuations were always the sum of two terms: one had just the form appropriate to fluctuations in a gas of independent particles; the other had the form appropriate to fluctuations produced by interfering waves. The first term would be the only one present in the Wien limit, (large values of $h\nu/kT$); the second term would be the only one present in the Rayleigh-Jeans limit, (large values of $kT/h\nu$). Planck's law included both aspects – the classical particle and the classical wave, and that was why Einstein was so sure that a "fusion" of the two was called for. He had stressed from the beginning that his proposal of independent quanta was only intended to apply in the region where the Wien distribution was valid.

12. By the time of the first Solvay Congress some physicists were beginning to feel that the idea of quanta had proved itself so well that it was time to reformulate it in a more general way. Arnold Sommerfeld

63. Further discussion and bibliography in M. J. Klein, *op. cit.* note 48.

asserted this point of view more explicitly and at greater length than any of his colleagues at the meeting in Brussels[64]. He began his review paper there by remarking that although it would be possible to construct a mechanics concerned exclusively with periodic circular motion, and although this mechanics would be useful and adequate for certain purposes, such a mechanics would necessarily give only a very incomplete picture of the system of laws known as Mechanics. The current status of the theory of quanta impressed Sommerfeld as resembling such a highly oversimplified mechanics. Restricted to dealing with simple harmonic oscillations, it might meet the needs of the theory of black-body radiation and some of the problems of the theory of specific heats, but it was very limited in what it could do. Sommerfeld was particularly interested in treating "non-periodic molecular phenomena" with the help of quanta, and he thought he had found an appropriate generalization of Planck's original theory.

Sommerfeld's new principle was to be a combination of the principle of least action with Planck's quantum of action. He proposed as his fundamental hypothesis that the action gained or lost by an atom during any process must be $h/2\pi$, where by action Sommerfeld meant the integral of the Lagrangian (kinetic energy minus potential energy) over the duration of the process. Although Sommerfeld had a variety of formal arguments with which to justify this hypothesis and a variety of physical applications to demonstrate its fruitfulness, it turned out to be untenable. He was, nevertheless, absolutely right in his view that an approach to the theory not limited from the outset to simple harmonic oscillations was the necessary next step in the quantum theory.

Both Sommerfeld and Planck[65] felt that the clue to a quantum theory of greater generality must lie in the idea of h as quantum of action, rather than $h\nu$ as quantum of energy. The constant h was a universal one, whereas $h\nu$ included the factor ν, peculiar to the particular system and characteristic of a simple oscillation.

Paul Ehrenfest was also deeply concerned with the problem of

---

64. A. Sommerfeld, "Application de la théorie de l'élément d'action aux phénomènes moléculaires non périodiques" in P. Langevin and M. de Broglie, *op. cit.* note 53, p. 313. Also see A. Sommerfeld, "Das Plancksche Wirkungsquantum und seine allgemeine Bedeutung für die Molekularphysik", Phys. Z. **12** (1911), p. 1057.

65. M. Planck, "La loi du rayonnement noir et l'hypothèse des quantités élémentaires d'action", in P. Langevin and M. de Broglie, *op. cit.* note 53, p. 113, *Planck Papers* II, p. 269.

generalizing the quantum theory to deal with systems other than the harmonic oscillator. His approach to the problem, however, was a very different one, growing naturally out of the studies described in his 1911 memoir. He had been especially struck by the way in which the special invariance property of the quantity E/ν, the ratio of oscillator energy to frequency, figured so crucially in his arguments. Because E/ν was invariant under slow parameter changes, the weight function in phase space had to be a function of just this invariant, if one were to maintain the statistical interpretation of the second law of thermodynamics. This was the true basis for Wien's displacement law, and Ehrenfest was sure that the method deserved further study.

A decade or so later, in 1923, he described the feelings with which he had approached this development[66]. "Let it be emphasized from the beginning: it was Boltzmann's radiation law and Wien's displacement law, or more precisely the mystery which was concealed behind the elegant electrodynamic and thermodynamic derivations of these laws, which baited the path that led to the Adiabatic Principle. Wien's displacement law had been derived on a purely *classical* basis. How could it, nevertheless, stand unshaken in the midst of the world of radiation phenomena whose *anti-classical* quantum character stood out ever more inexorably?...... One had côme upon the track of a *particular* type of 'pseudo-classical' behavior of a quantum system – even considered from today's standpoint – and by analysis of the derivation of the displacement law one should be able to learn how far one could go in obtaining further correct results in the midst of the quantum world with the help of classical mechanics and electrodynamics and of classical thermodynamics, and, therefore, even of Boltzmann statistics."

Ehrenfest's 1911 paper was, in part, just that "analysis of the derivation of the displacement law", and it left him convinced that there must be some sort of far-reaching extension of the analysis. Since the invariance of E/ν was a property peculiar to the harmonic oscillator, what was needed was an appropriate generalization of that invariance theorem. This problem was never far from Ehrenfest's thoughts throughout 1912, although he had had many other things to be concerned with during that year as we have seen: searching for and finding a job, moving across the European continent and establishing

66. P. Ehrenfest, "Adiabatische Transformationen in der Quantentheorie und ihre Behandlung durch Niels Bohr", Naturwissenschaften **11** (1923), p. 543. *Papers*, p. 463.

his family in a new country. Only a few days after his arrival in Leyden in October, Ehrenfest went back to his thoughts about more general "adiabatic" invariants, as he was already calling them in his notebooks. He worked out the case of particles which bounce back and forth between perfectly reflecting walls. He found that if the walls are made to approach each other very slowly (i.e. "adiabatically" – so that the work done moving the walls against the pressure of the gas all goes into kinetic energy of the particles), then the ratio of the kinetic energy of the molecules to the frequency of their collisions with the walls is constant. He realized that the result would have to be qualified in various ways, but found its "universality" to be "surprising, nevertheless". He was already sure that there must be an adiabatically invariant quantity for *any* periodic system, and wondered whether it might be the action during one period of the motion. Ehrenfest wrote himself a series of notes listing possibilities worth exploring. He wanted to go back and remind himself of the content of some of the mechanics he had studied years before which, he felt, might now be relevant – an old theorem due to Szily and some of Boltzmann's early work.

During November and much of December 1912 Ehrenfest was too busy with the infinite number of things that had to be done settling in a new place and starting on his new job (his inaugural lecture being not the least of these), to devote any serious thought to his problem. But on the evening of December 20th he sat down to it again. He began by exploring the possibility that E/ν was an adiabatic invariant for all one dimensional periodic systems, as suggested by his previous work on the harmonic oscillator, and the molecules reflected back and forth by slowly moving walls. (In the general periodic system, ν would still be the reciprocal of the period, θ, of the system, but it would depend upon the amplitude.) After trying this idea out on another case or two Ehrenfest noted another possibility to be considered: "Perhaps $\delta(\theta \bar{T}) = 0$?" The $\bar{T}$ here meant the average of the kinetic energy over the period θ. He also remembered that he wanted to check on the Boltzmann-Szily theorem, and tried to work it out again for himself, but not very successfully. For when he tried to check his version of this theorem on the example of the simple pendulum, Ehrenfest found an apparent contradiction. "Hmm, hmm!?!", he noted, and then "*Clarify.*"

The clarification came in the next entry, reporting his work on the following day. Ehrenfest had found what he wanted. He had reconsidered and perhaps done some reading, and now realized that the

Boltzmann-Clausius-Szily theorem, as he now called it, gave precisely what he was looking for—and even more: an adiabatic invariant for all periodic systems, no matter how many degrees of freedom they had. He recognized the importance of his result immediately, setting off the final equation in his notebook by drawing red and blue boxes around it.

Ehrenfest did not publish this result for almost a year, but he described it in a letter to Lorentz written on December 23rd[67]. More of Ehrenfest's excitement about his work comes through in the letter he wrote to Joffe[68], however, even though two months had already gone by. It is worth quoting at some length.

> "If you contract a reflecting cavity infinitely slowly, then the frequency ν and the energy ϵ of each proper vibration increase simultaneously in such a way that ϵ/ν remains invariant under this 'adiabatic' influence. You remember that I told you that this is really at the root of the fact that the *a priori* probability of an individual normal mode of frequency ν having energy between ϵ and $\epsilon + d\epsilon$ must have just the form $g(\epsilon/\nu)\, d\epsilon$.
>
> *The a priori probability must always depend on only those quantities which remain invariant under adiabatic influencing, or else the quantity* ln W *will fail to satisfy the condition, imposed by the second law on the entropy, of remaining invariant under adiabatic changes.*
>
> Consequently, it was of interest to answer the following question: if we go over from the special case of harmonic oscillations to general periodic systems (e.g. molecules moving back and forth between two walls, no forces acting; adiabatic influence: ∞ slow bringing together of the walls), *which function analogous to* ϵ/ν *then remains constant*?
>
> At Christmas I was able to prove the following remarkably simple and astonishingly general theorem. Given a conservative system of n degrees of freedom. The Lagrangian of the system may contain, in addition to the n coordinates q and velocities $\dot{q}$, any sort of parameters $a_1, a_2 \ldots, a_r$. (This means, for example, the coordinates of certain fixed centers of force, or quantities on which the moments of inertia of certain rotating masses can depend.)
>
> *The motion of the system is to be periodic for all values of the parameters* $a_1, a_2, \ldots a_r$ *and for arbitrary initial values of the*

67. P. Ehrenfest to H. A. Lorentz, 23 December 1912. See Chapter 1.
68. P. Ehrenfest to A. F. Joffe, 20 February 1913.

q's and $\dot{q}$'s. (The period θ may depend in any way on these initial values q_0, $\dot{q}_0$ and on the parameters $a_1, a_2, \ldots, a_r$.)

By an 'adiabatic influencing' of the system we will understand the following process: The system is carrying out one of its periodic motions, with period θ, for fixed values of the parameters $a_1, a_2, \ldots, a_r$. We vary the parameters *infinitely slowly* until they have assumed the values $a_1', a_2', \ldots, a_r'$ and then we hold them fixed again. The system will now carry out another periodic motion of period θ'.

Denote the 'frequency' $(1/\theta)$ by ν and $(1/\theta')$ by ν'. We also denote by $\bar{K}$ the *average kinetic energy* of the first motion and by $\bar{K}'$ that of the second. Then my theorem reads:

$$\bar{K}'/\nu' = \bar{K}/\nu. \tag{I}$$

The average kinetic energy of our system increases in the same proportion as the frequency under an adiabatic influencing.
Remarks:

1. $$\bar{K}/\nu = \theta\bar{K} = \theta\,\frac{\int_0^\theta K\,\mathrm{d}t}{\theta} = \int_0^\theta K\,\mathrm{d}t.$$

So the root of equation (I) is

$$\delta' \int_0^\theta K\,\mathrm{d}t = 0,$$

where δ' means an 'adiabatic' variation. You see, therefore, that Theorem I stands in close relationship to the Principle of Least Action (half more general, half more special – special case of the Boltzmann-Clausius-Szily principle).
2. In *special* cases (I) can be replaced by

$$\epsilon'/\nu' = \epsilon/\nu, \tag{II}$$

namely, whenever the average *potential* energy is in a fixed ratio to the average *kinetic* energy. Examples: (a) Harmonic oscillation. Here the average potential energy = average kinetic energy = $\frac{1}{2}$ total energy. Therefore ϵ/ν is constant for the cavity. (b) Force-free motion of a molecule between elastic walls, since here potential energy = 0, therefore $\bar{K} = \epsilon$.
A few more examples can be mentioned.
But in *general* ϵ/ν is *not* constant. It was already probable that

ϵ/ν couldn't mean anything special because ϵ contains an arbitrary additive constant. But $\bar{K}$ does not!!
3. The *a priori* probabilities are therefore to be chosen as functions of $\bar{K}/\nu$ and *not* ϵ/ν as soon as one wants to go over from the theory of radiation to more general cases.

In particular, Planck's energy level hypothesis must be generalized this way: $\bar{K}$ can only take on values which are integral multiples of $h\nu/2$. (h = Planck's h!!! $h/2$ here because, in Planck's case too, $\bar{K}$ changes only by amounts $h\nu/2$.)

You can imagine that it was by no means easy to guess at the correct generalization of ϵ/ν = constant. But I was firmly convinced that it had to succeed.

There are two things about my result that I don't like, and I don't know how to correct them:
1. That I have to limit myself to *periodic* systems. Periodicity undoubtedly has nothing to do with it.
2. That $\bar{K}/\nu = \int_0^\theta K\,dt$ is not invariant under Lorentz-Einstein transformations. This must be thoroughly clarified, since an infinitely slow increase in the translational velocity of an oscillating system is an adiabatic influencing. Unfortunately this is still completely unclear to me.

If you were here, I would surely get the whole business in order in a few days, stimulated by arguments with you. By the way, I essentially owe the discovery of the correct generalization to arguments with Tanya. She very justly criticized the appearance of the additive constant in ϵ.

If I can deepen this thing, it will throw the light of reason on Sommerfeld's generalization of the quantum hypothesis.

Don't judge my theorem too negatively. Naturally I don't know if this knowledge will amount to anything of value. But if you are just, you will have to admit that it essentially deepens our understanding of the roots of the $\epsilon/\nu = h$ assumption in the thermodynamics of radiation.

Don't you also get a little fun out of the fact that h is displaced by the cheaper $h/2$? If the quantum hypothesis is somehow carried over to ideal gases (e.g. to the specific heat of diatomic gases) the $h/2$ can play a role since the potential energy is zero there."

CHAPTER 11

The Adiabatic Principle

1. In December 1912, Paul Ehrenfest had found the result which he had long been searching for. This result had several aspects. It was in the first place, a rigorous theorem of classical mechanics, actually a special case of a theorem first proved by Boltzmann. This theorem asserted that, for any periodic system whose energy depends on certain parameters, the time integral of the kinetic energy taken over one period is invariant when the parameters are changed adiabatically (i.e. sufficiently slowly). Ehrenfest saw much more in it than that, however. He had strong reasons for believing that such adiabatically invariant quantities were the key to generalizing the statistical foundation of the second law of thermodynamics, so that it would withstand the undermining effect of the quantum theory. It was no accident that Planck had quantized the ratio of energy to frequency for his harmonic oscillators, since that ratio was the adiabatic invariant in this special case. And therefore, Ehrenfest conjectured, if one were to make a quantum theory for systems more general than Planck's oscillators, it would have to be done by quantizing the appropriate adiabatic invariants. The theorem he had found, that result in classical mechanics, was for him the clue to the generalization of the quantum theory.

All of this was clear in Ehrenfest's mind early in 1913 and was expressed in his letter to Abram Fedorovitch Joffe, quoted at length in the last chapter. He did not rush into print with these ideas; what seemed so clear might, after all, not be correct. What he needed was a test case, a suitable physical problem on which he could try out his ideas, and it did not take him long to find one. Ehrenfest applied his new hypothesis for the first time to the rotational motion of the hydrogen molecule, and used the quantized motions to calculate the specific heat of hydrogen gas.

2. Walther Nernst had undertaken a program of measuring the specific heats of solids at low temperature in order to test the "new heat theorem" he had proposed in 1905.[1] As his experimental results accumulated, they showed an apparently universal tendency for the specific heats of solids to vanish as the temperature was reduced to sufficiently low values. Nernst realized that this behavior had already been predicted by Einstein as a consequence of the quantum theory applied to atomic oscillations in the solid. As a result, Nernst became a staunch advocate of the quantum theory, and proved his devotion to the new ideas by trying to extend them. In a paper published in 1911,[2] Nernst argued that the quantum theory ought to affect the rotational motion of diatomic molecules, so that the contribution of this motion to the specific heat of a diatomic gas should also decrease as the temperature is reduced. Since one would generally expect such effects to set in at higher temperatures for lighter molecules, hydrogen should show this behavior most clearly. "A determination of the specific heat of hydrogen at low temperatures would, therefore, be especially interesting," wrote Nernst.

The measurements were carried out by A. Eucken[3] in Nernst's laboratory at Berlin, and reported early in 1912. As Nernst had predicted, the rotational contribution to the specific heat fell off from its value at room temperature, reaching an essentially constant value by about 60°K. The room temperature value of almost 5 cal/mole°K dropped with temperature to just about 3 cal/mole°K, so that from 60°K down hydrogen could be said to behave like a monatomic gas. (According to the equipartition theorem a monatomic gas would have a specific heat of $\frac{3}{2}R$, where R, the gas constant, is about 2 cal/mole °K.

1. W. Nernst, "Sur les chaleurs spécifiques aux basses températures et le développement de la thermodynamique," Journal de Physique **9** (1910), p. 721.

2. W. Nernst, "Zur Theorie der spezifischen Wärme und über die Anwendung der Lehre von den Energiequanten auf physikalisch-chemische Fragen überhaupt," Z. f. Elektrochemie **17** (1911), p. 265. Quoted passage on p. 271.

3. A. Eucken, "Die Molekularwärme des Wasserstoffs bei tiefen Temperaturen," Berliner Berichte (1912), p. 141.

The higher specific heat at room temperature reflected the contribution of two independent rotations.) While the curve representing the rotational contribution to the specific heat, c_r, as a function of temperature resembled the Einstein specific heat curve for a Planck oscillator, the agreement was evidently only qualitative. There was really no reason to expect an exact agreement since the rotation was an essentially different kind of motion.

Eucken's results posed a problem for the theorists, but Ehrenfest would probably not have attacked it if there had not been an earlier theoretical attempt to account for these results. He was not the sort of theorist whose work was closely tied to experimental data; very few of his papers refer directly to specific experimental results. Ehrenfest's interest was attracted to the problem by the earlier theoretical paper on it. He certainly would not ignore a paper that appeared under Einstein's name.[4]

Einstein, too, thought that Eucken's results might help to clarify a question of basic principle, but it was not the same question as Ehrenfest's. Einstein and his student Otto Stern saw a way of using Eucken's results to test the correctness of Planck's recent revision of his quantum theory.[5] Bothered by the difficulties of reconciling the quantized energy of his oscillators with the description of the electromagnetic field by Maxwell's equations, Planck had proposed a weaker version of his hypothesis of discontinuity. He now suggested that the oscillators absorb radiation continuously, exactly in accordance with classical theory. Only the emission process was to occur in discrete energy units, $h\nu$. This meant that the energy of the oscillator could take on all values, but emission would occur, following a statistical law, when the energy reached an integral multiple of $h\nu$. This revised theory led to exactly the same radiation distribution law that he had obtained in 1900 (not surprisingly), but it did introduce a change in the

4. A. Einstein and O. Stern, "Einige Argumente für die Annahme einer molekularen Agitation beim absoluten Nullpunkt," Ann. Phys. **40** (1913), p. 551. (The second part of this paper does not concern the rotational specific heat and is not discussed in the text.)

5. (a) M. Planck, "Eine neue Strahlungshypothese," Verh. d. Deutsch. Phys. Ges. **13** (1911), p. 138. *Planck Papers* II, p. 249.

(b) M. Planck, "Zur Hypothese der Quantenemission," Berliner Berichte (1911), p. 723. *Planck Papers* II, p. 260.

(c) This second version of the theory is also discussed in the second edition of Planck's book, the only one translated into English. M. Planck, *The Theory of Heat Radiation*, transl. M. Masius. Reprinted (New York: Dover Publications, 1959), pp. 135–175.

equation for the average energy of an oscillator. Where Planck's earlier theory had determined the average energy E of an oscillator as

$$E = \frac{h\nu}{\exp(h\nu/kT) - 1}, \tag{11.1}$$

the revised theory modified this into the form

$$E = \frac{h\nu}{\exp(h\nu/kT) - 1} + \frac{h\nu}{2}. \tag{11.2}$$

The only change was the extra energy $h\nu/2$ at all temperatures and in particular at absolute zero, so that it is always referred to as the zero point energy. The question was how to detect the presence of this extra term in the energy, which could not be measured directly and which did not contribute to the specific heat of an oscillator.

If, however, one had a system whose effective frequency varied with the temperature, then its specific heat would indeed depend on which form of the average energy, (11.1) or (11.2), was the correct one. Einstein proposed treating the rotating gas molecules as such a system, and gave reasons why their effective frequency should be determined by setting the kinetic energy of rotation equal to the oscillator energy. The rotational energy has the form $\frac{1}{2}I(2\pi\nu)^2$, where I is the moment of inertia and ν the rotational frequency. By using the two different expressions, (11.1) and (11.2), for the oscillator energy, Einstein and Stern could calculate the rotational contribution to the specific heat for both forms of the quantum theory. Their calculation did depend on the essential assumption that all molecules in the gas rotate with the same angular velocity. This failure to allow for the statistical distribution of velocities was a real limitation on the value of the theory, a limitation of which they were quite aware. When their two calculated curves for the rotational specific heat as a function of temperature were compared with Eucken's data, the conclusion was clear. Only that version of the theory which included the zero point energy could be used to account for the measurements.

Ehrenfest saw his own, very different, calculation of the specific heat as primarily "an extremely simple example of the application of my $h\nu/2$ philosophy of kinetic energy steps." That is how he described it in a letter to Joffe,[6] but in the paper[7] he sent off for publication a few

6. P. Ehrenfest to A. F. Joffe, 22 May 1913.

7. P. Ehrenfest, "Bemerkung betreffs der spezifischen Wärme zweiatomiger Gase," Verh. d. Deutsch. Phys. Ges. **15** (1913), p. 451. *Papers*, p. 333.

days later, at the end of May 1913, he put the emphasis differently. Here was a theoretical approach different from that used by Einstein and Stern: the statistical aspects of the problem were taken into account consistently and worked out completely. This alternate calculation accounted for the results at least as well, and without any appeal to the existence of a zero point energy. It was presented as a contribution to the theory of specific heats, and not as a proposal for generalizing the quantum theory. He had to state the assumptions on which the calculation was based, but these were not made the center of interest.

In the text of his article Ehrenfest stated his assumptions quite flatly. In his usual way, he simplified the problem into an almost schematic form – in this case, by treating "only one of the two rotations of the hydrogen molecule." This meant replacing the general rotation of the dumbbell shaped molecule by its rotation about one fixed axis, perpendicular to the dumbbell and passing through its center. The two assumptions which allowed Ehrenfest to apply the quantum theory to these simple rotations were the following. First, the only possible angular velocities of rotation, $2\pi\nu$, are those for which the kinetic energy is an integral multiple of $h\nu/2$, so that

$$\tfrac{1}{2}I(2\pi\nu)^2 = n\frac{h\nu}{2} \qquad (n = 0, 1, 2, \ldots). \tag{11.3}$$

This meant that in the phase plane of the molecule, the (q, p) plane where q is the angle of rotation and p is the corresponding angular momentum $I\dot{q}$, the only allowed regions are the origin and certain pairs of line segments. These segments are parallel to the horizontal q axis, and run from $q = -\pi$ to $q = \pi$ at heights $p = \pm h/2\pi, \pm 2h/2\pi, \pm 3h/2\pi, \ldots$. Ehrenfest's second stated assumption was that these "regions," the origin and the pairs of line segments just described, were to be treated as equally probable *a priori*, that is, as having equal statistical weights.

From here on, the normal methods of statistical mechanics could be applied. The possible energies ϵ_n could be written in the form $n^2h^2/8\pi^2I$, and one had to find the sum, $\Sigma_{n=0}^{\infty} \exp(-\epsilon_n/kT)$. If this sum is denoted by $Q(\sigma)$, where $\sigma = h^2/8\pi^2IkT$, and $Q(\sigma) = \Sigma_{n=0}^{\infty} \exp(-n^2\sigma)$, then the contribution, c_r, to the heat capacity of all rotational motions in the gas of N particles could be written in the form

$$c_r = 2N k\sigma^2 \frac{\mathrm{d}^2 \ln Q(\sigma)}{\mathrm{d}\sigma^2}. \tag{11.4}$$

The factor 2 was supplied to allow for the two independent rotational motions about mutually perpendicular axes (a procedure which did not do justice to the mechanics of rotation). The sum $Q(\sigma)$ could not be done in closed form, but Ehrenfest was able to handle it adequately with the help of a clever transformation. As already indicated, his results accounted reasonably for Eucken's data, with a plausible value for I, the molecular moment of inertia, which was the only adjustable parameter.

Ehrenfest restricted the arguments he gave to justify his two basic assumptions to two relatively brief footnotes. He said very little indeed about the first assumption. He simply took note of the fact that Lorentz[8] had made a passing remark at the Solvay Congress about quantizing the kinetic energy of rotational motion using units $h\nu$, and asserted that his own choice of the smaller kinetic energy units $h\nu/2$ "could be demonstrated from a very general point of view." The nature of that "very general point of view" was not explained, however. Ehrenfest was a little more explicit in his note on the second assumption. He described how the allowed regions in phase space for his rotator were related to the ellipses of Planck's oscillator. To see the relationship one had to imagine the molecular dumbbell as a dipole acted on by an external electric field. When the molecule's energy is small relative to the external field, the dipole oscillates about an equilibrium position. For this motion one has Planck's ellipses in the phase plane, and these were already treated as having equal statistical weights. As the energy of the dipole is increased, its motion changes character and the oscillation becomes a (non-uniform) rotation. The phase ellipses split into curved segments extending from $-\pi$ to π (and then repeating), and these become closer and closer to straight lines as the kinetic energy becomes more important relative to the dipole's energy in the orienting field. Ehrenfest justified his assignment of weights on the basis of the weights assigned to the oscillator: "By an infinitely slow reduction of the orienting field, all of the oscillating mole-

8. *La théorie du rayonnement et les quanta*, ed. P. Langevin and M. de Broglie (Paris: Gauthier-Villars, 1912), p. 447. Ehrenfest was not aware of this comment by Lorentz until it was pointed out to him when he lectured on his ideas in Leyden. See P. Ehrenfest to N. Bohr, 8 May 1922. Ehrenfest was equally unaware of the exchange between Lorentz and Einstein at the Solvay Congress in which Einstein remarked that the ratio of energy to frequency for a pendulum would be unaffected if the length of the string were slowly reduced. See p. 450 of the Solvay report and also the same letter from Ehrenfest to Bohr.

cules can be carried over 'adiabatically' into uniformly rotating ones, with the ellipses transforming into the pairs of straight line segments."

3. By November 1913 Ehrenfest was ready to say more about his ideas. The short paper[9] which Lorentz communicated on his behalf to the Academy at Amsterdam was in many ways typical of Ehrenfest's way of doing physics, much more so than his last paper had been. The new paper dealt directly with the fundamental issues that concerned Ehrenfest. He pointed out the existence of the adiabatic invariant, $\bar{T}/\nu$, the ratio of average kinetic energy to frequency, for any periodic system. Ehrenfest described it as both the generalization of the energy to frequency ratio which was the crucially important adiabatic invariant for the oscillator, and also as a special case of the "mechanical theorem of Boltzmann" to which he referred in the title of his paper. Having pointed out the existence of this general invariant, Ehrenfest posed the basic question: "How can it be applied heuristically, when Planck's assumption is extended to systems vibrating non-sinusoidally?" Ehrenfest was very cautious about trying to answer this question. "For the present I can only answer [this] question by giving an example." But he did point out at some length the difficulties involved in trying to go any further. The paper raised many more questions than it answered, and in this it was especially typical of Ehrenfest.

The example Ehrenfest discussed, as an illustration of how the adiabatic invariance of $\bar{T}/\nu$ might be used in extending the applications of the quantum theory, was the motion of a dipole, as already sketched in his previous paper. A dipole is suspended so that it can rotate freely about the z-axis, and it is subjected to an electric field along the x-axis. Suppose first that the field is held fixed at some large value. Then for low energies the motion of the dipole will be an oscillation about the field direction, a harmonic oscillation for sufficiently low energies. Planck's original quantization condition could be applied to these oscillations, and since $\bar{T}$ for a harmonic oscillator is just half the total energy, this condition could be put in the form,

$$(\bar{T}/\nu)_0 = nh/2, \qquad (n = 0, 1, 2, \ldots). \tag{11.5}$$

The subscript indicates that this condition holds for the original oscillatory motion.

9. P. Ehrenfest, "A Mechanical Theorem of Boltzmann and its Relation to the Theory of Energy Quanta," Proc. Acad. Amsterdam **16** (1913), p. 591. *Papers*, p. 340.

At higher energies the motion of the dipole changes its character, assuming the external field is held fixed; the oscillations increase their amplitude, becoming anharmonic, and eventually the dipole rotates around the z-axis, the rotation becoming more and more nearly uniform as the energy increases. Now this same change in the character of the motion could also be produced by a gradual (adiabatic) change in the field strength and dipole moment. In this way the rotational motion could be converted smoothly into harmonic oscillation by adiabatically increasing the electric field. Ehrenfest argued that his theorem could therefore be applied to find the allowed rotational motions. All this was essentially an elaboration of his previous paper, but he now pointed out a difficulty which he had not previously noticed. If one looked at the transition from oscillations of large amplitude to rotations, it appeared that one full oscillation corresponded to *two* complete rotations of the dipole, so that the appropriate frequency ν_1 for the rotational motion should be the angular velocity, $\dot{q}$, divided by 4π rather than by 2π. The adiabatic invariance condition would then read

$$\frac{\bar{T}_1}{\nu_1} = \frac{4\pi\bar{T}_1}{\dot{q}} = \frac{\bar{T}_0}{\nu_0} = \frac{nh}{2}, \qquad (n = 0, 1, 2, \ldots). \tag{11.6}$$

Since T_1 can be written as $\frac{1}{2}p\dot{q}$ (i.e. $\frac{1}{2}I\dot{q}^2$), this meant the allowed values of the angular momentum p were given by the equation

$$p = nh/4\pi, \qquad (n = 0, \pm1, \pm2, \ldots). \tag{11.7}$$

As Ehrenfest pointed out, this differed from his earlier work, where he had found a unit of angular momentum which was twice as large. (The specific heat calculation would be unaffected, if only the value of the moment of inertia were modified.) The question was, why was there an ambiguity?

Ehrenfest could not handle this question, but he could point to the source of the trouble, and illustrate it on the phase diagram. In going from the oscillatory motion to the rotational motion, or vice versa, one had to pass through a singular motion in which the period becomes infinite. (This is well known for the completely analogous case of the pendulum, where the oscillation of amplitude π has infinite period.) Although Ehrenfest did not point it out explicitly yet, the adiabatic assumption cannot hold in going through a singular motion: the parameter would have to be changed at a rate which is strictly zero if its change is to be small during one period, when that period is infinite.[10]

This difficulty was closely related to the problem of degeneracy. The single oscillatory motion can go into a rotation in either direction about the axis, with nothing in the problem to select one sense of rotation rather than the other.

The last difficulty Ehrenfest discussed (Einstein drew his attention to it), was of a rather different kind. Suppose one starts with an ensemble of N systems, all of which have the same value of the external parameter, the whole collection being in thermal equilibrium. This means that the various systems have the most probable distribution over their allowed motions, a distribution determined by the temperature and the statistical weights originally assigned. If the external parameter is now varied adiabatically, the ensemble of systems will automatically be distributed in a definite way over the new allowed motions, which are the adiabatic transforms of the original ones according to Ehrenfest's basic assumption. And this raises the question: is this new distribution the most probable one for the new value of the parameter? In other words, is the ensemble in equilibrium after the adiabatic transformation? Ehrenfest knew that there were cases in which the answer was yes. This is true, for example, for black-body radiation, which continues to be in equilibrium (i.e. it remains "black") when the reflecting enclosure that contains it is adiabatically expanded or contracted. This is true even if there is no mechanism, such as Planck's carbon speck, for allowing the exchange of energy among the independent normal modes. In the general case, however, if one starts with a distribution that is the most probable one and performs an adiabatic process, the resulting distribution will not be the most probable one for the new value of the parameter.

Ehrenfest made one remark in his paper which is particularly striking in view of its date. He was considering a possible objection to his whole idea, namely, that "there is no sense in combining a theorem derived from the equations of mechanics with the anti-mechanical hypothesis of energy quanta." To this he had a definite answer: "Wien's law holds out the hope to us that results which may be derived from classical mechanics and electrodynamics by the consideration of macroscopic adiabatic

10. This difficulty was straightened out by Niels Bohr in a paper that was not published until many years later. N. Bohr, *Abhandlungen über Atombau aus den Jahren 1913–1916*, transl. H. Stintzing (Braunschweig: Friedr. Vieweg & Sohn, 1921), footnote 3 on pp. 127–128. Ehrenfest's original choice of $nh/2$ had been correct, and Bohr showed how to justify it by arguing from the conical pendulum.

processes will continue to be valid in the future mechanics of energy quanta." From the beginning Ehrenfest saw his adiabatic hypothesis as a guide line in searching for the principles of the new physics, and not as a vestige of the past or as just a useful calculational device.

4. Earlier in 1913 a number of leading physicists had expressed their own ideas on various problems of the quantum theory at the Wolfskehl lectures in Göttingen. Ehrenfest had also attended, but he was not one of the invited speakers, as we have already seen. Some of the papers read in Göttingen make an interesting contrast to Ehrenfest's approach.[11]

Max Planck lectured on a possible way in which the quantum theory could be applied to the study of the ideal gas. He had been trying to minimize the use of quanta in the theory of radiation, as mentioned above. But by this time the evidence from specific heats seemed to be a much stronger support for the quantum theory than the radiation law. Planck pointed out that one could account for the sharp reduction in specific heats as the temperature was lowered, and still maintain the equipartition theorem, if the number of degrees of freedom of the solid were reduced at low temperatures. Each quadratic term in the energy would still contribute its equipartition amount, $k/2$, to the specific heat, but the number of such terms would be reduced by the formation of larger and larger absolutely rigid complexes of atoms at lower and lower temperatures. Planck reported that he had "tried for a while to discover as simple a law as possible for this complex formation, on the basis of the chemical law of mass action, which would be in agreement with the measurements, but with no success worth mentioning." He finally convinced himself that such an explanation had to be ruled out as incompatible with other properties of solids at low temperatures. Matter did not form these rigid complexes, but energy did – the quanta of energy. Planck's quantum theory of the ideal gas can be omitted as irrelevant here. What is relevant is that, as late as 1913, he was still ready to consider an alternative to the use of the quantum theory in a field where it had proved so successful, the theory of the specific heats of solids.

Peter Debye's Wolfskehl lecture demonstrated quite a different

11. *Vorträge über die kinetische Theorie der Materie und der Elektrizität*, M. Planck, P. Debye, W. Nernst, M. v. Smoluchowski, A. Sommerfeld, and H. A. Lorentz, (Leipzig: B. G. Teubner, 1914). The discussions that took place at Göttingen were not published. Also see Chapter 9.

point of view. He realized that what had been done so far with the idea of quanta was "only the first step." If one looked at the quantum theory as essentially a modification of statistical mechanics, then one had to admit that it simply did not supply the methods needed for dealing with some of the most important applications of that theory. Debye was particularly interested in determining the equation of state of solids at arbitrary temperatures. He showed how one could handle the nonlinear atomic vibrations when classical statistical mechanics applied by using a simple approximate method, and derived the important result that thermal expansion depended for its existence on asymmetric, nonlinear vibrations. The question was how to determine the equation of state at low temperatures where the quantum theory must come into play, as shown by the behavior of the specific heat. Although he had no general solution to offer, Debye had found "a trick" by means of which he could push through a first approximation.

Debye's idea was to make use of Planck's approach, in which h was taken to be the quantum of action. For the harmonic oscillator the allowed phase curves were those energy ellipses which contained areas in the phase plane equal to integral multiples of h. Debye proposed extending this concept to his anharmonic oscillators. The curves of constant energy in the phase plane would no longer be ellipses, but one could still restrict the motion to those curves, whatever their shape might be, which enclosed an area equal to an integral multiple of h. Debye admitted the arbitrariness of his procedure, but obviously thought it worth trying in the current unsettled state of the theory. It did lead to results and these results could be subjected to experimental test. (Debye's rather pragmatic attitude was, in that sense, justified by his results. His gift for finding the right physical simplification did not fail him, and his insights into the nature of thermal expansion and thermal conductivity have proven to be of lasting value.)

The Wolfskehl week took place in May 1913, several months too early for any discussion of the most significant new work on the quantum theory, the first of Niels Bohr's papers, "On the Constitution of Atoms and Molecules."[12] Bohr, whose boldness and originality were comparable only to Einstein's, had written an extremely rich and

12. N. Bohr, "On the Constitution of Atoms and Molecules," Phil. Mag. **26** (1913), p. 1. The other two papers in the series appeared in the same volume, at pages 476 and 857. The three papers were reprinted in a book under the same title, with an extremely interesting and informative introduction by L. Rosenfeld. (Copenhagen: Munksgaard 1963). The three papers appear in German translation in the book cited in note 10.

suggestive work. It was the first phase of an attempt to develop a theory of atomic and molecular structure and to use this theory to understand a whole range of physical and chemical properties of matter.[13] In this first paper Bohr had studied the binding of a single electron by a nucleus with a single positive charge to form a hydrogen atom. His work was based on Rutherford's model of the atom as a small massive nucleus surrounded by light electrons. Among other things Bohr worked out the frequencies of the spectral lines that could be emitted or absorbed by hydrogen on the basis of his theory. Although this was not a part of his original program, it turned out to be the most strikingly successful aspect of the theory. Bohr's results agreed remarkably well with experiment, not only for the general structure of the spectrum, but also for its single numerical constant. As Bohr pointed out, Planck's constant was just what was needed to complete the determination of the scale of atomic dimensions.

Since Bohr was working on a quantum theory of atomic systems, he had to face the same question that his fellow theorists were puzzling over: how was the quantum theory of Planck's oscillator to be extended to more general systems? It is characteristic of Bohr that he did not try to give a simple answer to this difficult question.[14] He saw that in Planck's theory of the oscillator the motions were restricted to certain discrete values of the energy. When the oscillator emitted or absorbed energy it was by means of a definite transition from one of these allowed states to another, and no radiation was emitted or absorbed while the oscillator remained in one of its allowed states. Planck's requirement that the energy be capable of only certain discrete values was, in effect, a way of achieving stability, a stability utterly foreign to classical electrodynamics in which an oscillating charge (or any accelerated charge) would necessarily radiate energy. It was apparently this feature of the quantum theory, the stability imposed by the existence of discrete stationary states, that impressed Bohr most strongly, for it was the stability of atomic systems which seemed both so characteristic and so hard to understand.

---

13. For general discussion of Bohr's papers see Rosenfeld's introduction, *op. cit.*, note 12. Also see Jammer's book, *op. cit.*, note 10e of Chapter 10, pp. 62–88, and K. M. Meyer-Abich, *Korrespondenz, Individualität, und Komplementarität*, (Wiesbaden: Franz Steiner Verlag, 1965), pp. 1–45.

14. See Bohr's early discussion of his work in a lecture "On the Spectrum of Hydrogen," given in Copenhagen on 20 December 1913. It is reprinted in N. Bohr, *The Theory of Spectra and Atomic Constitution*, (Cambridge: University Press, 1922), p. 1.

At the most general level Bohr postulated that atomic systems possessed a series of allowed stationary states, in which no radiation occurred. While the system remained in such a state one could apply ordinary mechanics to discuss its dynamical equilibrium, but the process of transition from one such state to another could not be treated that way. For the transition process Bohr assumed that the radiation emitted was "homogeneous," of a single frequency ν. This frequency ν was to be related to the energy E emitted, the energy difference between the initial and final states, by Planck's relationship, $E = h\nu$. The first of these assumptions amounted to a severe restriction on the validity of mechanics, to which the concept of stationary state is foreign. Bohr described the second as being "in obvious contrast to the ordinary ideas of electrodynamics," but he was notoriously mild in his language: it was absolutely heretical. He renounced any attempt to identify the frequency of the emitted light with the frequency of some motion of the sources of the electromagnetic field, the electrically charged particles in the atom.

Bohr also made a more specific assumption about the way in which the quantum of action must be used to determine the stationary states. He actually did this in several different ways in this first paper, beginning with the assumption that the frequency ν of the light emitted when a free electron at rest is bound into a stationary state, is just half the frequency ν' of the electron's revolution in its circular orbit in that state. Bohr then used "Planck's relationship" to set the energy E emitted equal to $nh\nu$ or $nh\nu'/2$, with n an integer. (Bohr shifted his interpretation of the equation $E = nh\nu'/2$ in the course of his paper, originally treating n as the number of quanta emitted, there being a different number for each stationary state, but then later treating E as one quantum of the appropriately higher frequency.) It was also in the course of his probing into the significance of this assumption that Bohr introduced a mode of argument he would later use more and more. This was the requirement that any result obtained from the quantum theory would have to agree with the result of a classical calculation "in the region of slow vibrations," where the classical theory was in agreement with experiment.

While Bohr did not arrive at his results this way, he found "a very simple interpretation" of his stationary states "by help of symbols taken from the ordinary mechanics." He warned, however, that "there obviously can be no question of a mechanical foundation of the calculations given in this paper." The "interpretation" Bohr referred to was the remark that the angular momentum of the electron in its *n*th circular orbit around the nucleus would be precisely $nh/2\pi$. This interpretation, in which Bohr was surely influenced by the prior work of J. W. Nicholson,[15] quickly took a central position in Bohr's work on atomic problems. It is, in fact, remarkable to see how the idea of quantized angular momentum, (to use an anachronistic term), which appears so guardedly on page 15 of Bohr's paper, after much discussion of other ways of interpreting his generalization of Planck's equation, becomes the key idea in his thinking about atomic and molecular problems throughout the rest of his series of three papers. His "main hypothesis" in studying "the permanent states" of atomic and molecular systems is that "the angular momentum of every electron round the centre of its orbit" will be $h/2\pi$ in this permanent state.

5. Bohr's contemporaries reacted to his work in a great variety of ways. Sommerfeld,[16] for example, wrote Bohr to thank him for his "extremely interesting work." He remarked that he had been thinking for some time about the problem of expressing the Rydberg constant in terms of Planck's *h* and had even "talked about it to Debye a few years ago." He was "for the present still rather skeptical about atom models in general," but felt "nevertheless" that "the calculation of this constant is indisputably a great achievement." Sommerfeld was interested enough to want to apply Bohr's methods to the Zeeman effect, and asked whether Bohr had any plans along this line.

This was one characteristic reaction – characteristic, that is, of Sommerfeld. Elsewhere there were doubts, as at Göttingen, where Richard Courant was "laughed at for taking such fantastic ideas seriously."[17] Ernest Rutherford, Bohr's great mentor, found his ideas "very ingenious," but thought that "the mixture of Planck's ideas with the old mechanics" made it "very difficult to form a physical idea of

15. For a discussion of Nicholson's work and its influence on Bohr see R. McCormmach, "The Atomic Theory of John William Nicholson," Archive for History of Exact Sciences **3** (1966), p. 160.
16. A. Sommerfeld to N. Bohr, 4 September 1913. Quoted in full by Rosenfeld, *op. cit.*, note 12, p. lii.
17. Quoted by L. Rosenfeld and E. Rüdinger in their article, "The Decisive Years 1911-1918," in *Niels Bohr. His Life and Work as Seen by his Friends and Colleagues*, ed. S. Rozental, (Amsterdam: North-Holland, 1967), p. 57.

what is the basis of it."[18] James Jeans presented Bohr's ideas in a favorable light to the British Association in September 1913. Having been convinced of the necessity of quanta by Poincaré's work, Jeans was ready to take the very large next step needed to accept Bohr's new theory.[19]

We have Einstein's reaction to Bohr's work in the words of George Hevesy. When Hevesy told him that new experiments confirmed Bohr's explanation of the origin of the Pickering lines (as produced by ionized helium), "the big eyes of Einstein looked still bigger and he told me 'Then it is one of the greatest discoveries.'"[20] In another letter Hevesy described Einstein as being particularly impressed by one point: "'Then the frequency of the light does not depand at all on the frequency of the electron' – (I understood him so??) And this is an *enormous achiewement.* The theory of Bohr must be then wright."[21] (The picturesque spelling is Hevesy's.) Just this point would be a major objection to Bohr's work for most physicists. Here is what Erwin Schrödinger had to say about it in 1926: "The frequency discrepancy in the Bohr model, on the other hand, seems to me, (and had indeed seemed to me since 1914), to be something so *monstrous*, that I should like to characterize the excitation of light in this way as really almost *inconceivable*."[22]

Schrödinger's words of 1926 are not terribly different from those that Paul Ehrenfest used in expressing his first reaction to Bohr's paper. In a letter to Lorentz written in August 1913, Ehrenfest commented: "Bohr's work on the quantum theory of the Balmer formula (in the *Phil. Mag.*), has driven me to despair. If this is the way to reach the goal, I must give up doing physics."[23] This was not just a casual remark. Ehrenfest's notebooks and correspondence clearly indicate that

---

18. E. Rutherford to N. Bohr 20 March 1913. Quoted in full in Bohr's Rutherford Memorial Lecture of 1958, "Reminiscences of the Founder of Nuclear Science and of Some Developments Based on his Work," Proc. Phys. Soc. **78** (1961), p. 1091. Reprinted in N. Bohr, *Essays 1958–1962 on Atomic Physics and Human Knowledge*, (New York: John Wiley, 1963), p. 41.
19. See J. H. Jeans, *op. cit.* note 58, Chapter 10, p. 50.
20. G. Hevesy to E. Rutherford, 14 October 1913. Quoted by Rosenfeld, *op. cit.*, p. xlii.
21. G. Hevesy to N. Bohr, 23 September 1913. Quoted by Rosenfeld, *op. cit.*, p. xlii.
22. E. Schrödinger to H. A. Lorentz, 6 June 1926. See *Letters on Wave Mechanics*, E. Schrödinger, M. Planck, A. Einstein, and H. A. Lorentz, ed. K. Przibram, transl. M. J. Klein, (New York: Philosophical Library, 1967), p. 61.
23. P. Ehrenfest to H. A. Lorentz, 25 August 1913.

he did not try to work himself into a full understanding and acceptance of Bohr's approach for several years. Once he met Bohr and absorbed the essence of his ideas and attitudes in direct conversation, Ehrenfest became an enthusiastic booster and close friend, but that had to wait for the end of the World War. In 1913 and 1914 Ehrenfest continued to work on his own approach to the problems of the quantum theory, uninfluenced by Bohr's contribution.

6. One of the questions that had troubled Ehrenfest for years was what the quantum theory might do to the statistical foundations of thermodynamics. Boltzmann's derivation of the second law of thermodynamics as a theorem in statistical mechanics had used as one of its essential links the assumption that equal volumes in phase space were to be assigned equal statistical weights. It was of the essence of the quantum theory, however, that this classical uniform weighting of phase space was to be abandoned, and only certain particular regions of phase space were to be allowed, that is, to be given nonzero weight. In the case of Planck's oscillators, for example, the allowed domains were those particular ellipses of constant energy whose enclosed areas were integral multiples of h.

Ehrenfest had already considered this question in detail for the quantum theory of harmonic oscillators in his 1911 paper.[24] He had shown that Boltzmann's relationship between entropy and the number of ways of achieving the most probable distribution remained valid precisely because Planck had quantized the oscillator's adiabatic invariant, the ratio of its energy to its frequency, Ehrenfest had proved more generally that if, and only if, the statistical weight function depended only on this adiabatic invariant, the statistical thermodynamics of the oscillator was secure.

In 1914 Ehrenfest was ready to discuss the problem again on a much more general basis, without any special reference to the oscillator. In a paper[25] sent off to the *Physikalische Zeitschrift* in May, he dealt systematically with one sharply defined question. For a general ideal molecular system, what restrictions must be placed on the statistical weight function in phase space so that Boltzmann's relationship between the thermodynamic properties of the system and the averages

24. See the reference in note 48 of Chapter 10, and the discussion in section 9 of Chapter 10.
25. P. Ehrenfest, "Zum Boltzmannschen Entropie-Wahrscheinlichkeits-Theorem," Phys. Z. **15** (1914), p. 657. *Papers*, p. 347.

of its molecular properties over the most probable distribution remains valid?

Ehrenfest considered a system of independent molecules, each of which was specified by its coordinates $q_1, \ldots, q_r$ and their corresponding conjugate momenta $p_1, \ldots, p_r$. The energy ϵ of the molecule was to be considered a function of the q's and p's, (giving the position of the molecule in its $2r$ dimensional phase space), and also of certain externally controllable parameters $a_1, a_2, \ldots$, such as the volume,

$$\epsilon = (q_1, \ldots, q_r, p_1, \ldots, p_r; a_1, a_2, \ldots), \tag{11.8}$$

or, in a more concise notation,

$$\epsilon = \epsilon(q, p; a). \tag{11.9}$$

The generalized force exerted by a molecule, corresponding to a change in one of the parameters a_i, would be given by the negative derivative, $-\partial\epsilon/\partial a_i$. If the stationary distribution function in phase space is denoted by $f(q, p; E, a)$, where E is the total energy of the system of N molecules, then the force A_i exerted by the system is given by the equation

$$A_i = -\int \mathrm{d}\tau f \frac{\partial\epsilon}{\partial a_i}, \tag{11.10}$$

where $\mathrm{d}\tau$ stands for $\mathrm{d}q_1 \ldots \mathrm{d}p_r$, and the integral is over all of the molecular phase space. The work done by the system in any process in which the parameters a_i and the energy E are changed can then be written as

$$\sum_i A_i \delta a_i = -\int \mathrm{d}\tau f \delta\epsilon \tag{11.11}$$

where $\delta\epsilon = \Sigma_i (\partial\epsilon/\partial a_i)\delta a_i$. Since the total energy E of the system is given by the equation

$$E = \int \mathrm{d}\tau f \epsilon \tag{11.12}$$

its change δE in this process can be written,

$$\delta E = \int \mathrm{d}\tau \, \epsilon \delta f + \int \mathrm{d}\tau f \delta\epsilon. \tag{11.13}$$

The heat δQ supplied to the system in such a process is defined as the sum of the system's energy change and the work done by the system, so one immediately obtains the result

$$\delta Q = \int d\tau\, \epsilon \delta f. \tag{11.14}$$

Ehrenfest now introduced the statistical weight function $G(q, p; a)$, where $G d\tau$ is the weight attached to the phase space volume $d\tau$ at the point (q, p), the external parameters having the value a. The probability W of a given distribution f can then be expressed in the form

$$W = \Pi\, (G\, d\tau)^{f d\tau} \frac{N!}{\Pi\, (f\, d\tau)!}, \tag{11.15}$$

where the products are taken over all volume elements in phase space. The most probable distribution is the one that maximizes W for given total energy E, and it is readily shown to be

$$f(q, p; a) = (N/Z) G(q, p; a) \exp\{-\mu\epsilon(q, p; a)\}. \tag{11.16}$$

Here $Z = \int d\tau G \exp(-\mu\epsilon)$, and μ is determined by the energy constraint, (11.12). With this explicit form for f, the expression (11.14) for the heat δQ can be reworked into the form

$$\mu\delta Q = \delta(\mu E + N \ln Z) - (N/Z) \int d\tau \exp(-\mu\epsilon)\delta G. \tag{11.17}$$

The quantity δG is the change in the weight function G due to the changes in the parameters; $\delta G = \Sigma_i (\partial G/\partial a_i)\delta a_i$.

On the other hand, the quantity $\ln W$ can be calculated for the most probable distribution using (11.15) and (11.16), and it is not hard to show that

$$\delta \ln W = \delta(\mu E + N \ln Z). \tag{11.18}$$

If we compare (11.17) and (11.18) we see that they differ only by the extra term on the right hand side of (11.17), the term involving the change in the weight function. Now the theorem that relates thermodynamics to statistical mechanics is that $\mu\delta Q$ and $\delta \ln W$ are equal. In other words μ is an integrating factor for the differential heat, and so, up to a scale factor, μ is the reciprocal of the absolute temperature and $\ln W$ is the entropy. If this theorem, proved by Boltzmann for the special case that $G(q,p;a)$ is everywhere constant, is to hold, the extra term involving δG must vanish. Because this must hold for all

μ and all a, Ehrenfest could re-express the necessary and sufficient condition for the validity of Boltzmann's theorem in the form

$$\int_{\epsilon=\epsilon_1}^{\epsilon=\epsilon_2} \mathrm{d}\tau \, \delta G = 0; \qquad (11.19)$$

the integral of δG between any two energy shells must vanish.

This condition, which Ehrenfest referred to as the "δG-condition," is a somewhat guarded way of stating that the weight function must be adiabatically invariant if the statistical foundations of the second law are to remain valid.[26]

Ehrenfest investigated one particular situation in some detail. Suppose one considers an energy surface in the $2r$ dimensional molecular phase space, that is, a $(2r-1)$ dimensional surface of the form $\epsilon(q_1, \ldots, p_r; a_1, a_2, \ldots) =$ constant. This surface will enclose a $2r$ dimensional phase volume which can be denoted as $i(q_1, \ldots, p_r; a_1, a_2, \ldots)$, since for given values of the parameters a, there will be only one energy surface passing through a definite point $q_1, \ldots, p_r$. Ehrenfest showed that any weight function $G(q, p; a)$ which had the form $\Gamma(i)$, (any weight function, that is, which depended only on the phase volume enclosed by the corresponding energy surface), would be adiabatically invariant and so would satisfy the δG-condition. Furthermore, he could prove for this case something that he could not prove in general, namely, that the distribution parameter μ now had *all* the thermodynamic properties of the temperature. This case, where the weight function had the form $\Gamma(i)$, actually included the classical Boltzmann weighting and all those generalizations of it already used in the quantum theory by Planck, Debye, and Ehrenfest himself. It was not clear, however, that all interesting and relevant weight functions would have to be of this particular form.

Ehrenfest's proof that statistical weights had to be adiabatically invariant did not draw very much attention. Few people realized that there was anything to be concerned about. Even Einstein, who was much taken with Ehrenfest's idea of the importance of adiabatic transformations, did not see at first what he was driving at in this

26. Ehrenfest was probably allowing for the possibility of discrete weighting, characteristic of quantization. The argument can be given in more condensed form if one assumes that there are only discrete states, but that would have been begging the question Ehrenfest asked. See N. Bohr, "On the Application of the Quantum Theory to Atomic Structure. Part I. The Fundamental Postulates," Proc. Cambr. Phil. Soc. (Supplement) (1924), footnote pp. 16–17.

calculation. Einstein thought for a while that the weight function could not depend on external parameters as a matter of principle.[27] But Ehrenfest soon convinced him that the situation was quite the reverse: if one imposed quantum conditions at all, the weight function would necessarily depend on parameters, and therefore one had to raise precisely the question Ehrenfest had considered.[28] Once Einstein saw the point he was thoroughly convinced. He helped to spread Ehrenfest's ideas by quoting and applying them in his next paper,[29] where the term "Ehrenfest's adiabatic hypothesis" appeared for the first time.

It must have been terribly frustrating for Ehrenfest that most of his contemporaries did not even see what he was driving at in this 1914 paper. He had not forgotten this four years later, as one can see from his remarks in a letter to Bohr.[30] "It is still extremely interesting that there is a pre-established harmony [31] between the 'δG-condition,' necessary for the second law to hold (see my 1914 note in the *Physikalische Zeitschrift*), and the adiabatic invariance of the quantum conditions. Oh, if you knew how much I had to fret before I succeeded in convincing anyone that there is a problem here: that we had lost the basis for Boltzmann's old proof of the second law *in principle*, as soon as we followed Planck in abandoning Boltzmann's old assumption that $G \equiv 1$."

7. Ehrenfest had not yet published a detailed and systematic treatment of his ideas on adiabatic invariance and its role in the quantum theory. His paper on rotational specific heats[7] was widely read, but it contained only a couple of rather cryptic footnotes on the principles underlying the calculation. The 1914 paper[25] dealt only with the adia-

27. A. Einstein to P. Ehrenfest, 18 May 1914.

28. A. Einstein to P. Ehrenfest, 25 May 1914.

29. A. Einstein, "Beiträge zur Quantentheorie," Verh. d. Deutsch. Phys. Ges. **16** (1914), p. 826.

30. P. Ehrenfest to N. Bohr, 10 May 1918.

31. It is interesting that Ehrenfest used the phrase "pre-established harmony" here. Precisely this phrase had been used by Poincaré in 1912 when he referred to the "mysterious, pre-established harmony" by which Planck's oscillators were "endowed with the only mechanical property which could be suitable" for their quantization. (See Section 10 of Chapter 10.) Ehrenfest never referred to Poincaré's paper in print, so far as I know, but he was apparently echoing it here, when dealing with a very similar point. One wonders whether this was done consciously, or whether it was an unconscious reflection of strongly negative feelings about Poincaré's paper. See Chapter 8.

batic invariance of statistical weights from the standpoint of one interested in the foundations of statistical mechanics. It did not discuss the behavior of quantum states under adiabatic transformations at all, and in any case it received very little notice, as we have just seen. Ehrenfest had discussed his general ideas in only one article, the short paper[9] presented to the Academy at Amsterdam in 1913. Even here he had avoided giving a sharp statement of the adiabatic hypothesis, and had limited himself to an illustrative example of how it worked. He was so scrupulous about pointing out the problems related to the adiabatic hypothesis which remained to be solved, that he failed to give his readers a sufficiently clear picture of the hypothesis itself. The Amsterdam *Proceedings* were also less widely read than Ehrenfest might have hoped.

As a result of these various factors Ehrenfest's contemporaries, including some of those most concerned with the same complex of problems, simply did not know what he had done. Max Planck, for example, wrote to Ehrenfest in November 1914.[32] He had been puzzled by the footnote in Ehrenfest's paper on specific heats, and wrote to ask the nature of the "very general" viewpoint from which Ehrenfest deduced that the rotational kinetic energy must be a multiple of $h\nu/2$. Planck was obviously unaware of the paper in the Amsterdam *Proceedings*, but he had become interested in the problem of the rotating dipole on other grounds and was about to make it the basis of several important studies of his own.[33] These studies, published late in 1915, constituted one detailed approach to the generalization of the quantum theory, by means of an analysis of what Planck called the structure of phase space.[34]

Planck's papers appeared at about the same time as another, very different, approach to the extension of quantum methods. Arnold Sommerfeld published several very long papers[35] on the quantum

32. M. Planck to P. Ehrenfest, 28 November 1914.

33. M. Planck to P. Ehrenfest, 22 December 1914, 29 March 1915, 27 April 1915, 23 May 1915.

34. See in particular M. Planck, "Die Quantenhypothese für Molekeln mit mehreren Freiheitsgraden," Verh. d. Deutsch, Phys. Ges. **17** (1915), pp. 407, 438, and M. Planck, "Die physikalische Struktur des Phasenraumes," Ann. Phys. **50** (1916), p. 385. *Planck Papers* II, pp. 349, 362, 386, respectively.

35. A. Sommerfeld, "Zur Theorie der Balmerschen Serie," Münchener Berichte, (1915), p. 425; "Die Feinstruktur der Wasserstoff und wasserstoffähnlichen Linien," *ibid.*, p. 459; "Zur Quantentheorie der Spektrallinien," Ann. Phys. **51** (1916), pp. 1, 125. See M. Jammer, *op. cit.*, pp. 89–109.

theory of spectra in which he generalized Bohr's methods and applied them to a thorough analysis of the spectrum of atomic hydrogen. By including the effects of the relativistic variation of the electron's mass, Sommerfeld was able to account in detail for the fine structure in the spectral lines of hydrogenic atoms. Bohr had already suggested that such an explanation might be possible,[36] but he had not attempted a real calculation since no precise experimental data were available for comparison. Sommerfeld, however, had the new fine structure measurements made by Paschen available to him, and he had worked the problem through completely, illustrating the power of his new method of determining the allowed quantum states.

Sommerfeld's work made a great impression, particularly on Bohr, who was in the best position to appreciate it. "I thank you so much for your most interesting and beautiful papers," Bohr wrote to Sommerfeld.[37] "I do not think that I ever have enjoyed the reading of anything more than I enjoyed the study of them, and I need not say that not only I but everybody here [at Manchester] has taken the greatest interest in your important and beautiful results." Bohr was so convinced of the significance of Sommerfeld's work for the future of the quantum theory that he withdrew from publication an important paper of his own after reading Sommerfeld's papers.[38] Bohr actually did not publish anything further until he had had a chance to review and reconstruct his whole approach after absorbing the work by Sommerfeld and his followers.

Paul Ehrenfest was also impressed by what Sommerfeld had done, and particularly by the way Sommerfeld's work seemed to bear out his own suggestion about adiabatic invariance. Sommerfeld had not mentioned this concept and seemed no more aware of it than Planck had been. Ehrenfest wrote to Sommerfeld in the spring of 1916, telling him how happy he was over Sommerfeld's great success. Ehrenfest was still not reconciled to Bohr's methods, however, as he

36. N. Bohr, "On the Series Spectrum of Hydrogen and the Structure of the Atom," Phil. Mag. **29** (1915), p. 332.

37. N. Bohr to A. Sommerfeld, 19 March 1916. Quoted by L. Rosenfeld and E. Rüdinger, *op. cit.*, note 17, p. 67. Einstein was equally impressed by Sommerfeld's work. See *Albert Einstein – Arnold Sommerfeld, Briefwechsel*, ed. A. Hermann (Basel/Stuttgart: Schwabe, 1968), pp. 40–41.

38. This paper, "On the Application of the Quantum Theory to Periodic Systems," was already in proof for its appearance in the April 1916 issue of the *Philosophical Magazine*. It was finally printed, in German translation, in 1921, (see *op. cit.*, note 10, p. 123), but only after it had been superceded by later papers by Bohr and others.

made clear in his next sentence. "Even though I consider it horrible that this success will help the preliminary, but still completely monstrous, Bohr model on to new triumphs, I nevertheless heartily wish physics at Munich further successes along this path!" Later in his letter Ehrenfest "took the opportunity" to call Sommerfeld's attention to the connection between his work and the "adiabatic hypothesis." He referred Sommerfeld to his paper in the Amsterdam *Proceedings*, defined what he meant by a reversible adiabatic change of the parameters, and then formulated his hypothesis: under an adiabatic change of a system's parameters those motions which are allowed (forbidden) by the quantum conditions are transformed into new motions which are allowed (forbidden). Ehrenfest mentioned that this was how he had quantized rotational motions for his specific heat calculation. He wished he had an opportunity to talk to Sommerfeld and tell him about "this whole adiabatic question and its relations to [his] poor unreadable Note in the *Physikalische Zeitschrift* for 1914, but to write or print it completely so it can be read – impossible!"[39]

Sommerfeld answered with a postcard thanking Ehrenfest for his good wishes. As for Ehrenfest's feeling that the Bohr model was monstrous, Sommerfeld thought it no more and no less so that Walther Ritz's combination principle for spectral lines. What he did not find monstrous was his own new way of introducing the quantum conditions. Sommerfeld advised Ehrenfest to be prepared for further quantum surprises coming from Munich. As for the adiabatic hypothesis which Ehrenfest had described so eagerly, Sommerfeld's only comments were that Bohr had praised it highly to him, and that Paul Epstein was going to give a lecture on it at Munich.[40]

Sommerfeld's postcard may well have convinced Ehrenfest that it was time to try to write up his ideas "so they could be read," however "impossible" the task might have appeared to be. In any event he had a paper ready for the meeting of the Academy in Amsterdam at the end of June.[41] This time Ehrenfest took pains to make sure that his

39. P. Ehrenfest to A. Sommerfeld, May 1916.

40. A. Sommerfeld to P. Ehrenfest, 30 May 1916.

41. P. Ehrenfest, "On Adiabatic Changes of a System in Connection with the Quantum Theory," Proc. Acad. Amsterdam **19** (1916), p. 576. *Papers*, p. 378. Also see P. Ehrenfest, "Adiabatische Invarianten und Quantentheorie," Ann. Phys. **51** (1916), p. 327; "Adiabatic Invariants and the Theory of Quanta," Phil. Mag. **33** (1917), p. 500. The German text, rather than either of the two English versions or the Dutch publication, (Versl. Akad. Amsterdam **25** (1916), p. 412) is almost surely Ehrenfest's own, and it is the one I have followed.

work would not be invisible outside Holland by having his paper printed also in the *Annalen der Physik* and the *Philosophical Magazine*, with only minor modifications.

8. Ehrenfest began his paper by making clear the basic concern that had moved him in this work. The theories in use for some fifteen years, applied to a wider and wider range of problems, made use of both the classical mechanics and electrodynamics, and also the hypothesis of quanta, a hypothesis that directly contradicted classical physics. Up to this time, only "tact and, so far as possible, experimental checks," could establish the extent to which each of the two fundamental bases of the theory should be used or ignored. One urgently wanted some more general point of view for establishing the boundary between "classical" and "quantal." Ehrenfest pointed to the curious case of Wien's displacement law as a possible guide. Here was a law derived by classical arguments about the behavior of radiation in a reversible adiabatic process, and yet it "stood unshaken amidst the breaking waves of the quantum theory." This noteworthy fact had been Ehrenfest's guide, as he now pointed out explicitly. It suggested that more general cases of reversible adiabatic changes might be treated by classical methods, even though other sorts of processes demanded the use of quanta. In following up this suggestion, Ehrenfest now explained, he had used the "adiabatic hypothesis," so named by Einstein, as he repeated carefully, and proudly too, no doubt. This hypothesis allowed one to extend the quantum conditions to a new situation, if this new situation could be reached by reversible adiabatic transformation from another in which the quantum conditions had already been successfully applied.

The purpose of his new paper was to give a precise formulation to this adiabatic hypothesis, to explain the nature and use of the adiabatic invariants, and to show how these ideas were related to some of the latest developments in the quantum theory. Ehrenfest also undertook to point out the difficulties that blocked any extension of these ideas to broader classes of motions, particularly nonperiodic motions.

To formulate the adiabatic hypothesis sharply, Ehrenfest considered a system specified by generalized coordinates $q_1, \ldots, q_n$ whose energy might also depend on parameters $a_1, a_2, \ldots$ which could be varied infinitely slowly. Such an infinitely slow variation of the parameters from values $a_1, a_2, \ldots$ to values $a_1', a_2', \ldots$ would change a given motion, $\beta(a)$, into another, $\beta(a')$, its adiabatic transform. Suppose now that the quantum theory had already been applied to the system for the particular parameter values a_{i0}, so that one knew the allowed motions for these values of the parameters. Ehrenfest's hypothesis was the following: "For general values $a_1, a_2, \ldots$ of the parameters, those and only those motions are allowed which are adiabatically related to the motions which were allowed for the special values $a_{10}, a_{20}, \ldots$ (i.e. which can be transformed into them, or may be derived from them in an adiabatic reversible way)." Although the hypothesis did not refer in so many words to periodic motions, it was implicitly limited to them because one could not properly define a reversible adiabatic transformation in any other case.

By an adiabatic invariant Ehrenfest meant any quantity whose value would be the same for the adiabatically transformed motion, $\beta(a')$, as it was for the original motion, $\beta(a)$. One example he could cite was the invariant, $\bar{T}/\nu$, the ratio of average kinetic energy to frequency for a periodic system. (Ehrenfest included a proof of the adiabatic invariance of $\bar{T}/\nu$ for periodic systems as an appendix to his paper. He also pointed out in some detail how this theorem was related to Boltzmann's original unsuccessful attempt to give a strictly mechanical interpretation of the second law of thermodynamics.) Ehrenfest could now indicate some new aspects of this adiabatic invariant. The kinetic energy T of any dynamical system can be expressed in terms of the generalized velocities, $\dot{q}_i$, and the corresponding canonical momenta, p_i, (which are defined by the equations $p_i = \partial T/\partial \dot{q}_i$), through the equation

$$2T = \sum_{i=1}^{n} p_i \dot{q}_i. \qquad (11.20)$$

(This result assumes only that T is a homogeneous quadratic function of the generalized velocities.) The adiabatic invariant can then readily be transformed as follows,

$$\frac{2\bar{T}}{\nu} = \int_0^P 2T\,\mathrm{d}t = \int_0^P \mathrm{d}t \sum_{i=1}^{n} p_i \dot{q}_i = \sum_{i=1}^{n} \int p_i \mathrm{d}q_i = \sum_{i=1}^{n} \int\!\!\int \mathrm{d}q_i \mathrm{d}p_i, \qquad (11.21)$$

where the last integrals are extended over the "projection" of the periodic motion onto the phase plane corresponding to the coordinate, q_i, and its conjugate momentum, p_i. (P is the period of the motion.)

For a one dimensional system this result simply said once more that the area enclosed by the phase space trajectory was adiabatically invariant. This is what Planck had quantized as an integral multiple of

h for the harmonic oscillator, where $2\overline{T} = E$, the energy of the oscillator. Debye's[11] quantization of this area for the anharmonic oscillator, also in integral multiples of h, could now be seen to be exactly appropriate according to the adiabatic hypothesis. Ehrenfest also considered the transformation from harmonic oscillator to free rotator again, this time determining the frequency of the rotational motion as the number of rotations per second. This led directly to a quantized angular momentum whose possible values were integral multiples of $h/2\pi$. The difficulty associated with the nonadiabatic passage through the singular motion of infinite period had not yet been overcome, although Ehrenfest suggested that the use of a conical pendulum might help here.[42]

Sommerfeld[35] had based his analysis of the hydrogen spectrum on the postulate that, with a suitable choice of the n generalized coordinates of a system, *each* of its coordinate-momentum pairs must satisfy an equation of the form

$$\int p_i \mathrm{d}q_i = n_i h \ (i = 1, \ldots, n; n_i = 0, 1, 2, \ldots) \qquad (11.22)$$

where the integral is extended over one period of the coordinate. (This has to be stated differently and more carefully for nonperiodic motions.) While this general formulation of the quantum conditions, n of them for a system of n degrees of freedom, obviously went further than the adiabatic hypothesis, Ehrenfest could show that in at least one important case, the two were equivalent. For a particle moving under the influence of a central force, the motion occurs in a plane. One of the two equations of motion, using polar coordinates, is just the constancy of the angular momentum. Since this equation contains no reference to the central force, the angular momentum, which Sommerfeld had quantized, is necessarily an invariant under adiabatic transformations of the parameters in the central force. The radial equation of motion can then be treated separately as if it described a one dimensional motion in which the apparent force is the actual force plus a "centrifugal repulsion," whose strength depends on the angular momentum. If the energy of the particle is such that the motion is bounded then this equation describes a one dimensional periodic motion, whose adiabatic invariant was again precisely the

---

42. Ehrenfest did not work this suggestion out, but he was correct, as shown by Bohr. See N. Bohr, *op. cit.* notes 10 and 38.

other quantity to which Sommerfeld had applied his quantum condition. Thus, Sommerfeld's quantization procedure was completely in accord with the adiabatic hypothesis for this case, although Ehrenfest could not extend his reasoning to cover Sommerfeld's further analysis including relativistic effects.

As already indicated, Ehrenfest had not made any progress with the problem of treating degenerate motion. He discussed it again, this time including a treatment of the transition from the nonperiodic (multiply periodic) two dimensional anisotropic oscillator to the isotropic (periodic) oscillator.

Ehrenfest also drew attention once more to the problem of adapting the foundations of statistical mechanics to the new situation introduced by the quantum theory. He emphasized again the difference between his reversible adiabatic processes and the adiabatic processes of thermodynamics: except for special cases, thermodynamic equilibrium would not be preserved by a "reversible adiabatic" transformation. The most probable distribution after adiabatic transformation would not be the transform of the initial most probable distribution.[43]

By the time Ehrenfest read the proof of his paper he had become aware of new papers by Epstein[44] and Schwarzschild[45] which carried the quantum theory still another step forward. Both men had shown how one could apply the quantum theory to multiply periodic systems, by separating the Hamilton-Jacobi equation using action and angle variables. The theory was immediately put to use to explain quantitatively the Stark effect in hydrogen, the splitting of the spectral lines by a strong electric field. Ehrenfest added a note to his paper, raising a question about the relationship between these successful new methods and his own ideas: "How far are the different parts into which these authors [Epstein and Schwarzchild] separate the integral of action . . . adiabatic invariants?" This question would be answered promptly, not by Ehrenfest but by his student, J. M. Burgers.[46] Ehrenfest was

43. For a recent discussion of this problem with references to some of the modern literature see R. D. Mountain, "Statistical Mechanics of Adiabatic Processes," Physica **30** (1964), p. 808 and M. J. Klein and R. D. Mountain, "Adiabatic Processes and the Equation of State," Physica **30** (1964), p. 818.

44. P. S. Epstein, "Zur Theorie des Starkeffektes," Ann. Phys. **50** (1916), p. 489; "Zur Quantentheorie," Ann. Phys. **51** (1916), p. 168.

45. K. Schwarzschild, "Zur Quantenhypothese," Berliner Berichte (1916), p. 548.

46. J. M. Burgers, "Die adiabatischen Invarianten bedingt periodischer Systeme," Ann. Phys. **52** (1917), p. 195. Also see his dissertation, *Het Atoomodel van Rutherford-Bohr* (Haarlem: De Erven Loosjes, 1918).

convinced that these "phase integrals" must be adiabatic invariants, but he never mastered the technique of the theory of contact transformations well enough to attempt a proof. At Ehrenfest's suggestion Burgers took up the question and succeeded in demonstrating this important result by several distinct methods. For Burgers, "it was somewhat like solving a puzzle,"[47] but Ehrenfest never lost the feeling that he himself was "horribly ignorant, and inept," and he had "fallen far behind when it came to the theory of integration for canonical systems."[48] Whether that was true or not, it was surely Ehrenfest who saw the importance of the problem, which he could view in the context of his whole approach to the quantum theory.

9. Ehrenfest's ideas continued to meet with mixed reactions. Sommerfeld finally responded[49] more fully to Ehrenfest's letter on the adiabatic hypothesis after he read the detailed discussion which appeared in the *Annalen*[41] at the end of October 1916. In his first paper Sommerfeld had referred to his form of the quantum conditions as "unproven and perhaps unprovable." He did not seem to think that Ehrenfest's adiabatic ideas threw any more light on the quantum conditions, and preferred to be guided by the spectral lines themselves, as he put it. Since *they* had nothing to say about such difficulties as those that appeared in going from an anisotropic to an isotropic oscillator, neither would he. What he really would have liked was some assistance from the adiabatic principle in performing specific calculations, and when he found it his respect for the principle went up sharply.[50]

(Sommerfeld ended his somewhat lukewarm letter with the remark that it was a shame Walther Ritz could not have been alive to take part in these new developments. His rich imagination, and his optimistic readiness to lay hold of a problem, unencumbered by too critical an approach, would have found just the right channel here. This remark was very likely made quite innocently, but it must have had a crushing effect on Paul Ehrenfest. Although Sommerfeld made no direct comparison between Ehrenfest and his brilliant friend who had died in 1909,

47. J. M. Burgers, *Autobiographical Notes* (Unpublished), p. 31.
48. P. Ehrenfest to J. M. Burgers, 13 February 1923.
49. A. Sommerfeld to P. Ehrenfest, 16 November 1916.
50. See, for example, Sommerfeld's book, *Atomic Structure and Spectral Lines*, (New York: E. P. Dutton, 1923) transl. H. L. Brose (from 3rd German edition), pp. 304–313 594–596.

Ehrenfest would have rushed to the comparison himself, as he always did. This was particularly so because Sommerfeld had praised Ritz for having just those qualities whose lack Ehrenfest felt so keenly.)

Whatever Ehrenfest may have felt about Sommerfeld's initial response to his paper would have been more than compensated for two years later when he read Niels Bohr's monograph, *On the Quantum Theory of Line-Spectra*.[51] Bohr, whose work had seemed so alien to Ehrenfest at first, had taken the adiabatic hypothesis to heart, and incorporated it into the complex structure of his own ideas on the quantum theory. He preferred to call it the "principle of mechanical transformability," and he did not view it exactly as Ehrenfest did. Nevertheless, Bohr bracketed it with Einstein's recent work on the theory of radiation as a sign of "great progress" in the attempt to solve the "many difficulties of fundamental nature" in trying to formulate a really "consistent and developed structure" for the quantum theory. The adiabatic principle achieved its full promise in Bohr's work in the years from 1918 on.

For Ehrenfest himself the adiabatic hypothesis would always be his most prized work. "I know that I have never discovered anything – and quite surely never will discover anything – that I can love so fervently as this train of thought which I found with so much joy."[52]

51. N. Bohr, "On the Quantum Theory of Line-Spectra," Det. Kgl. Danske Vidensk. Selsk. Skrifter, 8. Raekke, **4** No. 1 (1918–22). Quote from p. 4.
52. P. Ehrenfest to N. Bohr, 8 May 1922.

CHAPTER 12

Einstein and Ehrenfest

Before his appointment to the Leyden professorship, Paul Ehrenfest was ready to move his family to Zürich, even though he had no position there, just to be near Albert Einstein. His one visit to Einstein, in Prague early in 1912, had made that deep an impression on Ehrenfest. When he did move to Leyden, Ehrenfest wasted no time in trying to arrange another visit. A projected trip to Zürich during the Christmas holidays in 1912 did not work out, although Einstein wrote that he would be "always welcome".[1] But in June 1913 Paul and Tatyana were able to spend several weeks in Switzerland, leaving their daughters at home with Tatyana's mother who was visiting from Russia.

Einstein had been working very hard on the theory of gravitation. In a letter[1] to Ehrenfest the previous autumn he had reported "advances" in the theory, among other "bits of scientific news", but he felt his progress was "very little, after all, compared to the enormous efforts" he had been making. By the time he wrote to Ehrenfest at the end of May[2] things were looking better. Einstein, like most of Ehrenfest's correspondents, could not keep up with his friend's passion for writing long and frequent letters. "My heart pounds as I stare at this sheet of paper destined for you, conscious of how grievous are my letter writing debts. But my excuse is the actually superhuman exer-

1. A. Einstein to P. Ehrenfest, undated, late fall 1912.
2. A. Einstein to P. Ehrenfest, 28 May 1913.

tions with which I have devoted myself to the problem of gravitation. I am now inwardly convinced that I have found the right way, but at the same time I am also sure that a murmur of indignation will travel down the rows of our colleagues when the work appears, which will happen in a few weeks." Einstein invited Ehrenfest to stay at his home in Zürich, partly for the pleasure of having him there, but especially because it would prove that Ehrenfest had not taken offense at his "perpetual silence".

The Ehrenfests actually stayed at a pension in Zürich, but they spent a great deal of time with Einstein and his family.[3] Ehrenfest had the opportunity to meet Einstein's old college friend, the mathematician Marcel Grossmann, who had just been collaborating with Einstein on the mathematical aspects of his effort to generalize the theory of relativity. He also met Michele Besso, the Italian engineer with whom Einstein had been discussing his revolutionary scientific ideas since his days at the Swiss patent office in Bern. Einstein once referred to Besso as the best sounding-board in Europe,[4] and immortalized him by thanking him for his "loyal assistance" and for "several valuable suggestions" in the only acknowledgment contained in the 1905 paper on the special theory of relativity.[5]

The morning after the Ehrenfests' arrival at Zürich, Einstein called for Paul at their pension and the two men walked up into the hills together. It was a hazy hot day, and they rested under the trees overlooking a meadow while they talked about the theory of gravitation.[6] Einstein compared his own ideas with those of Gunnar Nordström, who was working on another approach to a relativistic theory. Nordström was also in Zürich at that time, and there were to be many more discussions of gravitation. (Nordström later spent a year at Leyden with Ehrenfest.)

One evening Ehrenfest attended a colloquium on Lawrence Bragg's new simplification of the theory of X-ray diffraction by crystals. After the colloquium a number of the participants continued the discussion in the garden of a nearby coffee house. Ehrenfest argued with Einstein and Max von Laue about Bragg's X-ray reflections, and then Einstein

3. Diary "I", 18 June–July 1913.
4. C. Seelig, *Albert Einstein. A Documentary Biography*, transl. M. Savill (London: Staples Press, 1956), p. 71.
5. A. Einstein, "Zur Elektrodynamik bewegter Körper", Ann. Phys. **17** (1905), p. 891.
6. Diary "I", 19 June 1913.

and Ernst Zermelo, Boltzmann's former antagonist, argued about the Brownian motion.[7]

While they were in Zürich, the Ehrenfests visited the mother and sister of their lamented friend, Walther Ritz.[8] Ten years had gone by since Paul had spent a week with Ritz in Zürich after his Italian holiday.[9] Ritz may already have known then that he had only a few years left, years that he spent in scientific work to the limit allowed by his failing health. "There may be only a little time left to me," he had written in 1906, "and I am firmly resolved to spend this time in scientific and intellectual surroundings, which are the only ones that can give me any satisfaction and make me feel alive, and so are perhaps needed for my health. I cannot hope for the joys of family life or the comforts of the old bachelor who has kept his health; all that is left for me is science and the intellectual life."[10] Ritz had been dead for four years but the Ehrenfests would never forget him.

Ehrenfest spent as much time as he could with Einstein talking physics, making music, taking photographs, and renewing his acquaintance with young Albert. Only once during the two weeks in Zürich did Ehrenfest record in his notebook, "A day without Einstein!"[11]

In the fall of 1913 Ehrenfest had surprising news from Zürich. Einstein had accepted an offer of a research professorship at the Kaiser Wilhelm Institute and the University of Berlin. Max Planck and Walther Nernst had persuaded the German authorities to appoint Einstein to a specially created chair and, at the same time, to make him a member of the Prussian Academy of Sciences.[12] It was an unprecedented honor for the 34 year old Einstein, as well as being a very attractive offer financially. Only six years earlier Einstein's application to become an unpaid *Privatdozent* at the University of Bern had been rather scornfully rejected by the faculty. Now he was invited to come to one of the world's scientific centers on terms which left him completely free to work on his own ideas. The offer was irresistible. As Einstein wrote to

---

7. Diary "I", 24 June 1913.
8. Diary "I", 25 June 1913.
9. See Chapter 3.
10. W. Ritz to P. Weiss, quoted in Weiss's preface to W. Ritz, *Oeuvres* (Paris: Gauthier-Villars, 1911), p. ix.
11. Diary "I", 25 June 1913.
12. C. Seelig, *op. cit.*, p. 143.

Ehrenfest,[13] after the usual apology for his "offensive silence": "First of all, thanks very much for your congratulations on my Berlinization. I accepted this peculiar sinecure because giving lectures grates so oddly on my nerves, and I don't have to lecture there at all."

Einstein was not very enthusiastic about his move to Berlin. He is quoted as having said: "The Germans are gambling on me as they would on a prize hen. I do not really know myself whether I shall ever really lay another egg."[14] He wrote to Besso that winter: "I am somewhat uneasy as I see the Berlin adventure approaching."[15] One reason can perhaps be found in his remark in the same letter, in connection with the very limited response to his new work on the theory of gravitation, that the German theorists were not receptive to general discussions based on fundamental principles. "A free unprejudiced look is generally not characteristic of (adult) Germans. It's as if they wore blinders." The one place where Einstein did look for a real and understanding interest in his work was Leyden, and he planned a trip there in March 1914, while his household goods were being moved to Berlin. His wife and children were going to Locarno for a few weeks, so that the younger son could convalesce after a recent attack of whooping cough.[16] Einstein was extremely eager to see Ehrenfest and to discuss his work with him, and especially with Lorentz.

Einstein spent a week in Leyden at the end of March, lecturing on his recent work, visiting Lorentz in Haarlem and being shown the sights of Holland. He made the trip to Leyden with Lorentz's former student, A. D. Fokker, who had been working with Einstein for some months in Zürich. Almost two weeks after his departure from Leyden, Einstein wrote to Ehrenfest from Berlin.[17] "Shame on me that I haven't written you until now. I spent an unforgettable week with you both in an unhurried, relaxed and comfortable way. My warmest thanks to you and your wife." One sign of the progress of their friendship is that Einstein switched for the first time to the informal "Du" in addressing Ehrenfest in this letter. The two men had obviously spent a great deal of time discussing the highly problematic quantum theory, and especially Ehrenfest's new ideas on a general principle that might

13. A. Einstein to P. Ehrenfest, undated, winter 1913-14.
14. C. Seelig, *op. cit.*, p. 148.
15. A. Einstein to M. Besso, undated, winter 1913-14.
16. A. Einstein to M. Besso, undated, March 1914.
17. A. Einstein to P. Ehrenfest, 10 April 1914.

guide future developments, the adiabatic principle. Einstein's letters through the next few months are full of questions and critical remarks on this subject, mostly expressed in his usual very physical way in terms of a simple system intended as a counter example. Ehrenfest gave as much as he got in the arguments. "You are completely right, you temperamental young fellow," Einstein wrote in one note.[18] "I had already noticed it before your threatening letter arrived." Einstein was deeply impressed by Ehrenfest's adiabatic transformations. They "would not leave him alone", and he thought they "might be the most valuable aid in the generally hopeless situation" of the quantum theory.[19]

In May 1914 Ehrenfest spent a week with the Einsteins in Berlin. Ehrenfest had for some time been struggling with the problem of how the developing quantum theory affected the foundations of statistical thermodynamics. He had just sent off a paper[20] on the subject to the *Physikalische Zeitschrift*, but it had not answered all the questions, and he was not satisfied with it. He had been arguing about aspects of this problem with Einstein by mail, but now it was time for more direct conversation. Ehrenfest went on with his calculations, even on the train taking him from Amsterdam to Berlin.

There was a lot to talk about in Berlin besides his own ideas on the quantum theory. Einstein also had some things to say on the quantum theory, as well as on his developing theory of gravitation. And Berlin was very much alive scientifically. As Einstein wrote just a few days before Ehrenfest's visit,[21] "It is enormously stimulating here, so that I recommend you to visit here just on that basis . . . . Franck and Hertz have discovered that electrons are elastically reflected from Hg atoms up to a velocity corresponding to 4.8 volts, and that they lose all their kinetic energy in collisions at this velocity, with the emission of monochromatic light such that the kinetic energy = $h\nu$ within a few percent. A wonderful inverse of the photoelectric phenomenon. A striking confirmation of the quantum hypothesis."

Ehrenfest also had a chance to renew his acquaintance with Einstein's wife, Mileva, who was not happy in Berlin and would have preferred

18. A. Einstein to P. Ehrenfest, 25 May 1914.
19. A. Einstein to P. Ehrenfest, undated, early 1914.
20. P. Ehrenfest, "Zum Boltzmannschen Entropie-Wahrscheinlichkeits-Theorem", Phys. Z. **15** (1914), p. 657. *Papers*, p. 347. See Chapter 11.
21. A. Einstein to P. Ehrenfest, 25 May 1914.

her husband to remain in Zürich. He enjoyed seeing young Albert again, and made it his business to take him to the zoo, and for walks in the park.[22]

On a Sunday morning late in July 1914, only a week after the Ehrenfests had moved into the new house on Witterozenstraat, Paul was outside, chasing a black cat out of his garden. A neighbor called over to tell him that Austria and Serbia had announced general mobilization. It was a month after the assassination at Sarajevo, and within days all of Europe was at war.[23]

The Netherlands had a century old policy of neutrality, but in the light of what happened to neighboring Belgium there was no assurance at any time during the war that Dutch neutrality would be respected. As it turned out, the country did maintain its neutral position during the four long years of destruction, even though many parts of the Netherlands were literally within earshot of the artillery fire. The Dutch borders were officially closed and the army was mobilized throughout the war.

Paul Ehrenfest belonged to a Home Guard unit during the war, along with other members of the faculty of the University of Leyden. They drilled at Katwijk, and he learned to handle a gun, but the unit was never called upon for any activity. The war also halted his plan to acquire Dutch citizenship as soon as possible. That had to wait until peace came.[24] But apart from the relatively minor hardships and restrictions of wartime life in neutral Holland, the greatest impact of the war on Ehrenfest was the shock to his picture of the civilized world. He himself had never had a blind national loyalty, neither to the Austria where he was born and grew up, nor to the Germany where he had studied and lived, nor even to the Russia where he had felt so much at home. What shocked Ehrenfest so deeply was the unexpected appearance of blatant chauvinism among the best educated, the most learned, the most deeply cultured of his scientific colleagues in the belligerent countries. Early in September 1914 Ehrenfest wrote to Lorentz on this theme.[25]

---

22. Ehrenfest's 1914 diary, entries of 29 May to 4 June.
23. Diary, 26 July 1914.
24. Personal communication from Mrs. T. Ehrenfest-Afanassjewa and Professor W. J. M. van Eysinga.
25. P. Ehrenfest to H. A. Lorentz, 4 September 1914.

"The typical newspaper item reporting that Haeckel, as a public gesture, has rejected his English academic honorary degrees has left me *very* depressed. Whatever one may think of Haeckel as a scientist and philosopher, he is at all events a man who is really true to his own conscience, or so it seems to me. Then how could he do such a thing? Now he should also throw away Darwin's books and portraits as another gesture. Will Ostwald and Klein now do something similar?

Isn't it possible that a few quiet and *warm* words from *distinguished* Dutch scholars might provide a 'condensation-nucleus' for the expression of sympathetic sentiments by scholars in the belligerent countries? I don't doubt for an instant that a large fraction of the *scientists* in all these countries know that the distinctive element of Haeckel's action (treating an honorary degree from Cambridge as if it were a military decoration) is completely irrational and perverse....

It is so terribly sad that even the most highly educated people don't clearly perceive that none of the great powers is completely 'free of guilt' in this catastrophe, and that now it's at best only a matter of each individual standing up for whatever is dear to him."

A week or so later Ehrenfest was glad to report to Lorentz that Felix Klein, who meant so much to both Paul and Tatyana, had written, and that he was keeping aloof from the "chauvinistic uproar".[26] (Ehrenfest had written Klein to ask whether, as residents of a neutral country, they could be of any help with the international correspondence in connection with the Encyclopedia.[27] He also commented that, considering the way honorary degrees were being treated, he feared that kinetic theorists in the belligerent countries would soon be ceremonially burning the papers of Boltzmann and Maxwell respectively.) The following month brought a new blow, the "Manifesto to the Civilized World".

Germany's invasion of neutral Belgium at the outbreak of the war, and the reports of what the German army had done in Belgium, shocked much of the world. In response to the widespread criticism that ensued, a public declaration was issued in Berlin in October 1914,

26. P. Ehrenfest to H. A. Lorentz, 14 September 1914.
27. P. Ehrenfest to F. Klein, 8 September 1914.

a "Manifesto to the Civilized World".[28] This document affirmed the wisdom and justice of German actions and ended with the flat assertion that German culture and German militarism were inseparable. Ninety three of Germany's leading intellectuals signed this remarkable Manifesto, including Ernst Haeckel, (the famous biologist whose actions had already depressed Ehrenfest), Wilhelm Röntgen, Felix Klein and Max Planck. One of the most distinguished non-signers of the Manifesto was Albert Einstein.

In the first month of the war Einstein had written to Ehrenfest:[29] "Europe, in her madness, has now started something incredible. At such a time we see what a wretched species of beast we belong to. I am going on quietly with my peaceful meditations, and feel only a mixture of pity and disgust." A few months later[30] Einstein, referring to himself and Ehrenfest as "international people", was wishing for an island somewhere for the thoughtful and those of good will. "I too would be an ardent patriot there." In Berlin, however, Einstein could not join the patriotic hysteria. He was one of the four men at the University who were willing to put their names to a document challenging the "Manifesto of the Ninety Three".[31] This was a calmly worded, rational plea for European unity prepared by the pacifist physician Georg Nicolai. This "Manifesto to Europeans" called on intellectuals to preserve the goal of "a universal world-wide civilization", to put aside all nationalistic passions and try to exert their influence for a peace that would unite Europe. It was meant to be a call to "good Europeans" everywhere, but it gained so few signatures at Berlin that it was not issued as a public document. "In those days even the best Germans were unwilling to be good Europeans, or dared not display their true feelings," as Nicolai described it.[31]

Both Einstein and Ehrenfest were glad to be able to turn to Lorentz for support in their internationalist views, and for his counsel. On one occasion, for example, Ehrenfest reported another upsetting incident in a letter to Haarlem.[32]

---

28. This document is quoted in part and described in O. Nathan and H. Norden eds., *Einstein on Peace* (New York: Schocken, 1968). See pp. 1–8 and the notes on pp. 649–650. There is also some discussion in P. Frank, *Einstein. His Life and Times*, transl. G. Rosen, ed. S. Kusaka (New York: Alfred A. Knopf, 1947), pp. 119–121.

29. A. Einstein to P. Ehrenfest, 19 August 1914.

30. A. Einstein to P. Ehrenfest, December 1914.

31. O. Nathan and H. Norden, *op. cit.*, pp. 4–6.

32. P. Ehrenfest to H. A. Lorentz, undated.

"I received a letter today from a young German physicist, (one of the well known theorists), with the urgent request that I forward it to Michelson in America. Since it contains no communication whose circulation would be against the laws or the interests of Holland I felt obliged to forward it immediately, though of course I accompanied it with a letter of my own in which I explained that I shared neither the opinions nor the feelings of the otherwise acute author. This letter, (whose author is prevented from taking part in the fighting for reasons of health, as he writes), contains the following words, along with other mad invective: 'You know me personally as a quiet peaceful person, but I must tell you this. This nation of pirates must be wiped off the earth before we Germans can rest, and the day that the English fleet is sunk and London is levelled to the ground will be the finest day of my life; and every single German thinks and feels this way.' (I am quoting from memory, but I would swear to the correctness of the characteristic words.)"

Ehrenfest wondered if he had done the right thing in forwarding the letter. He felt he had no right to act as a censor, but he thought perhaps he should have waited for an answer to his own reply to the German author before taking action. And then he added these words to Lorentz.

"Why am I writing you all this? Because my mistake distresses me very much. And, in general, in the chaos that is now taking place in the world, one has to have a very special temperament and maturity to know with absolute certainty on every occasion what one's conscience demands."

As the war dragged on, Ehrenfest tried to get Einstein to visit Leyden again. Einstein was living alone in Berlin, separated from his wife who was in Switzerland with their two sons. He was still trying to do what he could to preserve the bonds among scientists of different countries. "Isn't the small group of scholars and intellectuals the only 'fatherland' which is worthy of serious concern to people like ourselves? Should *their* convictions be determined solely by where they happen to live?"[33] In December 1915 he wrote[34] to Ehrenfest, in

33. A. Einstein to P. Ehrenfest, 23 August 1915.
34. A. Einstein to P. Ehrenfest, 26 December 1915.

response to another invitation to come to Holland: "Every fibre of my being itches to get away from here." Unfortunately it was impossible. His mother was visiting in Berlin, and at the Easter vacation he had promised to go to Switzerland to see his son, Albert. "But if I am already tied up and cannot travel to visit you, I will at least prove my desire to do so in writing by being exceptionally talkative."

Einstein was intensely busy again, developing the covariant form of the gravitational field equations. He had extremely interested readers in Leyden, and his letters of the winter of 1915–16 to Ehrenfest and Lorentz are full of his pleased answers to their detailed questions. On New Year's Day 1916 he wrote Ehrenfest,[35] "If the trip weren't so difficult now I would come immediately despite my lectures and the Academy." Einstein was delighted with the real interest his Leyden friends took in the general theory of relativity, and he regularly sent Ehrenfest copies of the proof sheets of his articles. Leyden began to seem like an oasis to him. On his return from Switzerland in April he wrote,[36] "I have a constant desire to show up suddenly at your home." And a month later,[37] "I have such a need to see you all again that I can hardly wait for peace to come without at least making an attempt to visit you. I will try during the summer vacation."

Arranging a trip to Holland from Berlin for a German-born Swiss citizen was not so easy under wartime restrictions. On August 25th Einstein reported that he had been to the Ministry of Foreign Affairs and had learned that some sort of official invitation from the University of Leyden would make it much easier for him to get permission for the journey. A few days later[38] he was ready to give up. "I am not going to Holland," he wrote to Besso, "because the trip has become so horribly difficult. I intend to wait until the war is over." But the official invitation came, with Lorentz's help, and Einstein went on with his plans. He had to get the original of his Swiss naturalization papers from Zürich, and then expected "a series of further obstacles, still obscured by fog."[39] He was looking forward to the trip "with indescribable pleasure", but feared there might still be further delays. Finally, on September 27, 1916, he left for Leyden, to stay for two weeks.

The day after his arrival Einstein and Ehrenfest went to Haarlem

35. A. Einstein to P. Ehrenfest, 1 January 1916.
36. A. Einstein to P. Ehrenfest, 29 April 1916.
37. A. Einstein to P. Ehrenfest, 24 May 1916.
38. A. Einstein to M. Besso, 27 August 1916.
39. A. Einstein to P. Ehrenfest, 6 September 1916.

to visit Lorentz. Einstein did not always share Ehrenfest's enthusiastic and impulsive reactions to people, but he was completely with him in his love and veneration for Lorentz, whom he called "the greatest and noblest man of our times."[40] "For me personally", Einstein would write shortly before his own death, "he meant more than all the others I have met on my life's journey."[41] Einstein particularly looked forward to seeing Lorentz in 1916, after his decade of successful struggle which had just culminated in the general theory of relativity. Lorentz had been one of his most eager, receptive and thoughtful readers, and there was no one with whom Einstein would rather have discussed his work. Ehrenfest was delighted to be able to bring them together, and described the meeting some years later.[42]

> "In his usual way, Lorentz saw to it first at dinner that Einstein felt himself enveloped in a warm and cheerful atmosphere of human sympathy. Later, without any hurry, we went up to Lorentz's cozy and simple study. The best easy chair was carefully pushed in place next to the large work table for the esteemed guest. Calmly, and to forestall any impatience, a cigar was provided for the guest, and only then did Lorentz begin quietly to formulate a finely polished question concerning Einstein's theory of the bending of light in a gravitational field. Einstein listened to the exposition, sitting comfortably in the easy chair and smoking, nodding happily, taking pleasure in the masterly way Lorentz had rediscovered, by studying his works, all the enormous difficulties that Einstein had had to overcome before he could lead his readers to their destination, as he did in his papers, by a more direct and less troublesome route. But as Lorentz spoke on and on, Einstein began to puff less frequently on his cigar, and he sat up straighter and more intently in his armchair. And when Lorentz had finished, Einstein sat bent over the slip of paper on which Lorentz had written mathematical formulas to accompany his words as he spoke. The cigar was out, and Einstein pensively twisted his finger in a lock of hair over his right ear. Lorentz, however, sat smiling at an Einstein completely lost in medi-

---

40. A. Einstein, *The World as I See It*, (New York: Covici Friede, 1934), p. 250.
41. A. Einstein in G. L. de Haas-Lorentz, ed., *H. A. Lorentz. Impressions of His Life and Work*, (Amsterdam: North-Holland, 1957), p. 8.
42. P. Ehrenfest. Fragment, presumably a draft of his funeral oration for H. A. Lorentz, 9 February 1928.

tation, exactly the way a father looks at a particularly beloved son – full of secure confidence that the youngster will crack the nut he has given him, but eager to see how. It took quite a while, but suddenly Einstein's head shot up joyfully; he 'had it'. Still a bit of give and take, interrupting one another, a partial disagreement, very quick clarification and a complete mutual understanding, and then both men with beaming eyes skimming over the shining riches of the new theory."

Soon after Einstein's return to Berlin, he wrote to the Ehrenfests ("Dear P and T") to thank them. "The reviving days spent with you have blended into a beautiful dream, which I do not weary of dreaming over and over."[43] "The trip to Holland was indescribably good for me, physically and spiritually; I am much fresher and happier. One can bear solitude only up to a certain limit."[44] And a week later, Einstein wrote to Besso about his wonderful days in Holland, where there was such a lively interest in relativity. "I spent unforgettable hours with Ehrenfest and especially with Lorentz, not only stimulating but also refreshing. I feel in general that I am incomparably closer to these people."[45]

This particular visit from Einstein left Ehrenfest with even more than his usual renewed spirits, scientific invigoration, and warm feelings. The two men always played violin and piano duets when they met, since music meant so much to both. Until this visit Ehrenfest's favorite composer had been Beethoven, though he also played and enjoyed music from Haydn to Brahms, at least. But he had never particularly liked the music of Bach. Einstein opened the world of Bach's music to his friend during his stay in Leyden, and Ehrenfest was completely captured. Einstein referred to this in his first letter. "I see that you are beginning to be one of the new admirers of these magnificent things [Bach chorales]. T will also join in, despite the alarming rectilinearity of her intellectual world line (an exception to the law of motion?)"[43] Ehrenfest must have written describing the pleasure he was getting from his playing of Bach, a pleasure his wife did not especially share. (His letters from this period were not preserved, but his notebooks indicate that for several months he devoted much more time to Bach

43. A. Einstein to P. Ehrenfest, 18 October 1916.
44. A. Einstein to P. Ehrenfest, 24 October 1916.
45. A. Einstein to M. Besso, 31 October 1916.

Fig. 11 *Standing:* Albert Einstein, Paul Ehrenfest, Willem de Sitter

Seated: Arthur S. Eddington, H. A. Lorentz.

Fig. 12 Albert Einstein and H. A. Lorentz.

Fig. 13 Paul Ehrenfest, Paul Ehrenfest Jr., and Albert Einstein in Ehrenfest's home, 1920.

than to physics.) "You are well off in being able to play Bach yourself and not having to wait for someone to play it for you," Einstein wrote at the beginning of his next letter.[44] "But I am well off too, because I can travel to Leyden where I can have my share not only of Bach but also of all other choice pleasures."

Ehrenfest did not see Einstein again until after the war was over. He invited Einstein again and again to come to Leyden, but there were too many factors working against it. Travel became even more difficult as time went on and Germany's position in the war grew worse. These were difficult years for Einstein. His health was often poor, and he had to keep to a strict diet and maintain a regular routine in his life. He was separated from his wife, and the marriage was soon to end in divorce. When he could travel at all, he went to Switzerland to see his children. Both his younger son and his wife were also ill. "Troubles and more troubles," as he wrote to Ehrenfest.[46] He could still console Ehrenfest, however. "You are complaining about yourself again, and are dissatisfied with yourself. Just think how little difference it will make in twenty years how one has loitered about on this earth, just so long as one has done nothing base. Whether you write this or that article yourself, or whether someone else writes it, makes very little difference. Stupid you certainly are not, except insofar as you keep thinking about whether or not you are stupid. So away with the hypochondria! Rejoice with your family in the beautiful land of life!"[47]

In March 1918 Ehrenfest wrote Einstein about his latest passion. "In the last few months I have become completely fascinated by some very definite problems of *theoretical* economics, and I have been reading deeply and thinking much about this whole complex of questions. Naturally this is in violation of my professional obligations, but it has taken too strong a hold on me – it's no use; it just corresponds to the structure of my thinking machinery."[48] Ehrenfest had been interested in economics for several years. Some of his earlier notebooks list "political economy" as a field to investigate, and W. Stanley Jevons is listed in 1914 among the authors Ehrenfest was reading, or intended to read. In the fall of 1917, however, he was struck by the possibility of developing an analogy between thermodynamics and economics.[49]

---

46. A. Einstein to P. Ehrenfest, 25 May 1917.
47. A. Einstein to P. Ehrenfest, 3 June 1917.
48. P. Ehrenfest to A. Einstein, 27 March 1918.
49. Ehrenfest's Notebooks, XXIII, XXIV, entries from October 1917 to May 1918.

He plunged into the works of Böhm-Bawerk, Schumpeter, Marx, Taussig and other economists. For the next six or eight months economic questions take by far the largest place in the notebook record of Ehrenfest's ideas. Most of his notes try to formulate economic concepts as parallels of thermodynamic concepts, with the concept of equilibrium occupying the central position in both theories. He hoped to be able to use the highly developed formalism of thermodynamics to gain new insight into economic problems. Ehrenfest never published any of his work in economics, but he always maintained his interest in the subject. It turned out a few years later that his work in economics, far from being a violation of his obligations as professor of physics, allowed him to guide one of his students, Jan Tinbergen, wisely and well from theoretical physics to mathematical economics.

When he wrote to Einstein in May,[50] Ehrenfest was still "miles away from all of physics," still working away at his economic ideas, "with the shoulder-shrugging blessing of Papa Lorentz". Ehrenfest wondered when he would return to physics, "full of remorse and ready to mend [his] ways". For the moment he found it "unspeakably detestable". He was also ill, and had been in bed with jaundice for ten days when he wrote. (His full recovery during the summer did bring a renewed interest in physics, and Ehrenfest was soon as deeply involved as ever in the quantum theory and statistical mechanics.)

Einstein was sympathetic to Ehrenfest's illness: "Now you are sick, too, you poor fellow." He was also interested and encouraging about his work in economics. "I am curious about what you are doing in economics. I always have the feeling that everything gets very long winded in this field just because of a lack of simple and vigorous thinking. But I find it intensely interesting, and understand how these problems excite you." He was feeling better but still had to take care of himself, because "my belly revolts at the slightest provocation; why should it have fewer rights than the brain?" He was enjoying sailing, his favorite amusement, and recommended that Ehrenfest "exchange wisdom for a period of harmless vegetation in Nature".[51]

Early in 1918 Ehrenfest received an invitation to attend another Wolfskehl week at Göttingen. The invitation came in the names of David Hilbert and Max Planck. Ehrenfest declined, partly because he

50. P. Ehrenfest to A. Einstein, 8 May 1918.
51. A. Einstein to P. Ehrenfest, 5 June 1918.

would not easily get his passport as an Austrian citizen, due to a conflict with the Austrian consulate in Holland. More important, however, was the state of his feelings about the war: "My opposition to the continuing oppression of Belgium, my sympathy with the views of my closest Russian friends, and my opposition to the general views in Austria and Germany are so passionate that I would constantly have to be afraid of violating my obligations as a guest." Ehrenfest also wrote, however, about his "warm gratitude for all he had received in Göttingen from teachers and fellow students", including his experiences at the Wolfskehl lectures in 1913. He was also grateful for the invitation and for the warm words with which Hilbert had accompanied it.[52]

Hilbert felt that Ehrenfest was being unreasonable in his views, and Einstein had to agree. "I believe it is wrong to vent your feelings on such people," he wrote to Ehrenfest. "Just imagine that you had been called to Göttingen instead of Leyden a few years ago, and were now to be held responsible for the whole evil machinery! I said nothing to Planck about this business. He is like a child in these public matters, and would understand you about as much as a cat understands the Lord's Prayer."[53] Ehrenfest insisted that he had not "vented his feelings" but expressed his affectionate memories of his student days in Göttingen, and for the rest, "I did what I had to."[50]

The previous year, in a letter to Klein,[54] he had commented on how painful it was to both Ehrenfests that Klein, who had done so much for international understanding, had put his name to the Manifesto of the 93, that "disastrous, devastating manifesto." Ehrenfest thought that some of the bitterness the manifesto had produced might be removed if several courageous men among the original signers would publicly withdraw their support. It did not happen.

Einstein's own feelings had tempered somewhat as he saw the horrors of the war going on and on. "I have grown much more tolerant, though my basic views have not changed in the slightest. On the political side I see that those with the least restraint and the greatest thirst for power are often the very ones who, in private life, would not harm a fly. We are dealing with an epidemic delusion which, having caused infinite suffering, will one day vanish and become a

52. P. Ehrenfest to D. Hilbert, 11 March 1918.
53. A. Einstein to P. Ehrenfest, 1 May 1918.
54. P. Ehrenfest to F. Klein, undated, June 1917.

monstrous and incomprehensible source of wonderment to a later generation."[55]

The war years were full of personal problems for Einstein – the end of his marriage and the serious illness of almost every member of his family. It was during these years in Berlin, though, that he renewed a childhood acquaintance with his cousin Elsa, now a widow with two daughters, and eventually made a happy second marriage with her.

For Ehrenfest, on the other hand, these years were good ones as far as his family life was concerned. In May 1915 his third child was born, a son who was named Paul. To the Ehrenfest family he was always Pavlik. In August 1918 the fourth and last child, Vassily (Vassik), was born. Einstein wrote to congratulate the Ehrenfests on the birth of their son, saying how much he admired their will to live. "I too am full of hope for the future again, even if I don't express it so concretely."[56]

The Ehrenfest home was indeed full of life. The children did not go to school. This was a matter of principle with Paul and Tatyana. They were convinced that the active inquiring minds of their children should be given the opportunity to develop freely at home rather than being regimented and dulled at school. Tanya, their older daughter, had a tutor – a young schoolteacher, Miss S. W. Hellema, who began giving her lessons soon after the Ehrenfests came to Leyden. Although Miss Hellema had some reservations about this "solitary" instruction, she was impressed by the freedom given to her by the Ehrenfests and also by their obvious deep concern with their daughter's intellectual development. Miss Hellema was encouraged to work according to her own best judgement and not to feel bound by the school curriculum. Contrary to her own expectations, the Ehrenfests did not interfere or try to take part in the lessons. They had decided to put their faith in her, and she was given a free hand. But Paul Ehrenfest always made himself available when she wanted to discuss her pupil, and there were many conversations in the study-living room or walking along the canals of Leyden.

It was somewhat lonely for a child to have lessons at home, but the Ehrenfests were more concerned with encouraging her intellectual

---

55. A. Einstein to P. Ehrenfest, 3 June 1917. Quoted in O. Nathan and H. Norden, *op. cit.*, p. 19.

56. A. Einstein to P. Ehrenfest, 4 September 1918.

growth and originality than with the possible hazards of loneliness. After a few years Miss Hellema was able to combine Tanya's lessons with those of another little girl whose health did not permit her to go to school, and both teacher and pupil found this a relief.[57]

Ehrenfest's home was always the center of his activities, and his children were aware at an early age of scientific meetings and colloquia. The girls even had a "colloquium" game and gave "lectures" of their own.[58]

The fall of 1918 brought another pleasure to Ehrenfest. He received news from Russia for the first time since the revolution. "I received a detailed letter today from Krutkow in Petersburg dated 14 September 1918(!!!!)," he wrote to Jan Burgers early in October. "They are often starving and freezing, but all my friends are still alive and working and even publishing!... 'My' colloquium [the one Ehrenfest had started during his Russian years] is still constantly active, even though the people don't know in all seriousness whether they will be able to survive this winter. You know – the enthusiasm of Russian youth is *really* unique. I can't tell you how this news has warmed my heart. *Home*!!"[59]

On November 7th Jan Burgers received his doctorate as Ehrenfest's first Ph.D. student, something of a milestone. A few days later the war was over. Ehrenfest, who had not seen Einstein for two years, immediately renewed his invitations. Einstein had made other plans, however. He intended to go to Paris to plead with the Allies for food to avert starvation in Germany during the coming winter, hoping that his word would be believed as the bitter truth, coming as it would from an uncompromised source. This trip did not work out, but Einstein spent several months in Zürich in 1919, lecturing at the University. He planned to give a brief cycle of lectures twice a year in Zürich. "I am doing this," he wrote Ehrenfest, "to show my gratitude for an appointment which I did not accept. I could not make up my mind to leave here under the prevailing conditions. But who knows whether sheer necessity won't finally require me to do so?"[60]

---

57. Personal communication from Mrs. T. van Aardenne-Ehrenfest and Mrs. A. van Bommel-Ehrenfest. Also a reminisçence by Mrs. S. W. Kits van Heijningen-Hellema obtained for me by Mrs. A. van Bommel-Ehrenfest.
58. The "Red Book", undated entry for 1916.
59. P. Ehrenfest to J. M. Burgers, 2 October 1918.
60. A. Einstein to P. Ehrenfest, 12 December 1918.

During the summer of 1919 Paul Ehrenfest decided that he really wanted more than just infrequent, brief, and hard to arrange visits from Albert Einstein. It was now almost three years since he had seen his friend, and he was ready to attempt something more ambitious. Here is the letter he wrote to Einstein on September 2nd,[61] after sounding out his Dutch colleagues on his plans.

"In connection with other circumstances, *all* of us are now in complete agreement that we must undertake to get you to Leyden. The business is extremely simple: if you just say *yes* to me, it will be possible – at least according to all human expectations – to arrange things extremely quickly according to your wishes.

1. You will receive as much salary as you need for yourself, your children, and whatever else is necessary to you. You have to compute that yourself. (Our *maximum* salary of 7500 guilders is *your minimum*.)
2. No lecturing duties will be required of you; the only real obligation will be to choose Leyden, or someplace in its environs, as your principal residence.
3. You can spend as much time as you want in Switzerland, or elsewhere, working, giving lectures, traveling, etc., provided only that one can say 'Einstein is in Leyden – in Leyden is Einstein'.

You understand: I talk as though I already have everything in the bag – of course that isn't so. But trial discussions have given me the greatest imaginable hope that it will be possible immediately to arrange everything exactly according to your wishes, if you just tell me you are ready in principle to negotiate with us. You might be interested in hearing that Onnes, in particular, (to say nothing of all the rest of us!), is very warmly for this idea!

You will never have to lecture in Dutch. What Dutch you need for daily life you will learn in two weeks.

You will have here in your immediate neighborhood, or in good contact with you: Lorentz, de Sitter, me, Kuenen, Droste, de Haas, his wife, Fokker, Burgers, Julius, Zeeman, and, continually, a group of very intelligent young fellows, noble and pleasant people. Also guests like Nordström, Bohr, etc.

---

61. P. Ehrenfest to A. Einstein, 2 September 1919.

By living further out of town you will be able to isolate yourself as much as you like.

Dear, dear Einstein, don't say no to me, but instead come as soon as you can to stay with us for two weeks, or at least a week. But in any case answer me directly: 'Well, what you are proposing to me here is not so stupid.'

Dear, dear Einstein! I need that and nothing more to set everything going directly. And listen: don't refuse right away, because even if later on you can't make up your mind to come, it would, in any case, be of *very* great value to our activity in physics here for a whole set of people to have organized themselves with the special goal of obtaining *you*.

Einstein – bear in mind that you would be taken in here by a group of people who are really fond of you personally, and not just of the brain drippings that ooze out of you! No one here expects results from you; everyone simply wants to have you here close to us.

I assure you that you will have such *extraordinarily* extensive freedom here, that, without any more ado, you will be able to fulfill all the moral obligations you feel with respect to Switzerland. You will also be able to do a great deal for Berlin, which you regard as desirable.

Dear, dear Einstein! Don't destroy all my dreams and hopes. Help me in my efforts by immediately sending me an answer favorable enough for me to put everything in motion directly."

There is also a remarkable, Ehrenfestian postscript: "It is really very annoying that you have something to say, too, about a matter that we can survey much better than you!!"

Einstein answered within a few days.[62]

"Your offer is so fabulous and your words are so friendly and so full of affection that you can hardly imagine how confused I have been as a result of your letter. You know, of course, how delighted I am with Leyden. And you know how much I like all of you. But my situation is not so simple that I can do the right thing by just following the inclination of my own feelings. I am sending you a letter that Planck wrote me while I was in Zürich. After receiving it I promised him not to turn my back on Berlin

62. A. Einstein to P. Ehrenfest, 12 September 1919.

unless conditions were such that he would regard such a step as natural and proper. You have hardly any idea of the sacrifices that have been made here, with the general financial situation so difficult, to make it possible for me to stay and also to support my family in Zürich. It would be doubly base of me if, just when my political hopes are being realized, I were to walk out unnecessarily, and perhaps *in part* for my material advantage, on the very people who have surrounded me with love and friendship, and to whom my departure would be doubly painful at this time of supposed humiliation. You have no idea with what affection I am surrounded here; they are not all out to catch the drops of oil my brain sweats out.

So you see how things stand with me. I can leave here only if there is a turn of events that makes it impossible for me to stay on. Such a turn of events could occur. But if it does not, my departure would be tantamount to a despicable breach of my word to Planck. I would be breaking faith and would certainly come to reproach myself later on. (I feel like some relic in an old cathedral – one doesn't quite know what to do with the old bones, but –).

Well, whatever may come of it, this is a fine excuse to come and visit you, especially since it is a terribly long time since we've seen each other. If my tyrannical belly permits it, I will visit you this fall. I wonder whether one can get a travel permit right away? Have you perhaps heard anything there about the English solar eclipse expedition? The rest of the shop talk can be postponed until we meet. I don't have much to report anyway."

Einstein's letter gave Ehrenfest a feeling of "warm, proud joy", even though it cancelled the plans over which he had been dreaming and scheming for months.[63] He was still upset, however, by the thought that monetary considerations (such as the exchange rate between Switzerland and Germany) could play a part in determining what happened to a man like Einstein. Ehrenfest copied Einstein's letter and sent it to Lorentz, remarking on his own reactions of shame and then pride in his friend, and adding: "If at least one could take away his monetary worries!! The Nobel Prize?!"[64] (Einstein had not received

63. P. Ehrenfest to A. Einstein, 21 September 1919.
64. P. Ehrenfest to H. A. Lorentz, 21 September 1919.

the Nobel Prize yet, despite his achievements, and would not receive it for another two years.)

In arranging for Einstein's forthcoming trip to Leyden, which involved special efforts by Lorentz and Kamerlingh Onnes to get his Dutch travel permit promptly, Ehrenfest was very solicitous of his friend's health and welfare. He advised Einstein to travel by the best available train accommodations, since all his travel expenses would be taken care of whether or not he gave a lecture during his Leyden visit. Ehrenfest offered to do whatever was necessary to keep Einstein from becoming overtired, and asked for a "power of attorney" to forbid visitors and obtain complete isolation for Einstein whenever he wanted it.[65] This would be the last occasion when such isolation would be relatively easy to attain.

Two weeks before leaving Berlin for Leyden, Einstein received the answer to the question he had casually asked at the end of his last letter to Ehrenfest. Word had reached Lorentz that the observations of the solar eclipse of March 29 had finally been analyzed by Eddington and his co-workers: the results confirmed the predictions of the general theory of relativity. "Joyful news today!" Einstein wrote on a postcard to his mother dated September 27, 1919. "Lorentz has telegraphed me that the English expedition has really proved the deflection of light by the sun."[66] The news was not made public until November, so the great public uproar it created came after Einstein's trip to Holland, which took place during the last two weeks of October.

Ehrenfest had always reacted enthusiastically to seeing Einstein, and this time was no exception. "You have left waves of joy and harmony behind you here, and they will ripple back and forth undamped among all of us for a long time to come. You have really made us all completely happy and left us full of boundless optimism."[67]

It was Einstein, however, who wrote with unusual emotion and warmth after this visit. "Yes, that was really a beautiful and tranquil time that we spent together." He appreciated all the thoughtful care for his health and comfort, which even kept his "clanking old body"

65. P. Ehrenfest to A. Einstein, 5 October 1919. Also see O. Nathan and H. Norden, *op, cit.*, p. 650.

66. A. Einstein to Frau Pauline Einstein, 27 September 1919. Reproduced in facsimile in O. Nathan and H. Norden, *op. cit.*, p. 28, and also in B. Kuznetsov, *Einstein* (Moscow: Progress Publishers, 1965), p. 224.

67. P. Ehrenfest to A. Einstein, 3 November 1919.

in a good mood. Einstein sent a warm kiss of thanks to each individual member of the Ehrenfest family, from the toddler Vassik to Tatyana's aunt, each kiss to be suited to the "sex, tradition and temperament" of its intended recipient. "I have never before taken part in such a happy family life. It proceeds smoothly from two independent people, who are not united just by compromises. I have come to feel that all of you are a part of me and that I belong to you." And to Paul Ehrenfest in particular he added, "We will keep in close personal contact with each other from now on. I know that it is good for both of us, and that each of us feels less of a stranger in this world because of the other."[68]

One can easily imagine Ehrenfest's reactions to these words from Einstein. He, who was so easily moved to emotional utterance, could not find words this time.[69] "I have struggled and struggled for a long time, trying to answer your letter and to thank you for your photographs which have pleased us no end. But now I know that I shall *not* answer your letter. For your answer you must simply imagine the children's faces, full of love for you."

Only a few days after Einstein's return to Berlin, the great news that his general relativity theory had been confirmed by the eclipse data was made public. The Royal Society of London heard the announcement at its meeting on November 6th, and learned that "a great adventure in thought had at length come safe to shore," to quote the words Alfred North Whitehead used in describing the "supreme incident" at which he was present.[70] The newspapers picked up the story immediately, sensing the drama: a challenge to Newton's concepts offered by a lonely theorist, the test of his prediction of the bending of light rays during an eclipse, (always mysterious, especially when observed by expeditions to Brazil and West Africa), and his triumphant justification by the results – a "scientific revolution" indeed.

It was also a revolution in Einstein's life. His name became a household word almost overnight as a result of all the newspaper "cackling", as he referred to it. Physicists everywhere were asked to give popular lectures explaining relativity, or to write popular articles on the subject for the newspapers. Einstein himself wrote a brief article for the *Times* of London, and took the opportunity to make a pointed

68. A. Einstein to P. Ehrenfest, 9 November 1919.
69. P. Ehrenfest to A. Einstein, 24 November 1919.
70. Whitehead is quoted in P. Frank, *op. cit.*, p. 140–141.

joke. "Here is yet another application of the principle of relativity for the delectation of the reader: today I am described in Germany as a 'German savant' and in England as a 'Swiss Jew'. Should it ever be my fate to be represented as a *bête noire*, I should, on the contrary, become a 'Swiss Jew' for the Germans and a 'German savant' for the English."[71] Ehrenfest was, of course, delighted with this relativity joke, and also happy to report that Lorentz, too, had enjoyed it.[72]

Ehrenfest had agreed to write a sketch of relativity for one of the principal Amsterdam newspapers, and came to regret his decision. He had made himself ill worrying over the assignment. This was partly because he felt that his talent for giving lectures at a popular level did not carry over to writing popular articles. More important was his inability to reconcile the various viewpoints that Einstein had adopted in the successive stages of his work. As he put it, the special theory required the equivalence of observers in uniform relative motion because there was no reason for them to be inequivalent, since they moved with respect to nothing; i.e. there was no aether, no absolute frame of reference. Motion with respect to nothing could have no physical significance. In the same way, the general theory required the equivalence of all observers because acceleration with respect to nothing could also have no physical significance. *But*, if one adopted Mach's principle that inertia (mass) had its origin in the gravitational effects of distant masses, the system of fixed stars, then what? "Now one can no longer say that they move with respect to nothing, for now they move with respect to an enormous something!" And Ehrenfest went on, "Einstein, my upset stomach hates your theory – it almost hates you yourself! How am I to provide for my students? What am I to answer to the philosophers?!!"[69]

Einstein answered that he understood Ehrenfest's difficulty in accounting for the development of relativity. It arose because Ehrenfest was trying to give the 1905 theory an *epistemological* justification through the nonexistence of the aether, whereas Einstein had, in fact, based his arguments on the *empirical* equivalence of all inertial systems with respect to light. He had not introduced epistemological considerations until 1907, when he asked: "Why should relativity govern only uniform motion, especially since the equality of inertial and gravitational masses suggests an extension of the principle of

71. A. Einstein, *The World as I See It*, p. 81.
72. P. Ehrenfest to A. Einstein, 9 December 1919.

relativity?" This led, after years of effort, to the principle of general covariance and then to the new field equations. It was only in 1917 that Einstein showed how the metric field was completely determined by matter, a step one did not need to accept, if one were willing to assume that the mean density of matter in the universe is zero. Einstein, however, found this alternative "detestable", as he wrote to Ehrenfest.[73]

Ehrenfest's relativistically irritated stomach was presumably soothed by this answer, which he found extremely valuable. He even suggested an exchange with Einstein on these subjects in the pages of one of the physics journals, for the illumination of all the others who must have been bothered by the same questions.[72] This exchange, unfortunately, never took place.

When Einstein declined to leave Berlin and accept a professorship at Leyden, Ehrenfest immediately began to speculate about the possibility of somehow attaching his friend to Leyden without separating him from Berlin. By the end of November 1919 Ehrenfest was writing to Einstein about the tentative plans to offer him an appointment as Special Professor at Leyden. This would bring him to Leyden regularly for three or four weeks a year. The plans were expected to go through smoothly, but Ehrenfest asked Einstein to say nothing yet "because my super-solid colleagues could be annoyed with me for disturbing you before everything is *absolutely* settled."[69] Einstein immediately liked the idea of such a "comet-like existence in Leyden", and told Ehrenfest that he would surely accept such an offer.[73] Ehrenfest enjoyed just thinking about "the Einstein-present we are giving to ourselves", feeling "as happy as a child about it – and so it's hard to hold my tongue."[72]

The official approval for Einstein's special professorship did not come as quickly as Ehrenfest would have liked. During the first few months of 1920 Ehrenfest sent Einstein all sorts of detailed advice about the form and circumstances of the inaugural lecture with which Einstein would begin his professorial duties when the official appointment came. He pointed out to Einstein that one could not use a blackboard, slides, or any other aids; one couldn't even talk with his hands, since they would be hidden by an academic robe![74] Ehrenfest described the Dutch custom of adding words of thanks to appropriate

73. A. Einstein to P. Ehrenfest, 4 December 1919.
74. P. Ehrenfest to A. Einstein, 21 January 1920.

groups and individuals at the end of the lecture, sketched some samples, and suggested that Einstein pause before these extra remarks long enough to drink a glass of water, so that the old gentlemen in the audience could wake up after the lecture. He even gave Einstein the proper Dutch pronunciation for the sentence ("Ik heb gezegd") which traditionally ends such a speech; (but only when it is given in Dutch.)[75]

Einstein planned to talk on "The Aether and the Theory of Relativity", a subject closely related to his recent correspondence with Ehrenfest. Ehrenfest advised him to have lots of copies of the talk printed because he was sure they would go "like hot cakes". The date for the talk was set tentatively for May 5, 1920. But in mid-April, Ehrenfest wrote that the matter was still dragging in the government bureaus, even though the funds had been made available early in February by the (private) Foundation for the Leyden University Fund.[76] On May 4th, however, Einstein was still in Berlin waiting for the red tape to be untied.[77] He did get to Leyden for a few weeks later in May, but his appointment was not yet official so the inaugural lecture had to be postponed. It was not until July that everything was finally settled and Einstein wrote that he would come in October to give the lecture he had written months earlier.[78] Three curators were appointed for Einstein's special chair – an esteemed Leyden notary, a high official from the Ministry of Foreign Affairs and the physicist Pieter Zeeman. "In other words," Ehrenfest wrote, "Mr. Einstein is recognized as a man whom even 'respectable' people can trust not to cut capers. I hope you will not prove yourself unworthy of this 'trust placed in you by the solid citizens'. (If you do, a lot of people dear to you will suffer for it directly.) It's like marrying a countess, but we are the mothers-in-law."[79]

One of the less weighty problems in connection with Einstein's forthcoming visit concerned the tailcoat he would have to wear for the ceremonial reception. Einstein owned one, but was not anxious to take it with him since he was not traveling directly to Leyden. "Naturally there are plenty of people here who would lend you a tailcoat," Ehrenfest wrote, "but whether or not one of them will fit you is a

75. P. Ehrenfest to A. Einstein, 10 March 1920.
76. P. Ehrenfest to A. Einstein, 13 April 1920.
77. A. Einstein to P. Ehrenfest, 4 May 1920.
78. A. Einstein to P. Ehrenfest, 19 July 1920.
79. P. Ehrenfest to A. Einstein, 6 August 1920.

geometrical problem whose solvability has not been demonstrated."[80] Einstein took his own coat with him when he left Berlin.

To celebrate the beginning of Einstein's official relationship with the University of Leyden, Ehrenfest and Kamerlingh Onnes arranged to have several other guests present at the same time for scientific discussions. Paul Langevin of the Collège de France and Pierre Weiss of Strasbourg were to be the other guests, and the subject for discussion was to be the theory of magnetism. Langevin and Weiss had made extremely significant contributions to the theory of the magnetic properties of matter, Onnes's laboratory was engaged in the study of magnetic phenomena at low temperatures, and Einstein had recently participated in an experiment with de Haas on the relation between magnetism and rotation. As for Paul Ehrenfest, he was working on the theory of magnetism during that summer of 1920, puzzling, as he so often did, over a paradox that had struck him. He wrote Einstein in July that he was looking forward to these discussions on magnetism, since he had prepared carefully for them by a thorough study of the literature. "I can clear away an enormous amount of rubbish, and I am burning with impatience for these discussions."[81]

Einstein looked forward to seeing Langevin whom he liked very much, and wrote that he too was happy with the idea of discussing magnetism at low temperatures, "a subject that is now completely ripe for theoretical treatment."[82] Ehrenfest was surprised by Einstein's "theoretical optimism" on the subject of paramagnetism. "Yes – the paramagnetism of gases – good. But the paramagnetism of solids?!"[79] This was just the paradox that Ehrenfest had been puzzling over: how could one understand the fact that some solids obeyed the Curie law for paramagnetism, just like gases, even down to extremely low temperatures, although the elementary magnets in a crystal could hardly be considered free to rotate like those in a gas? Ehrenfest had some suggestions to make on this subject, but he was correct in thinking that the mystery was not yet fully ripe for solution; nor would it be solved until after the development of the new quantum mechanics.[83]

---

80. P. Ehrenfest to A. Einstein, 16 August 1920.

81. P. Ehrenfest to A. Einstein, 24 July 1920.

82. A. Einstein to P. Ehrenfest, 30 July 1920.

83. P. Ehrenfest, "Note on the Parmagnetism of Solids", Proc. Acad. Amst. **23** (1921), p. 989. *Papers*, p. 443. Also see J. H. Van Vleck, *The Theory of Electric and Magnetic Susceptibilities*, (Oxford: Oxford University Press, 1932), esp. Chapters 4–6, 9, 11.

Ehrenfest was not at home in French, and he decided to prepare for Langevin's visit by reading French so that he would be able to understand at least some of what was said. He chose the writings of Henri Bergson, and found him "wonderful", and "a fine fellow". "You will be astonished at what a pleasure it is," he wrote Einstein, "when I show you a few pages selected from his books."[79]

Ehrenfest was very hopeful that this visit of Langevin and Weiss with Einstein would have some real scientific value. "I really believe that if we all spend a week in *ordered* discussion – first on the paramagnetism of gases, then solid (non-conducting) crystals (Curie and anti-Curie behavior at low temperatures), liquid oxygen, and *finally* on ferro-magnetism – that something *must* come out of it."[84] But despite all his enthusiasm and preparation, in the same letter that contained his new ideas on the paramagnetism of crystals, Ehrenfest still expressed his recurrent doubts and disgust with himself. "I am very, very depressed – partly because of the eternal (minor!!!) worries about money, partly because I am not working at all. What I *can* do is not science but only a bit of entertaining salon conversation or promenade conversation about physics – the physics done by others." And after his enthusiastic remarks about Bergson: "How extraordinarily happy I could be, if only I were not so slack and so *unproductively ambitious*. It is all completely clear to me, but that doesn't help at all. I can obtain as much pleasure as I want out of *everything*, but it's all like confectionery and marmalade – the bread is work that succeeds after exertion, and *there* I fail completely."[79]

Einstein counseled resignation. "Don't complain and don't vex yourself. *We* too may make use of the human law that one always gets lazier and stupider as one grows older. In that way we acquire the merit of easing the consciences of others."[85] Ehrenfest's reaction to this is revealing; it recalls his childhood situation as the very little brother of four older, stronger and seemingly wiser boys: "Don't be impatient with me. Bear in mind that I hop around among all of you big beasts like a harmless and helpless frog who is afraid of being squashed."[80]

Paul Ehrenfest lived an essentially private life, as did almost all of his colleagues and friends. Among these friends Albert Einstein could

84. P. Ehrenfest to A. Einstein, 2 September 1920.
85. A. Einstein to P. Ehrenfest, 13 August 1920.

have served as the perfect example of a private man, whose real satisfactions were completely independent of the public at large, and hardly even demanded the existence of a few intimate friends. (It is hard to imagine any other scientist saying with unquestionable sincerity, as Einstein did, that a job as lighthouse keeper might be the ideal position for him.) But it was Einstein's fate to become a public man, and it was often a harsh fate. One of the worst periods for Einstein came in the first postwar years. Much of the bitterness and hatred that swept through Berlin in the aftermath of Germany's defeat found a convenient target in Einstein and his theory of relativity. Einstein was a Jew, a pacifist, and an outspoken believer in democratic internationalism. This collection of obviously undesirable traits, combined with his notoriety as the creator of a mysterious, incomprehensible, and fundamentally radical scientific theory, were enough to provoke a stream of abuse. The abuse reached a peak in the summer of 1920 when an organization to oppose Einstein's theories was set up, offering sizeable sums of money to authors who would write or speak against Einstein and his work.

On August 24, 1920, this group, the Anti-Einstein League as it was known, held a widely publicized public meeting at the Berlin Philharmonic Hall. One of the speakers was Paul Weyland, an engineer who led the League. The other was the experimental physicist, E. Gehrcke. Einstein attended, accompanied by his colleague Walther Nernst, and heard himself denounced as a publicity hound, a plagiarist, a charlatan, and a scientific Dadaist. All this took place in a poisonous atmosphere, with swastikas and anti-semitic pamphlets on sale at the entrance to the hall.[86] The scientific reporter for the *Berliner Tageblatt* made it clear in his account of the meeting that no serious scientific objection had been raised, and a day later Einstein's reputation was staunchly defended in a letter to the paper from his colleagues Nernst, Rubens, and von Laue. Nevertheless, Einstein felt it necessary to write his own reply, and the August 27 issue of the *Berliner Tageblatt* carried on its front page an article by Einstein entitled "My Answer to Anti-Relativity Theory, Inc.".

The next day Ehrenfest read an account of the whole affair, including a mention of Einstein's answer, in the Dutch papers, and he immediately sent off a letter to his beleaguered friend. "I am awfully distressed that you have been dragged into the mud. Please, just spit on *all* these

86. P. Frank, *op. cit.*, pp. 158–166.

attacks." Even though he had not been able to reach Lorentz and Kamerlingh Onnes, Ehrenfest was ready to assure Einstein, on his own responsibility, that he could come to Holland and take up a Dutch professorship (with no particular duties). "You must not misunderstand me: I know how definitely you declined to leave a defeated Germany a year ago, especially not to hurt Planck. And *really* (please believe me) it is not that I am trying to entice you to Holland and away from Germany by using this 'favorable opportunity'. No – it's only that if the situation there embitters you so that you cannot work, you can count on the fact that we will exert all our efforts to get you here." He advised Einstein not to write a single word of reply in "these filthy newspapers", but to let his friends do it for him, particularly friends like Planck and Lorentz.[87]

A few days later, when he had finally seen the August 27 issue of the *Berliner Tageblatt* for himself, Ehrenfest wrote again, in a rather different tone. After telling Einstein that he and his wife had read through the newspaper article four or five times, Ehrenfest went on this way.[84] "Forgive me – this is the first time I have allowed myself to meddle in your affairs without being asked; I do it very apprehensively, but I feel obliged to do it *just because* we feel so warmly toward you." And if anything in what he was about to write annoyed Einstein, Ehrenfest urged him to recall how much he meant to the Leydeners, and that would remove any trace of his annoyance. "With a heavy heart" Ehrenfest then gave Einstein his reactions to what he had read.

> "My wife and I absolutely cannot believe that you yourself wrote down at least some of the phrases in this article, 'My Answer'. We don't forget for a minute that you have certainly been provoked in an especially vulgar way, and we also don't forget in what an abnormal moral climate you live there, but nevertheless this answer contains particular reactions that are completely non-Einsteinian. We could underline them one by one in pencil. If you really did write them down with your own hand, it proves that these damned pigs have finally succeeded in touching your soul which means so terribly much to us.
>
> You must understand me correctly: *I* might have sinned 100 times worse, but it is a question of you and not of me. And you

87. P. Ehrenfest to A. Einstein, 28 August 1920.

are of such a nature that this 'My Answer' doesn't correspond to you, but rather sounds like an *echo* of the filthy attacks on you.

Why am I writing this letter? 1) To say what my wife and I feel so strongly. 2) To help prevent this poisonous filth from really penetrating to you. 3) In any case to urge you as strongly as I can not to throw one more word on this subject to that voracious beast, the 'public'.

And now please don't be angry with me. Whatever else may happen never forget how faithfully all of us here are attached to you – from Pavlik to Lorentz."

Ehrenfest was right: Einstein's "My Answer" does sound as though he had been hurt badly enough by the wretched attacks on him to reply in an unbecoming style. He had answered his attackers reluctantly. "I am very well aware of the fact that both speakers are unworthy of an answer from my pen, since I have good grounds for believing that there are motives other than the search for truth behind this undertaking. (If I were a German nationalist with or without a swastika instead of a Jew with internationalist views, then . . .) I am answering only because I have been repeatedly asked to do so by well-meaning people so that my interpretation becomes known."[88] He cited a list of prominent theoretical physicists from half a dozen countries who were all active workers in the theory of relativity, whereas the only physicist "with an international reputation" who opposed it was Philipp Lenard. Lenard was a distinguished experimentalist, (Nobel Laureate in 1905), but he had contributed nothing significant to theoretical physics and his objections were so superficial that Einstein did not feel the need to refute them in detail. Einstein passed over Weyland's remarks as being without substance and consisting of nothing but "rude insolence and vulgar accusations" from someone without professional competence in physics. He answered Gehrcke's arguments in detail even though they were a "deliberate attempt to mislead the lay public". And Einstein ended by saying: "It will make a peculiar impression abroad, and particularly on my Dutch and English colleagues, H. A. Lorentz and Eddington, both of whom have worked extensively in the theory of relativity and repeatedly lectured on it, when they see that the theory as well as its originator are defamed this way in Germany itself."[88]

88. A. Einstein, "Meine Antwort über die anti-relativitätstheoretische G. m. b. H.", *Berliner Tageblatt*, 27 August 1920, p. 1.

When Einstein received Ehrenfest's reproaches he answered simply that he had indeed written the article himself, uninfluenced by anyone else, "all in one stroke, one morning". In his own defense Einstein pointed out that "there was nothing left except to defend myself publicly against the charges of charlatanism, self-advertisement, plagiarism, etc., which have been made repeatedly and publicly. I had to do this if I wanted to remain in Berlin, where every child recognizes me from the photographs. If one is a democrat, one also has to acknowledge the claims of publicity."[89]

Einstein finally began his official duties as Special Professor at Leyden by giving his long-delayed inaugural lecture on October 27, 1920.[90] He had had a hard year, whose first few months were dominated by the painful and protracted final illness of his mother. Pauline Einstein had come to Berlin from Switzerland to spend the last few months of her life at her son's home, "a frightful time", as he described it to Ehrenfest. His visits to Leyden in May and October were the brightest periods of the year.

"My thoughts are still often with you," he had written in June, saying how he missed the Ehrenfest children and how much he had enjoyed his discussions and music making with Ehrenfest himself. He couldn't account for the way he was "pampered and overestimated" in Leyden, but he was grateful for it with all his heart. And Einstein had some especially golden words for Paul Ehrenfest: "It's so remarkably good for both of us to be together more often, because it's just as though Nature had made us for each other."[91]

---

89. A. Einstein to P. Ehrenfest, 10 September 1920.

90. The lecture "Ether and Relativity" is reprinted in A. Einstein, *Sidelights on Relativity*, transl. G. B. Jeffery and W. Perrett (New York: E. P. Dutton and Company, 1922), p. 3. (It is incorrectly described as having been delivered on May 5, 1920, at Leyden.)

91. A. Einstein to P. Ehrenfest, 6 June 1920.

Index